1

QUADERNI

Highlights in the quantum theory of condensed matter

A symposium to honour Mario Tosi on his 72nd birthday

edited by
Fabio Beltram

EDIZIONI
DELLA
NORMALE

We aknowledge the generous
support provided by MPS

ISBN: 88-7642-170-X

Participants to the symposium
Highlights in the quantum theory of condensed matter
A symposium to honour Mario Tosi on his 72nd birthday
Scuola Normale Superiore, Pisa – September 10-11, 2004

Contents

Preface

The birth of condensed matter physics in Italy is linked to a small number of very distinguished scientists. Mario Tosi is unquestionably among the leading figures, a true founder of the theoretical activity in the country and a true catalyst of novel research directions internationally.

Mario Tosi is Professor of Physics of Matter at the Scuola Normale Superiore and a symposium to honour him on the occasion of his 72nd birthday was organised in Pisa on September 10-11, 2004. This volume collects the proceedings of this symposium designed to show Mario Tosi's broad, deep influence in very diverse areas of the quantum theory of condensed matter.

The breadth of his interests is well represented by the topics covered in the different sessions:

- Liquids
- Electronic states in complex structures
- Quantum degenerate gases
- Many-body physics

Many of the speakers and contributors are actually students and collaborators of Mario Tosi's and witness his achievements as teacher in different Italian universities and at the International Centre of Theoretical Physics, and his ability to inspire scientists towards exciting, innovative fields of research.

We wish to thank all participants to the symposium for making it an enjoyable and scientifically stimulating event. The Organiser would like to thank Zehra Akdeniz, Marco Giordano, Gaetano Senatore, Erio Tosatti for their help in the definition of the program. Thanks also to all those that helped in the production of this volume: first of all the contributors, but also thanks to Professor Michele Ciliberto, Director of the Edizioni della Normale who accepted to host this volume in the Quaderni Collection, and to Dr. Marina Berton that edited the text and ultimately made this work a reality. Last but not least, we wish to acknowledge the generous support provided by the Monte dei Paschi di Siena Bank.

Fabio Beltram
Symposium Organiser
Salvatore Settis
Director of Scuola Normale Superiore
Fulvio Ricci
Dean of the Classe di Scienze

Authors' addresses

ZEHRA AKDENIZ – Department of Physics, Istanbul University, Istanbul (Turkey)

LAURA ANDREOZZI – Dipartimento di Fisica and INFM, Università di Pisa, Largo B. Pontecorvo 3, 56127 Pisa (Italy)

MAURIZIO ARTONI – Dipartimento di Chimica e Fisica per l'Ingegneria e per i Materiali, Università di Brescia, Via Valotti 9, 25133 Brescia (Italy)

FRANCO BASSANI – Scuola Normale Superiore, Piazza dei Cavalieri 7, 56126 Pisa (Italy)

FEDERICO BECCA – INFM-Democritos, National Simulation Centre, and SISSA, 34014 Trieste (Italy)

FRANCESCO BUDA – Leiden Institute of Chemistry, Gorlaeus Laboratory, Leiden University, Einsteinweg 55, P.O. Box 9502, NL-2300 RA Leiden (The Netherlands)

LUCA CAPRIOTTI – Credit Suisse First Boston (Europe) Ltd. One Cabot Square, London E14 4QJ (UK)

PABLO CAPUZZI – NEST-INFM and Scuola Normale Superiore, 56126 Pisa (Italy)

DAVIDE CERESOLI – International School for Advanced Studies (SISSA), Via Beirut 2, 34014 Trieste (Italy)
– INFM-Democritos National Simulation Center, Via Beirut 2, 34014 Trieste (Italy)

CHRISTIAN CHERUBINI – Facoltà di Ingegneria, Università Campus Biomedico di Roma, Via Longoni 83, 00155 Roma (Italy)

STEFANO CHESI – Department of Physics, Purdue University, West Lafayette, IN 47907 (USA)

ALESSANDRO CRESTI – NEST and Dipartimento di Fisica 'E. Fermi', Università di Pisa, Largo B. Pontecorvo 3, 56127 Pisa (Italy)

VALENTINA DE GRANDIS – Dipartimento di Fisica, Università "Roma Tre", Democritos National Simulation Center, Via della Vasca Navale 84, 00146 Roma (Italy)

MASSIMO FAETTI – Dipartimento di Fisica and INFM, Università di Pisa, Largo B. Pontecorvo 3, 56127 Pisa (Italy)

ROSARIO FAZIO – NEST-INFM and Scuola Normale Superiore, Piazza dei Cavalieri 7, 56126 Pisa (Italy)

FRANCESCA FEDERICI – NEST-INFM and Scuola Normale Superiore, Piazza dei Cavalieri 7, 56126 Pisa (Italy)

CHIARA FORT – LENS and Dipartimento di Fisica, Università di Firenze, and INFM, Via Nello Carrara 1, 50019 Sesto Fiorentino FI (Italy)

PAOLA GALLO – Dipartimento di Fisica, Università "Roma Tre", Democritos National Simulation Center, Via della Vasca Navale 84, 00146 Roma (Italy)

GIANCARLO GALLI – Dipartimento di Chimica e Chimica Industriale and INSTM, Università di Pisa, Via Risorgimento 35, 56126 Pisa (Italy)

D. J. WALLY GELDART – Department of Physics, Dalhousie University, Halifax NS B3H3J5 (Canada)
– School of Physics, University of New South Wales, Sydney 2052 (Australia)
– Dipartimento di Fisica, Università di Camerino, 62032 Camerino (Italy)

PAOLO V. GIAQUINTA – Università degli Studi di Messina, Dipartimento di Fisica, Contrada Papardo, Salita Sperone 31, 98166 S. Agata ME (Italy)

Marco Giordano – Dipartimento di Fisica and INFM, Università di Pisa, Largo B. Pontecorvo 3, 56127 Pisa (Italy)

Gabriele F. Giuliani – Department of Physics, Purdue University, West Lafayette, IN 47907 (USA)

Vincenzo Grasso – Dipartimento di Fisica della Materia e Tecnologie Fisiche Avanzate, Università di Messina, and INFM, Unità di Messina, Salita Sperone 31, 98166 Messina (Italy)

Giuseppe Grosso – Dipartimento di Fisica 'E. Fermi', Università di Pisa, Largo B. Pontecorvo 3, 56127 Pisa (Italy)

Massimo Inguscio – LENS and Dipartimento di Fisica, Università di Firenze, and INFM, Via Nello Carrara 1, 50019 Sesto Fiorentino FI (Italy)

Alexander A. Klyachko – Faculty of Science, Bilkent University, Bilkent, 06800, Ankara (Turkey)

Giuseppe C. La Rocca – Scuola Normale Superiore, Piazza dei Cavalieri 7, 56126 Pisa (Italy)

Giovanni Modugno – LENS and Dipartimento di Fisica, Università di Firenze, and INFM, Via Nello Carrara 1, 50019 Sesto Fiorentino FI (Italy)

David Neilson – Dipartimento di Fisica, Università di Camerino, 62032 Camerino MC (Italy)
– School of Physics, University of New South Wales, Sydney 2052 (Australia)

Riccardo Nifosì – NEST-INFM and Scuola Normale Superiore, Piazza dei Cavalieri 7, 56126 Pisa (Italy)

Alberto Parola – Dipartimento di Scienze Chimiche, Fisiche e Matematiche, Università dell'Insubria, Via Valleggio 11, 22100 Como (Italy)

Giuseppe Pastori Parravicini – NEST and Dipartimento di Fisica 'A. Volta', Università di Pavia, Via A. Bassi 6, 27100 Pavia (Italy)

Marco Polini – NEST-INFM and Scuola Normale Superiore, 56126 Pisa (Italy)

ATTILIO RIGAMONTI – Dipartimento di Fisica 'A. Volta', Università di Pavia, Via A. Bassi 6, 27100 Pavia (Italy)

EMANUELE RIMINI – Istituto per la Microelettronica e Microsistemi, CNR, Stradale Primosole 50, 95121 Catania (Italy)

MAURO ROVERE – Dipartimento di Fisica, Università "Roma Tre", Democritos National Simulation Center, Via della Vasca Navale 84, 00146 Roma (Italy)

ALEXANDER A. SHUMOVSKY – Faculty of Science, Bilkent University, 06800 Ankara (Turkey)

LETTERIA SILIPIGNI – Dipartimento di Fisica della Materia e Tecnologie Fisiche Avanzate, Università di Messina, INFM, Unità di Messina, Salita Sperone 31, 98166 Messina (Italy)

SANDRO SORELLA – INFM-Democritos, National Simulation Centre, and SISSA, 34014 Trieste (Italy)

CORRADO SPINELLA – Istituto per la Microelettronica e Microsistemi, CNR, Stradale Primosole 50, 95121 Catania (Italy)

ROBIN STINCHCOMBE – Rudolf Peierls Centre for Theoretical Physics, University of Oxford, 1 Keble Road, Oxford OX1 3NP (UK)

SAURO SUCCI – Istituto Applicazioni Calcolo, CNR, Viale del Policlinico 137, 00161 Roma (Italy)

BILAL TANATAR – Bilkent University, Department of Physics, 06533 Ankara (Turkey)

UGO TARTAGLINO – INFM-Democritos National Simulation Center, Via Beirut 2, 34014 Trieste (Italy)

ERIO TOSATTI – International School for Advanced Studies (SISSA), Via Beirut 2, 34014 Trieste (Italy)
– INFM-Democritos National Simulation Center, Via Beirut 2, 34014 Trieste (Italy)
– International Center for Theoretical Physics (ICTP), Strada Costiera 11, 34014 Trieste (Italy)

VALENTINA TOZZINI – NEST-INFM and Scuola Normale Superiore, Piazza dei Cavalieri 7, 56126 Pisa (Italy)

PATRIZIA VIGNOLO – NEST-INFM and Scuola Normale Superiore, Piazza dei Cavalieri 7, 56126 Pisa (Italy)

FABIO ZULLI – Dipartimento di Fisica and INFM, Università di Pisa, Largo B. Pontecorvo 3, 56127 Pisa (Italy)

TANYA ZYKOVA-TIMAN – International School for Advanced Studies (SISSA), Via Beirut 2, 34014 Trieste (Italy)
– INFM-Democritos National Simulation Center, Via Beirut 2, 34014 Trieste (Italy)

1

LIQUIDS

Session 1
Chair
Zehra Akdeniz

Session 2
Chair
John E. Enderby

The phase diagram of confined fluids

V. De Grandis, P. Gallo, and M. Rovere

The phase diagram of a simple fluid confined in a quenched disordered structure is studied by grand canonical Monte Carlo simulation. The matrix is modeled to represent a fractal and highly porous silica aerogel and is generated from a diffusion limited cluster-cluster aggregation procedure. The multicanonical ensemble sampling technique combined with the histogram reweighting method makes possible to build the liquid-vapor coexistence curve which is narrower than in the bulk. A shouldering on the liquid side is observed which indicates the possibility of a liquid-liquid transition smeared out by finite size effects.

The properties of fluids confined in porous materials can be drastically changed from those observed in bulk. Understanding how the phase diagram of fluids is modified by confinement is a fundamental problem in technological applications such as catalysis and phase separation of fluid mixtures [1]. Statistical mechanical and computer simulation methods have been extended to study phase transitions, in particular the liquid-gas transition, in confined fluids. The case of confinement in a single pore with well defined geometry has been studied for long time but real porous materials are generally more complex and fluids are confined in a disordered structure of an interconnected network of pores. Vycor glass and silica gels are important examples. The liquid-vapor coexistence curve of ^4He [2] and N_2 [3] confined in very dilute aerogels results to be strongly distorted with respect to bulk while the critical temperature is lower than in the bulk. The effect of quenched disorder on the phase transitions has been considered from a more general point of view referring to the so called random field Ising model [4–6], an equivalent off lattice model has been widely studied by computer simulation and theoretical approach [6–12]. This "quenched-annealed" system is based on a fluid confined in a random matrix of quenched spheres. Several integral equation theories and computer simulation studies have been performed on such model, which is particularly appropriate to describe high porosity materials, like silica xerogels. We studied by grand canonical Monte Carlo simulation the phase diagram of a Lennard-Jones fluid confined in a random porous matrix generated by a Diffusion Limited Cluster-Cluster Aggregation procedure (DLCA). By the use of the DLCA method it is possible to build a disordered structure which reproduces quite well the main experimental features of silica aerogels [13].

The DLCA algorithm starts with a collection of N identical spherical particles of diameter σ_a randomly placed in a cubic box of side L with periodic boundary conditions applied. In our simulation we took $N = 515$ particles and $L = 15\sigma_a$ in order to have a volume fraction $\eta = \frac{\pi}{6}\sigma_a^3 \frac{N}{L^3}$ of 0.08, corresponding to a porosity of 92%. Aggregation proceeds via a diffusion motion of the particles. When two clusters collide they stick together and form a new larger cluster. Finally a single cluster is generated across the simulation cell.

By restarting the procedure more times we calculated the distribution function $g(r)$ and the corresponding structure factor $S(q)$. In Figure 1 it is evident the power law behaviour of $S(q)$ in the intermediate q-range $S(q) \sim q^{-D}$ with $D \sim 1.74$.

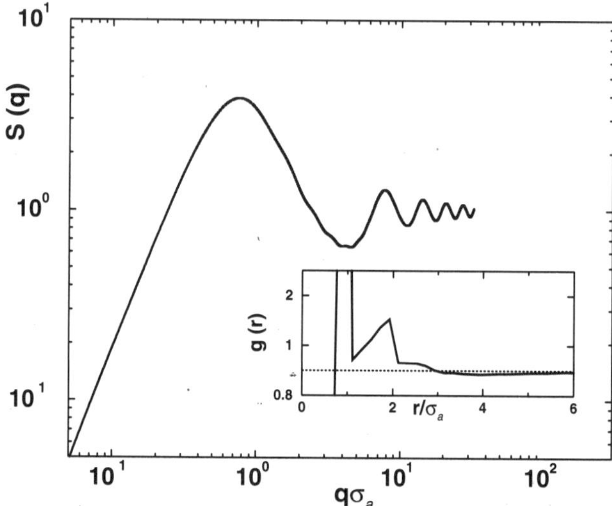

Figure 1. Log-Log plot of $S(q)$ versus $q\sigma_a$ of the aerogel generated by the LCDA algorithm (see text) containing 515 particles in a box of edge $L = 15\sigma_a$. In the inset the corresponding $g(r)$ versus r/σ_a

The cluster generated by the algorithm exhibits a fractal structure in the intermediate q-range with a fractal dimension $D = 1.74$ consistent with the value expected for silica aerogels. The radial distribution function g(r), shown in the inset, presents a strong peak at $r = \sigma_a$, associated with bonds between contacting particles and before the asymptotic limit of 1, g(r) has a minimum. The location of this minimum gives an estimate of the mean clusters size which is in our case about $4\sigma_a$.

In our simulations the confined system is a Lennard-Jones (LJ) fluid. The particles of the cluster are quenched and interact with the fluid particles by means of a purely repulsive hard sphere potential. We assume that the LJ diameter σ of the fluid particles has the same size as the aerogel hard core $\sigma = \sigma_a$. The length of the simulation box is $L = 15\sigma$. In the following Lennard-Jones units will be used: σ for lengths, ϵ for energies and ϵ/k_B for temperatures.

Monte Carlo simulation is performed in the grand canonical ensemble employing the algorithm used by Wilding [14, 15]. The main quantity to calculate is the number distribution function $P(N)$. The chemical potential is tuned at fixed temperature until a bimodal shape of $P(N)$ is obtained. The liquid-gas point of coexistence can be located in the (μ,T) by applying the equal peak weight criterion [15] (See Figure 2).

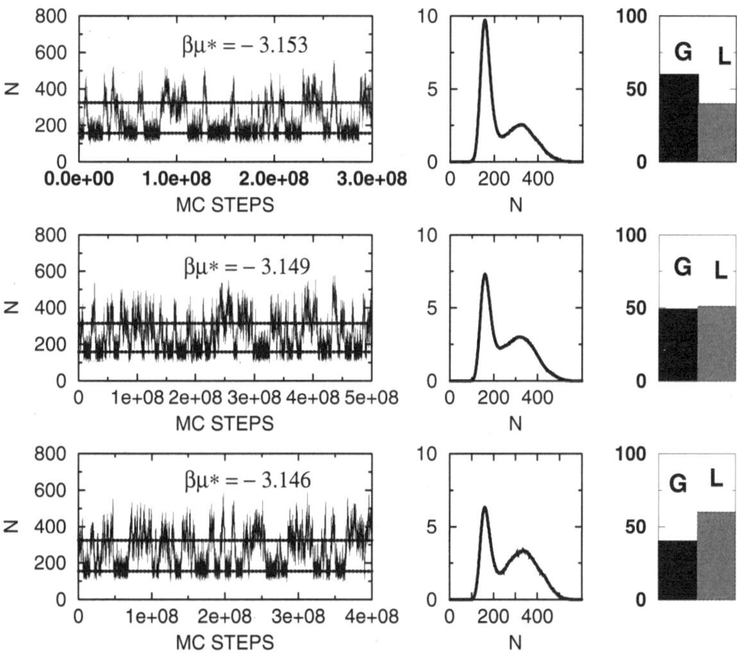

Figure 2. Fluctuations of the particle number in the subcritical region at $T = 0.95$ for three different values of $\beta\mu$ (column on the left); corresponding distribution function (column in the center) and areas below the liquid/gas (column on the right). The central row represents an almost equal area case.

In the subcritical region, where the large free energy barrier between the two phases makes difficult to sampling the phase space, we have used the Multicanonical Ensemble Sampling (MES) technique [14, 15]. This technique belongs to the class of the non Boltzmann distribution methods, similar to the umbrella sampling. A biased sampling function is introduced in order to sample with approximately the same probability the gas and the liquid configuration. The biased distribution can be estimated making use of the histogram reweighting technique [16]. Starting from the joint probability distribution of system energy and particles number at a temperature T_0 and a chemical potential μ_0

$$P(N, E|T_0, \mu_0) = \frac{e^{-\frac{H_0}{kT_0}}\mathcal{D}}{\mathcal{Z}_0} \tag{1}$$

where $\mathcal{D}(N, E)$ is the density of states and $\mathcal{Z}_0(T_0, \mu_0)$ is the grand partition function, the equivalent quantity $\tilde{P}$ for another thermodynamical state (T_1, μ_1) can be calculated as

$$\tilde{P}(N, E|T_1, \mu_1) = \frac{\mathcal{Z}_0}{\mathcal{Z}_1} e^{-(\frac{H_1}{kT_1} - \frac{H_0}{kT_0})} P(N, E|T_0, \mu_0) \tag{2}$$

The thermodynamical point (T_1, μ_1) has to be sufficiently close to (T_0, μ_0) in such a way that the statistical weight of the new configuration is not too different from the previous one. Then the bias function is obtained by

$$\tilde{P}(N|T_1, \mu_1) = \int dE\, \tilde{P}(N, E|T_1, \mu_1) \tag{3}$$

Extensive MES simulations of the order of $1 \cdot 10^9$-$2 \cdot 10^9$ are necessary to calculate the confined fluid phase diagram for a range of subcritical temperatures from $T = 0.96$ to $T = 0.79$.

In Figure 3 the distribution functions calculated with the MES method are reported for different temperatures as function of the density. On the same plot are reported the points of the coexistence located from the positions of the peaks of the $P(N)$ at each temperature. In this way the phase diagram (T vs. ρ) of the confined liquid is obtained and in the inset it is compared with the liquid-gas coexistence curve of the LJ bulk fluid.

The confined fluid phase diagram is greatly modified by the confinement in the aerogel system. The critical temperature and density are lower than in the bulk and the range of the vapor-liquid coexistence curve is much less extended. We recall that in the bulk the critical parameters are: $T_c = 1.1876$, $\rho_c = 0.3197$ and $\mu_c = -2.778$ [14].

Our findings about the gas-liquid coexistence properties are in qualitative agreement with previous computer simulation studies on Lennard-Jones fluids confined in random spheres matrices with purely repulsive adsorbent-adsorbate interactions [8, 11].

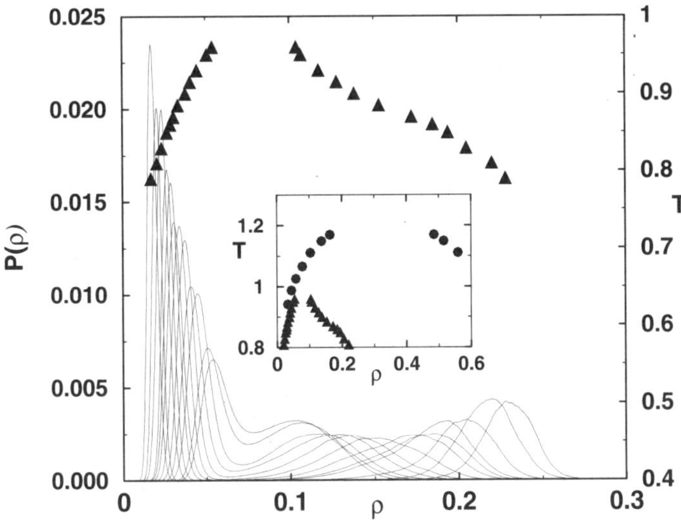

Figure 3. Distribution functions for temperatures ranging from $T = 0.79$ to $T = 0.96$ versus density (curves). On the same plot the triangles are the points of the coexistence curve of the confined fluid, reported as T versus density, obtained from the peak locations of the distribution functions. In the inset the phase diagram for the confined fluid (triangles) is compared with that of the bulk fluid [14](circles). All quantities are in Lennard-Jones units.

The phase diagram reported in Figure 3 does not present a liquid-liquid coexistence region, which has been sometimes reported, depending on the particular configuration generated for the adsorbent, random matrices of spheres [8, 11]. Our results, however, show a well defined shoulder on the liquid side boundary.

From a carefully observation of the behaviour of the distribution function in the region of the shoulder we find an asymmetry in the liquid peaks of the $P(N)$ with the signature of a crossover to the coexistence between a low density liquid and a slightly more dense liquid phase [18].

In spite of very sophisticated technique that we used in this study we cannot exclude that finite size effects be present. In particular the liquid-liquid coexistence could be smeared out by those effects. Extensive analysis with increasing sizes of the simulation box can give deeper insight on how sensitive the phase diagram is to the matrix realization.

References

[1] L. D. GELB, K. E. GUBBINS, R. RADHAKRISHNAN and M. SLIWINSKA-BARTKOWIAK, Rep. Prog. Phys. **62** (1999), 1573.

[2] A. P. Y. WONG and M. H. W. CHAN, Phys. Rev. Lett. **65** (1990), 2567.

[3] A. P. Y. WONG, S. B. KIM, W. I. GOLDBURG and M.H.W. CHAN, Phys. Rev. Lett. **70** (1993), 954.

[4] F. BROCHARD and P. G. DE GENNES, J. Phys. (France) Lett. **44** (1983), 785.

[5] P. G. DE GENNES, J. Phys. Chem. **88** (1984), 6469.

[6] M. L. ROSINBERG, In: "New Approaches to Problems in Liquid State Theory", C. Caccamo, J. P. Hansen and G. Stell (eds.), NATO ASI Series, C: Mathematical and Physical Science, **529**, Kluwer Academic Publishers, 1999, 245.

[7] W. G. MADDEN and E. D. GLANDT, J. Stat. Phys. **51** (1988), 537; W. G. MADDEN J. Chem. Phys. **96** (1992), 5422.

[8] K. S. PAGE and P. A. MONSON, Phys. Rev. E **54** (1996), R29; K. S. PAGE and P. A. MONSON, Phys. Rev. E **54** (1996), 6557.

[9] E. KIERLIK, M. L. ROSINBERG, G. TARJUS and P. A. MONSON, J. Chem. Phys. **106** (1997), 264.

[10] V. KRAKOVIACK, E. KIERLIK, M. L. ROSINBERG and G. TARJUS, J. Chem. Phys. **115** (2001), 11289.

[11] M. ALVAREZ, D. LEVESQUE and J. J. WEIS, Phys. Rev. E **60** (1999), 5495.

[12] L. SARKISOV and P. A. MONSON, Phys. Rev. E **61** (2000), 7231.

[13] A. HASMY, E. ANGLARET, M. FORET, J. PELOUS and R. JULLIEN, Phys. Rev. B **50** (1994), 6006.

[14] N. B. WILDING, Phys. Rev. E **52** (1995), 602.

[15] N. B. WILDING, Am. J. Phys. **69** (2001), 1147.

[16] A. M. FERRENBERG and R. H. SWENDSEN, Phys. Rev. Lett. **61** (1988), 2635; A. M. FERRENBERG and R. H. SWENDSEN, Phys. Rev. Lett. **63** (1989), 1195; A. M. FERRENBERG and R. H. SWENDSEN, Computers in Physics **3** (5) (1989), 101.

[17] E. KIERLIK, P. A. MONSON, M. L. ROSINBERG, L. SARKISOV and G. TARJUS, Phys. Rev. Lett. **87** (2001), 055701.

[18] V. DE GRANDIS, P. GALLO and M. ROVERE, Phys. Rev. E **70**, (2004), 061505.

Entropy revisited: The interplay between ordering and correlations

P. V. Giaquinta

I review the predictions of an entropy-based ordering criterion, exploiting an integrated measure of multiparticle density correlations, on the phase diagram of some reference models for the liquid state.

The connection between molecular interactions, correlations and phase transitions lies at the very heart of statistical-mechanical theories of classical and quantum systems. However, notwithstanding the impressive body of literature that has accumulated on this subject since the early foundations of statistical mechanics, some aspects of the overall theoretical framework are not yet altogether satisfactory. In particular, the origin and theoretical justification of first-order phase transitions at a more basic level as that provided, for instance, by the surmised relationship with the topological properties of the configuration space explored by the system still are an open question [1, 2]. Indeed, it is quite reasonable to expect some more or less direct link between topology and thermodynamics to be rooted in the entropy of the system. However, establishing such a relationship on rather general grounds is a formidable task [2]. A less ambitious program can be pursued if one turns to a statistical description of the system in terms of multiparticle density distribution functions. A useful starting point is the expansion of the excess (*i.e.*, solely configurational) entropy, S_{ex}, of a classical fluid:

$$S_{\text{ex}} = \sum_{n=2}^{\infty} S_n \, , \tag{1}$$

where the quantities S_n are partial entropies that can be calculated through a suitable re-summation of spatial density correlations involving n-particle multiplets. This expansion was first obtained for a closed system – *viz.*, in the canonical ensemble – by H. S. Green [3]. The grand-canonical version of this result is due to R. E. Nettleton and M. S. Green [4], while the formal equivalence between the canonical and grand-canonical expressions was later demonstrated by A. Baranyai and D. J. Evans [5]. Since its first appearance in the scientific literature, the mathematical formalism leading to the entropy expansion as well as its applications in the framework of statistical-mechanical theories of the liquid state have been the object of detailed analysis by a number of authors. I just mention here the generalization to quantum systems carried out by C. De Dominicis [6], and refer the reader to a few recent contributions which can help to trace the historical thread of the subject, both on the methodological and on the computational side, up to present times. [7–11].

The ensemble-invariant expression for the "pair" entropy showing up in Eq. (1) reads as:

$$S_2/N = -\frac{1}{2}k_B\rho \int [g(r)\ln g(r) - g(r) + 1]\, d\mathbf{r}\,, \tag{2}$$

where N is the (average) number of particles, k_B is Boltzmann's constant, and $g(r)$ is the radial distribution function (RDF). Truncating the expansion at the pair level already yields a very good estimate of the total configurational entropy of a liquid which can be computed once the RDF is available via theory, numerical simulation or experiment. And this has actually been the most frequent use of the entropy expansion in the past. However, it turns out that the residual part

$$\Delta S \equiv S_{ex} - S_2, \tag{3}$$

despite its minor quantitative relevance in the overall balance, conceals significant indications on the statistical thermodynamics of the system that are intimately related to the role played by higher-order ($n \geqq 3$) density correlations [12]. It is rather natural to expect the importance of this cumulative contribution to grow when the system undergoes any sort of structural transformation that may eventually end up in a real thermodynamic phase transition. In fact, the partial or full ordering of a fluid is an intrinsically cooperative phenomenon which becomes manifest through the emergence of some correlated pattern over an extended range. Maybe, the unexpected feature, at least *a priori*, is the way the corresponding information is conveyed by Δs, the so-called "residual multiparticle entropy" (RMPE). The pair entropy is, by definition, a negative quantity: the more structured the RDF (*i.e.*, the denser or cooler the fluid), the larger is its absolute value and, correspondingly, the reduction of the average number of configurational states effectively available to the system, as compared to the condition of an equally dense gas of non-interacting particles at the same temperature. The RMPE has a similar trend at low densities and at high temperatures. However, as soon as the system approaches a phase transition, its behaviour changes in a non-monotonic way, until this quantity becomes positive. As a result, resummed density correlations involving at least three particles no longer contribute to the overall decrease of accessible states that is largely set by the pair entropy. The crossover from one regime to the other has been interpreted as the underlying signature of a new condition that is being built up by the system, "forced" by compelling structural and thermodynamic constraints to exploit a more efficient arrangement on a local scale.

The vanishing of the RMPE has been found to herald in a very sensitive way the occurrence – in a proximate range of densities and/or temperatures – of a phase transition leading the system from a disordered (or, even, partially ordered) phase into a more structured one. Perhaps, the most surprising aspect is the existence of such an indication even for discontinuous phase transitions. In this respect, a paradigmatic example is offered by the freezing of hard spheres, *viz.*, point particles whose only interaction is an infinite repulsion for separations equal or smaller than a relative distance σ, the hard-core diameter. It is the entropy that drives the crystallization when hard spheres, whose internal energy is purely kinetic, occupy about 50% of the total available volume. As a matter of principle, no specific indications of the impending first-order transition need to be present in the thermodynamic properties of the fluid phase. And, actually, no anomaly whatsoever shows up in the compressibility factor of hard spheres along the fluid branch: the equation of state is found to continue smoothly into the metastable region, well beyond the freezing point, up to the random close-packing threshold. The finding that the RMPE of hard spheres vanishes almost precisely at the freezing density was, in many respects, an unexpected and provocative hint [12]. However, one could reasonably argue that it may well be a fortuitous happenstance occurring in a rather specific model. Since then, the surmise that the condition $\Delta S = 0$ does actually qualify as a general *one-phase* ordering criterion has been successfully tested against many diverse thermodynamic phenomena such as the freezing of one-component interacting fluids (hard spheres with an attractive Yukawa tail, Lennard-Jonesium, repulsive Gaussian particles) [12–17], also in two dimensions [18]; the gas-liquid phase transition [15]; the freezing and fluid-fluid phase separation undergone by binary (even non-additive) hard-core mixtures in three [19–22] as well as in two dimensions [23]; the formation of nematic and smectic mesophases in model liquid crystals [24–26]; the Ising and metal-insulator transitions in Coulomb lattice gases [27,28]; the density maximum anomaly in liquid water [29].

Indeed, a number of empirical one-phase criteria have been proposed in the past to estimate the location of the phase boundaries of liquid and solid phases [30,31]. Such criteria can be quite useful when it is not easy or straightforward to evaluate the free energies of the competing phases. However, these phenomenological rules are usually based on rather *ad hoc* assumptions which typically need some "external" input. In this respect, the zero-RMPE criterion is more general in that (i) it is a self-contained statement, based on an intrinsic feature of the configurational entropy of a fluid; (ii) it applies, without modifications, to a variety of structural and thermodynamic phenomena. Enlightening in this latter respect is the indication given by the criterion for a rather complex model of water, the four-site transferable intermolecular potential (TIP4P). Upon cooling TIP4P water at ambient pressure below 350K, one finds that the ordering threshold detected by the RMPE actually anticipates the freezing point of the model, falling very close to the temperature of maximum density of the liquid [29]. Indeed, it is below this temperature that the local structure of water starts expanding because of the formation of a statistically ordered pattern, the hydrogen-bond network, which becomes more and more persistent and extended. As is well known, this phenomenon preludes to the freezing of water into a rather open structure.

The implementation of the RMPE-based ordering criterion is straightforward since it requires just two properties of the fluid: the excess entropy and the RDF. However, it should be stressed that the criterion is by no means an alternative to the standard thermodynamic rules which state the conditions for the coexistence, at equilibrium, of distinct macroscopic phases. In fact, *a priori*, its indications on the thermodynamic conditions that herald the emergence of local order in a fluid need not coincide with the independently defined phase boundaries. Still, the criterion has proved to be a sensitive and reliable (theoretical) probe of the properties of a fluid at an intermediate level between thermodynamics and structure. The analyses carried out so far do actually show that the agreement with the phase diagram is definitely more than qualitative for dense fluids whose local structure is essentially determined by the repulsive component of the potential. This is certainly the case of hard-core models, at least in so far as the attractive interactions, if present, do not play a dominant role [32,33]. Moreover, a soft repulsion poses much less severe constraints on the nature of the spatial arrangements that can be exploited, on average, by N particles moving in a volume V at a given temperature. As a result, the predictions of the criterion for such potentials may be less successful on the quantitative side while maintaining a sound qualitative significance in the overall thermodynamic picture [17].

Upon looking retrospectively at the line of investigations on the RMPE synthetically retraced above, I believe that it is fair to conclude that the original finding on hard spheres [12] now rests on a more solid and credible basis, being supported by independent and coherent evidence on a number of different models that have been exploited in different thermodynamic scenarios. The very existence of an underlying signature in the properties of the fluid phase that can be unambiguously associated with the emergence of first-order phase transitions is, in my opinion, the most interesting and still challenging aspect of this finding.

References

[1] R. FRANZOSI and M. PETTINI, Phys. Rev. Lett. **92** (2004), 060601.

[2] L. ANGELANI, L. CASETTI, M. PETTINI, G. RUOCCO and F. ZAMPONI, arXiv:cond-mat/ET9144.

[3] H. S. GREEN, "The Molecular Theory of Fluids", North-Holland, Amsterdam, 1952.

[4] R. E. NETTLETON and M. S. GREEN, J. Chem. Phys. **29** (1958), 1365-1370.

[5] A. BARANYAI and D. J. EVANS, Phys. Rev. A **40** (1989), 3817-3822.

[6] C. DE DOMINICIS, J. Math. Phys. **5** (1962), 983-1002.

[7] S. PRESTIPINO and P. V. GIAQUINTA, J. Stat. Phys. **96** (1999), 135-167; **98** (2000), 507-509.

[8] M. PUOSKARI, Physica A **272** (1999), 509-544.

[9] J. A. HERNANDO and L. BLUM, Phys. Rev. E **62** (2000), 6577-6583.

[10] N. JAKSE and I. CHARPENTIER, Phys. Rev. E **67** (2003), 061203.

[11] S. PRESTIPINO and P. V. GIAQUINTA, J. Stat. Mech.: Theory and Exp. (2004), P09008.

[12] P. V. GIAQUINTA and G. GIUNTA, Physica A **187** (1992), 145-158.

[13] P. V. GIAQUINTA, G. GIUNTA and S. PRESTIPINO GIARRITTA, Phys. Rev. A **45** (1992), R6966-R6968.

[14] C. CACCAMO, P. V. GIAQUINTA and G. GIUNTA, J. Phys.: Condens. Matter **5** (1993), B75-B82.

[15] P. V. GIAQUINTA, G. GIUNTA and G. MALESCIO, Physica A **250** (1998), 91-102.

[16] F. SAIJA, S. PRESTIPINO and P. V. GIAQUINTA, J. Chem. Phys. **115** (2001), 7586-7591.

[17] P. V. GIAQUINTA and F. SAIJA, ChemPhysChem **6** (2005), 1768–1771; S. PRESTIPINO, F. SAIJA and P. V. GIAQUINTA, Phys. Rev. E **71** (2005), 050102 (R).

[18] F. SAIJA, S. PRESTIPINO and P. V. GIAQUINTA, J. Chem. Phys. **113** (2000), 2806-2813.

[19] F. SAIJA, P. V. GIAQUINTA, G. GIUNTA and S. PRESTIPINO GIARRITTA, J. Phys.: Condens. Matter **6** (1994), 9853-9865.

[20] F. SAIJA and P. V. GIAQUINTA, J. Phys.: Condens. Matter **8** (1996), 8137-8144.
[21] F. SAIJA, G. PASTORE and P. V. GIAQUINTA, J. Phys. Chem. **102** (1998), 10368-10371.
[22] F. SAIJA and P. V. GIAQUINTA, J. Phys. Chem. B **106** (2002), 2035-2040.
[23] F. SAIJA and P. V. GIAQUINTA, J. Chem. Phys. **117** (2002), 5780-5784.
[24] D. COSTA, F. SAIJA and P. V. GIAQUINTA, Chem. Phys. Lett. **283** (1998), 86-90; **299** (1999), 252.
[25] D. COSTA, F. MICALI, F. SAIJA and P. V. GIAQUINTA, J. Phys. Chem. B **106** (2002), 12297-12306.
[26] D. COSTA, F. SAIJA and P. V. GIAQUINTA, J. Phys. Chem. B **107** (2003), 9514-9519.
[27] M. G. DONATO, S. PRESTIPINO and P. V. GIAQUINTA, Eur. Phys. J. B **11** (1999), 621-627.
[28] S. PRESTIPINO, Phys. Rev. E **66** (2002), 021602.
[29] F. SAIJA, A. M. SAITTA and P. V. GIAQUINTA, J. Chem. Phys. **119** (2003), 3587-3589.
[30] H. LÖWEN, Phys. Rep. **237** (1994), 249-324.
[31] P. A. MONSON and D. A. KOFKE, Adv. Chem. Phys. **115** (2000), 113-179.
[32] A. A. LOUIS, Phil. Trans. R. Soc. Lon. A **359** (2001), 939-960.
[33] G. FOFFI, G. D. MCCULLAGH, A. LAWLOR, E. ZACCARELLI, K. A. DAWSON, F. SCIORTINO, P. TARTAGLIA, D. PINI and G. STELL, Phys. Rev. E **65** (2002), 031407.

The NaCl (100) surface: why does it not melt?

T. Zykova-Timan, D. Ceresoli, U. Tartaglino, and E. Tosatti

The high temperature surface properties of alkali halide crystals are very unusual. Through molecular dynamics simulations based on Tosi-Fumi potentials, we predict that crystalline NaCl (100) should remain stable without any precursor signals of melting up to and even above the bulk melting point T_m. In a metastable state, it should even be possible to overheat NaCl (100) by at least 50 K. The reasons leading to this lack of surface self-wetting are investigated. We will briefly discuss the results of calculations of the solid-vapor and liquid-vapor interface free energies, showing that the former is unusually low and the latter unusually high, and explaining why. Due to that the mutual interaction among solid-liquid and liquid-vapor interfaces, otherwise unknown, must be strongly attractive at short distance, leading to the collapse of any liquid film attempting to nucleate at the solid surface. This scenario naturally explains the large incomplete wetting angle of a drop of melt on NaCl (100).

1. Introduction

Interest is increasing toward adhesion, the structure and physics of solid-liquid interfaces, and the structure of liquid surfaces, particularly of complex and molecular systems. In order to gain more insight into these problems, there is a strong need for good case studies, to use as well-understood starting points.

One of the easiest starting points may be to study the relationship and contact of a liquid with its own solid, a clear situation where there will be no ambiguity of physical description, no uncertainty in chemical composition, no segregation phenomena, all of them complications present in the study of adhesion between different substances.

Adhesion of a liquid onto the surface of its own solid usually materializes spontaneously with temperature. Most solid surfaces are known to wet themselves spontaneously with an atomically thin film of melt, when their temperature T is brought close enough to the melting point T_m of the bulk solid. The phenomenon whereby the thickness $l(T)$ of the liquid film diverges continuously and critically as $T \to T_m$ is commonly referred to as (complete) surface melting [1,2].

There are actually a number of exceptions to this behavior. Some solid surfaces in particular remain fully crystalline as $T \to T_m$. This *surface non-melting* phenomenon, originally discovered in molecular dynamics simulations of Au(111) [3] and independently observed experimentally in Pb(111) [4], is known for the close-packed face of other metals too, such as Al(111) [5].

Here we are concerned with the surface of alkali halides, crystals well known for their unusually stable neutral (100) faces. Addressing a long time ago the NaCl (100) surface, bubble experiments by Mutaftschiev and coworkers revealed incomplete wetting of the solid surfaces by their own melt [6,7], moreover with an extraordinarily large partial wetting angle of 48°. This kind of incomplete wetting, as is physically clear, and as was demonstrated on metals surfaces [2,8], is associated with non-melting of the crystal surface.

In this paper we will review our recent theory work, where we showed by direct simulation the surface non-melting of NaCl (100) [9,10]. The reasons leading to this kind of surface non-melting are investigated. First, we will show the solid surface free energy calculated by thermodynamic integration, and see that at high temperature it drops due to a larger anharmonicity than in most other solid surfaces. Next, we will examine the surface tension of the liquid NaCl surface, we will find unusually large, owing to a surface entropy deficit, connected with the local surface short range molecular order. The solid-liquid interface free energy will finally be argued to be large, due to a 26% density difference. We will also discuss qualitatively – were hypothetically the solid-vapor interface to split into a pair of solid-liquid and liquid-vapor interfaces, with a thickness l of liquid in between – the interaction free energy $V(l)$ expected between them. This interaction is here strongly attractive at short range, leading to the collapse of any liquid film attempting to nucleate at the solid-vapor interface, and causing surface non-melting.

All quantitative results reviewed here were derived by means of molecular dynamics simulations, carried out extensively for NaCl (100) slabs. These simulations, as will be detailed below, are entirely based on interatomic potentials that Mario Tosi refined and published, together with Fausto Fumi, just over 40 years ago [12]. It is a fitting tribute to Mario's scientific perseverance, thoroughness, and general dependability, that these potentials still turn out to be so incredibly accurate, even well outside the range of temperatures for which they were constructed and tested, such a long time ago.

2. Simulations with Tosi-Fumi potentials

The high temperature properties of solid NaCl bulk and NaCl (100) slabs were studied by classical molecular dynamics (MD) simulations. NaCl was described with the potential of Tosi and Fumi [12] who accurately parametrized a Born-Mayer-Huggins form. The Coulomb long-range interactions were treated by the standard three dimensional (3D) Ewald method applied to a geometry consisting of infinitely repeated identical crystal slabs. Simulated systems generally comprised about 2000-5000 molecular units, with a time-step of 1 fs, and a typical simulation time of 200 ps. Long-range forces severely limit size and time in these simulations by comparison with the order-of-magnitude larger sizes and longer times typically affordable for systems with short range forces [8, 13]. We took explicit care to ensure that all our results are not crucially affected by small sizes, and that full equilibration was achieved in all cases. We checked that 80 Å of vacuum between repeated slabs are sufficient to prevent the interaction of a liquid slab with its own replicas [10]. Calculations were done at constant cell size and with periodic boundary conditions. Thermal expansion was taken care of by readjusting the (x, y) size of the cell at each temperature so as to cancel the (x, y) stress in the bulk solid. The theoretical thermal expansion was $4.05\times10^{-5}\,K^{-1}$, compared with $3.83\times10^{-5}\,K^{-1}$ in experiment. The theoretical bulk melting temperature T_M of NaCl was calculated by two phase coexistence to be 1066 ± 20 K. Remarkably, this Tosi-Fumi melting temperature is extremely close to the experimental melting temperature of 1073.8 K. The volume expansion at melting is about $(27\pm2)\%$, also in excellent agreement with the experimental value of 26%.

Tab. 1 lists some of the calculated thermodynamical quantities at high temperature, close to the melting point. These results provide an independent confirmation of the outstanding quality of the Tosi-Fumi description of thermodynamics of NaCl, even at very high temperatures and as we shall see even at surfaces, where it was by no way guaranteed.

3. Surface free energies and non-melting of NaCl(100)

High temperature simulations of crystalline NaCl(100) slabs directly showed the full stability of the dry, solid surface up to T_M. Moreover, a well pronounced metastability of the slab solid faces above the melting point T_M indicated a clear surface non-melting behavior. We found that a much higher ("surface spinodal") temperature $T_S \approx T_M + 150$ K needs to be reached before the crystalline NaCl (100) surface spontaneously melts. $T_S - T_M$ thus represents the maximum ideal overheating that a defect free NaCl (100) surface can theoretically sustain without becoming sponta-

	Simulation	Experiment
$T_{\rm M}$(K)	1066±20	1074
ΔV	27%	26%
L (eV/molecule)	0.29	0.29
ΔS_m (k_B)	6.32	6.38
dP/dT (kbar/K)	0.0311	0.0357
RMDS (ave.) (Å)	0.60	0.49
δ	20–24%	17–20% [11]

Table 1. High temperature properties of NaCl. $T_{\rm M}$is the melting temperature; ΔV is the volume jump at the melting point; L is the latent heat of melting; ΔS_m is the entropy variation at the melting point; dP/dT is the resulting Clausius-Clapeyron ratio at the melting point. RMSD is the averaged root mean square displacement of atoms at the melting point; δ is the RMSD over the Na–Cl distance, for the Lindemann melting criterion.

neously unstable against melting. This large overheating is quite similar in magnitude to that predicted for, e.g., Au(111) and Al (111) [8, 13].

At any temperature between $T_{\rm M}$ and $T_{\rm S}$, bulk melting can only originate through nucleation. Even though nucleation of the melt is in reality likely to proceed from a localized surface droplet or defact, it is instructive to consider a very idealized nucleus consisting of a uniform liquid film of thickness l. As a function of temperature, there will be a critical nucleation thickness $l_{\rm crit}$ decreasing from ∞ to zero between $T_{\rm M}$ and $T_{\rm S}$ [3, 13]. The free energy difference per unit area between a surface with a liquid film of thickness l and the same surface in its full crystalline state is

$$G(l) = -\rho\lambda l\left(\frac{T}{T_{\rm M}} - 1\right) + (\gamma_{\rm SL} + \gamma_{\rm LV} - \gamma_{\rm SV}) + V(l) \qquad (3.1)$$

The first term is the gain due to the melting of the solid at $T > T_{\rm M}$. Here λ is the latent heat per unit mass and ρ is the liquid mass density. The second term $\Delta\gamma_\infty \equiv (\gamma_{\rm SL} + \gamma_{\rm LV} - \gamma_{\rm SV})$, is the free energy imbalance caused by replacing the SV interface with the SL+LV pair of interfaces, supposed to be non-interacting. The last term $V(l)$ is an interface interaction, representing the correction to $\gamma_{\rm SL} + \gamma_{\rm LV}$ when the two interfaces are at close distance. This definition implies $V(+\infty) = 0$ and $V(0) = -\Delta\gamma_\infty$. At very large distance the interaction disappears. The solid-vapor crystal surface is instead recovered when the SL and LV interfaces collapse, and the liquid film disappears at $l = 0$. The non-melting condition $\gamma_{SL} + \gamma_{LV} > \gamma_{SV}$, or $\Delta\gamma_\infty > 0$ implies that here the interaction $V(l)$ is mainly attractive.

This formulation indicates three possible origins for non-melting: an exceptionally low free energy γ_{SV} of the solid surface; an unusually large free energy γ_{SL} of the solid-liquid interface; a relatively high surface tension γ_{LV} of liquid NaCl. As detailed elsewhere [9, 10], all three mechanisms are actually relevant to NaCl(100).

The solid-vapor interface free energy at the melting point was calculated through standard thermodynamic integration, using

$$\left(\frac{\partial(F/T)}{\partial(1/T)}\right)_{N,V} = E, \tag{3.2}$$

where E is the surface internal energy, extracted from simulation of the crystalline NaCl(100) slab and the corresponding bulk at increasing T from 50 K to 1200 K. The surface free energy in Figure 1 shows a large drop at high temperatures, with an increasing deviation from an effective harmonic behavior above 600 K, indicating very strong surface anharmonicity in this regime. The main source of this anharmonicity is connected with large root mean square thermal fluctuations of the surface Cl and Na ions above 20% of the Na–Cl distance, largely exceeding the canonical Lindemann values [10].

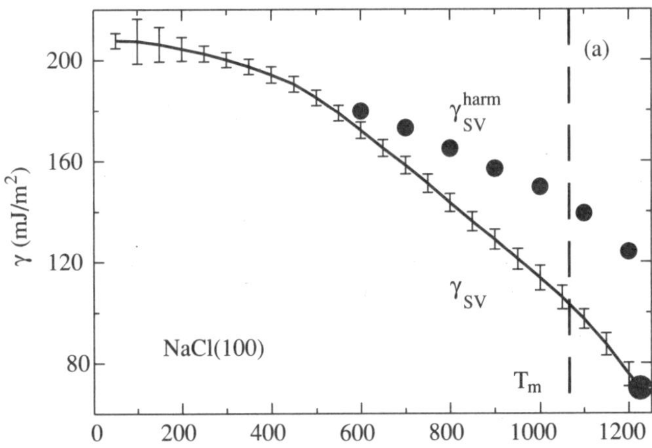

Figure 1. The solid surface free energy of NaCl(100) calculated from thermodynamic integration (circles: effective harmonic approximation).

The liquid-vapor free energy, equal to the liquid surface tension, was evaluated from simulations of liquid NaCl slabs via the standard Kirkwood-Buff formula [14]. The first thing we note in the result, shown in Figure 2

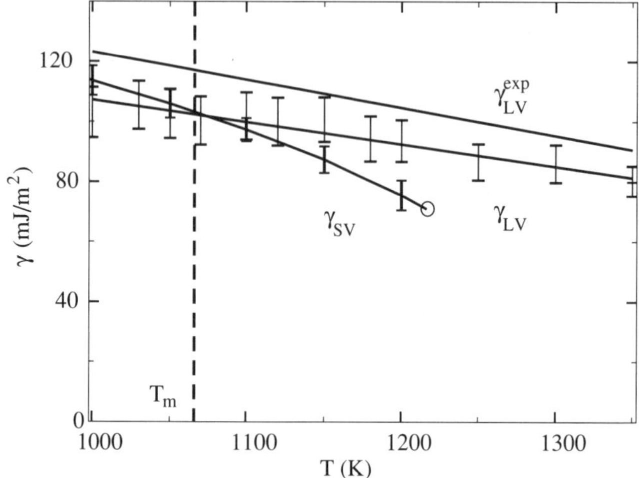

Figure 2. The calculated liquid surface free energy calculated. The NaCl (100) surface free energy is also shown in the temperature range from 1000 K to 1250 K.

is that right at T_m the solid and liquid surface free energies are essentially identical, 103±4 and 104±8 mJ/m^2 respectively. This is very unusual, and implies directly surface non-melting, because clearly $\Delta\gamma_\infty \equiv (\gamma_{SL} + \gamma_{LV} - \gamma_{SV}) > 0$. In fact, even though we did not calculate γ_{SL}, this interface free energy has no reason to be very small, owing to the large solid-liquid density difference. We independently estimated a lower bound for γ_{SL} to be 36±6 mJ/m^2 [10].

The question that remains to be explained is therefore the physical reason why the liquid surface tension is so relatively high. The liquid surface density profile, in particular, is very smooth, with none of the layering phenomena displayed by the metal surfaces (see Figure 3)

An important clue is provided by surface entropy (per unit area) $S_{\text{surf}} = -d\gamma/dT$. The temperature dependence of two surface free energies of Figure 2 shows a factor 2.6 *lower* surface entropy S_{LV} of the liquid surface compared with that of the solid surface. This liquid surface entropy deficit (SED) strongly suggests some underlying surface short range order. Short range order can in turn also explain why the surface tension is here as high as the solid surface free energy. The surface profile indicates that the order is clearly not layering: so what is it instead?

The answer we find is that charge order, already very important in bulk, plays a newer and enhanced role at the molecular liquid surface. If surface thermal fluctuations are indeed very large, we find them revealingly *correlated*. For a Na^+ ion that instantaneously moves e.g., out

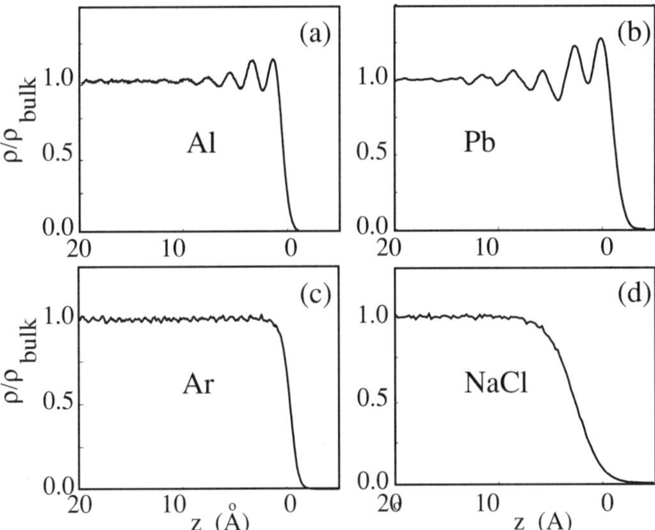

Figure 3. Surface density profile of liquid NaCl, compared with that simulated for liquid Ar, and for two liquid metals. Note that NaCl does not show layering as the metals do, and has an even smoother profile than Ar

of the surface, there is at least one accompanying Cl^-, also moving out; and vice versa. So on one hand the large fluctuations smear the average liquid vapor density profile, bridging very gently between the liquid and essentially zero in the vapor, (Figure 3). On the other hand the two body correlations, described e.g., by the the Na-Cl pair correlation function, or by its integral, the ion coordination number N, do not vanish identically in the vapor, but tend to a typical value corresponding to the NaCl molecule, (plus in fact a large concentration of dimers, Na_2Cl_2). The lack of freedom implied by the incipient molecular bond explains the entropy deficit, and the consequent large surface tension of the molten salt surface. To confirm if this is true, we recalculated γ_{SL} modifying the forces in the Kirkwood-Buff formula by removing all contributions from surface ions whose coordination is between zero and 1.3, the average NaCl vapor value, which amounts to suppress the molecular order at the liquid-vapor interface [9]. This construction, meant to provide a qualitative estimate of where would the surface tension drop if surface molecular order were absent, gives a surface tension of about 50 mJ/m^2 [9]. With a surface tension this low, surface non-melting would in fact disappear, and wetting of the solid surface by the molten salt would be complete. Hence the high surface tension of liquid NaCl can indeed be ascribed to surface molecular short range order, ultimately due to charge neutrality. This result confirms an early surmise by Goodisman and Pastor [16].

4. Interface interaction

In this short speculative section we further rationalize the results above within the phenomenological framework of Eq.(1), where besides the bare interface free energies just calculated, an interaction $V(l)$ appears. We will not present a calculation of $V(l)$, but simply discuss it on physical grounds, in the light of our new microscopic understanding gained through simulations, and calculations of interface free energies just reviewed [9]. The definition of interface interaction $V(l)$ given earlier implied $V(+\infty) = 0$ and $V(0) = -\Delta\gamma_\infty$. At very large distance the interaction disappears. At the opposite limit, when the SL and LV interfaces approach each other and merge at very close quarters, they will eventually yield the SV interface upon their collapse, when the liquid film disappears altogether at $l = 0$. The non-melting condition $\gamma_{SL} + \gamma_{LV} > \gamma_{SV}$, or $\Delta\gamma_\infty > 0$ implies here that the interaction $V(l)$ is attractive at short range.

In non-melting metal surfaces, a source of finite-range attraction was described as the result of a constructive interference between two equal-period damped density oscillations, one entering the liquid film from the solid side, the other, due to surface layering, from the vacuum side. Here, one of the two oscillations, namely that on the vacuum side, is missing, because there is no layering at the molten salt surface. At large distance the main interaction between the SL and the LV NaCl interfaces will essentially be due to electrostatic forces and to dispersion forces. The latter in particular give rise to an additional long-range interface interaction $V_{dis}(l) = H\,l^{-2}$ which is dominant at large distances. Here H is the Hamaker constant, that can be estimated through the formula $H = (\pi/12)\,C_6(n_s - n_l)(n_l - n_v) = 0.00119\,\mathrm{eV}$, where $C_6 = 72.5\,\mathrm{eV\,\AA^6}$ is the coefficient of the Lennard-Jones interaction between chlorine ions, n_s, n_l and n_v are the number of Cl^- ions per unit volume respectively in the solid, liquid and vapor phases [17]. Since the liquid density is only about 79% that of the solid, this constant is positive which implies a long range repulsion of the SL and LV interfaces. A simple estimate indicates however that $V_{dis} < H/a^2 = 0.51\,\mathrm{mJ/m^2}$, a value that makes it irrelevant in practice.

Therefore we expect the effective interaction $V(l)$ to be very weak in NaCl, everywhere except very close to zero range, $l \approx a$. Here it will suddenly turn strongly attractive, $V(0) \approx -\Delta\gamma_\infty$. The physics of this short range attraction has already been described, because it amounts to the free energy gained by replacing the two costly LV and SL interfaces, with the single and less costly SV interface.

5. Conclusions

Summarizing, the NaCl (100) surface is predicted to show non-melting and to sustain overheating up to a theoretical maximum of about 150 K above the bulk melting point. The thermodynamics of surface non-melting in alkali halides is shown to differ from that of metal surfaces, e.g. Al (111), Pb (111) or Au (111). Unlike metals, non-melting in alkali halides is not connected with liquid layering, but to molecular short range order raising the liquid surface tension, as well as to strong anharmonicty that lowers the free energy of the solid surface. It is argued moreover that the thermodynamical SL-LV interface interaction should consist mainly of a strong short-range attraction. Fresh microscopic experimental work, absent so far, is called for to check these predictions on the high temperature behavior of NaCl(100) and other alkali halide surfaces.

ACKNOWLEDGMENTS. We wrote this paper to honor Mario Tosi on his 72th birthday. The high accuracy and dependability which we found for his potentials is a tribute to his long-lasting work. This project was sponsored by the Italian Ministry of University and Research, through COFIN2003, COFIN2004 and FIRB RBAU01LX5H; and by INFM, through the "Iniziativa Trasversale Calcolo Parallelo". Calculations were performed on the IBM-SP4 at CINECA, Casalecchio (Bologna). We are grateful to E. A. Jagla for his help and discussions about NaCl (100).

References

[1] J. F. VAN DER VEEN, B. PLUIS and A. W. DENIER VAN DER GON, In: "Chemistry and Physics of Solid Surfaces VII", R. Vanselow, R. F. Howe (eds.), Springer, Heidelberg, 1988.

[2] U. TARTAGLINO, T. ZYKOVA-TIMAN, F. ERCOLESSI and E. TOSATTI, Phys. Reps. **411** (2005), 291.

[3] P. CARNEVALI, F. ERCOLESSI and E. TOSATTI, Phys. Rev. B **36** (1987), 6701.

[4] B. PLUIS, A. W. DENIER VAN DER GON, J. W. M. FRENKEN and J. F. VAN DER VEEN, Phys. Rev. Lett. **59** (1987), 2678.

[5] A. W. DENIER VAN DER GON, R. J. SMITH, J. M. GAY, D. J. O'CONNOR and J. F. VAN DER VEEN, Surf. Sci. 227, Issues 1-2, 1 March 1990, 143-149

[6] G. GRANGE and B. MUTAFTSCHIEV, Surf. Sci. **47** (1975), 723.

[7] L. KOMUNJER, D. CLAUSSE and B. MUTAFTSCHIEV, J. Cryst. Growth **182** (1997), 205.

[8] F. D. DI TOLLA, F. ERCOLESSI and E. TOSATTI, Phys. Rev. Lett. **74** (1995), 3201.

[9] T. Zykova-Timan, D. Ceresoli, U. Tartaglino and E. Tosatti, Phys. Rev. Lett. **94** (2005), 176105.
[10] T. Zykova-Timan, U. Tartaglino, D. Ceresoli and E. Tosatti, J. Chem. Phys. **123** (2005).
[11] M. A. Viswamitra and K. Jayalakshmi, Acta Crystllogr. A **28** (1972), S189.
[12] M. P. Tosi and F. G. Fumi, J. Phys. Chem. Solids **25** (1964), 45.
[13] F. di Tolla, E. Tosatti and F. Ercolessi, *Interplay of melting, wetting, overheating and faceting on metal surfaces: theory and simulation*, In: "Monte Carlo and Molecular Dynamics of Condensed Matter Systems", K. Binder, G. Ciccotti (eds.), Italian Physical Society, Bologna, 1996, 345.
[14] N. H. March and M. P. Tosi, "Atomic dynamics in liquids", Macmillan, London, 1976.
[15] T. Zykova-Timan, U. Tartaglino, D. Ceresoli, W. Sekkal-Zaoui and E. Tosatti, Surf. Sci. **566-568** (2004), 794.
[16] R. W. Pastor and J. Goodisman, J. Chem. Phys. **68** (1978), 3654.
[17] J. N. Israelachvili, "Intermolecular and Surface Forces", Academic Press, San Diego, 2nd edition, 1991, 177.

2

ELECTRONIC STATES

Session 1
Chair
Emanuele Rimini

Session 2
Chair
Alfonso Baldereschi

Chemical and morphological characterization of annealed sub-stochiometric silicon oxide layer by energy filtered transmission electron microscopy

C. Spinella and E. Rimini

Annealed sub-stoichiometric silicon oxide layers, deposited by plasma enhanced chemical vapor deposition, was analyzed by electron energy loss spectroscopy and energy filtered transmission electron microscopy. The composition of the layer was evaluated by the Si and O electron energy ionization losses. The Si clusters, formed by the high temperature anneal and embedded in a oxide host, were detected using energy selected bright field imaging with an energy loss tuned to the value of the Si bulk plasmon. The clustered Si volume fraction in the central region of the layer was deduced from a fit to the experimental electron energy loss spectrum using a theoretical approach to calculate the dielectric function of a system of spherical particles of equal radii, located at random in a host material. The adopted methodology demonstrates that the clustered Si concentration is only one half of the initial excess Si present in the layer. The depth clustered Si concentration profile is uniform on the central region of the layer whilst it significantly reduces close to the interfaces with the silicon substrate and with the thin surface oxide layer.

1. Introduction

Si nanodots (ND) have been extensively investigated over the last decade, due to their peculiar physical properties, of relevance also for application in microelectronics [1, 2] and photonics [3]. Specifically, in the non-volatile memory (NVM) devices technology, the use of Si nanocrystals embedded in a SiO_2 layer as storage nodes has emerged, over the last years, as a very important alternative to conventional floating gates, because of the high reliability associated with the discrete-trap structures and for the opportunity to realize multibit devices [4, 5]. In the field of optoelectronics, Si ND have attracted in the last years the interest of the scientific community as promising materials for the fabrication of a Si-based light source. Indeed, due to quantum confinement effects, Si ND are characterized by an energy band gap which is enlarged with respect to bulk Si, and an intense room temperature photoluminescence can be obtained in the visible-near infrared range [6, 7]. For both applications (nonvolatile memory devices or silicon based light source) the control of the Si ND mean radius and density play an important role in the final device performance. Among the several methods proposed to synthesize the Si dots the one based on high temperature anneal of sub-stoichiometric silicon oxide (SiO_x) films, prepared by plasma enhanced chemical vapor deposition, is demonstrating particularly effective. It allows the control of the Si dot size by playing on the annealing temperature as well as on the silicon excess in the SiO_x film [7]. A key point for a full understanding of the electrical and optical properties of this system is the availability of a clear picture of its structure and morphology and their dependence on deposition parameters and thermal annealing, i.e. the knowledge of the phase diagram of the system. Under this respect, several techniques have been employed to characterize Si ND, namely transmission electron microscopy (TEM) [7–10], X-ray diffraction [10], and Raman spectroscopy [11]. All of these techniques provide some estimations of the Si ND mean radius. In particular, very reliable determinations of the Si ND size have been obtained by TEM analysis (from dark field or high resolution measurements), but this technique is unable to give a full quantitative picture of the system, as for instance the Si ND density. Furthermore all of these techniques are almost blind to the presence of amorphous Si clusters. Very reliable determinations of the cluster size distribution have been recently obtained using energy filtered transmission electron microscopy (EFTEM) bright field images, with electron energy loss tuned to the value of the Si bulk plasmon (17 eV). The method overcomes the problem related to the visibility of amorphous silicon dots [12]. The quantitative determination of clustered silicon amount from dot size distributions, however, requires the accurate measurement of the local thickness of the analyzed sample and moreover an extremely thin portion of the layer is analyzed [12]. These problems can be solved by coupling the analysis of the EFTEM images with the acquisition and interpretation of the electron energy loss spectrum (EELS) taken on the same region. Indeed many theoretical models have been developed and refined to describe the different contributions to the energy losses of energetic electrons interacting with nanometer solid spheres, in vacuum or embedded in a dielectric host, regarding to the different cluster geometry, their density, and to the electron probe configuration (quantum probe or parallel beam) [13–19]. In this paper we illustrate the methodology we used to quantify the volume fraction of clustered silicon in sub-stoichiometric oxide layer based on the combination between EFTEM imaging and electron energy loss spectrum (EELS) analysis in cross-sectional configuration. The methodology allows us to determine the average radius of the Si dots and the SiO_x stoichiometry from the EFTEM micrographs, and the clustered Si concentration from the EELS analysis based on the theoretical approach proposed by Barrera and Fuchs [18].

2. Experimental

A sub-stoichiometric silicon oxide film, 70 nm thick, was deposited by plasma enhanced chemical vapor deposition (PECVD) on (100) silicon wafer at a substrate temperature of 400 °C by using a mixture of SiH_4 and N_2O at a pressure of 2.8 Torr. The SiH_4 and the N_2O flows were kept constant to the values of 60 sccm and 155 sccm, respectively. After deposition, the sample was annealed at 1100 °C for half hour, under N_2 flux, and thinned in cross-sectional configuration, by mechanical lapping and argon ion milling. The sample was then analyzed using a JEOL JEM 2010F electron microscope, operating at an acceleration voltage of 200 kV and equipped with a *Gatan* electron energy loss imaging filter. The apparatus allowed us to get electron energy loss spectrum with an energy resolution of about 0.7 eV and energy filtered TEM (EFTEM) micrograph of the analyzed sample.

The EFTEM chemical analysis of elemental Si and O employed the three windows background subtraction method [20]. The thinned specimen was analyzed in regions whose thickness varied only weakly (less than 8%) moving from the bulk silicon to the film surface. During analysis the sample was tilted $\sim 10°$ off the [110] silicon zone axis, around the $< 001 >$ direction, in order to avoid electron axial channeling through the substrate, while maintaining a vertically aligned layer surface. To obtain the energy filtered Si maps, the Si_L 99 eV electron energy loss edge was employed, with a post-edge 13 eV window placed at 109 eV and pre-edge windows placed at 76 and 89 eV, respectively. For the energy filtered O maps we used the O_K 532 eV electron energy loss edge, with a post-edge 15 eV window placed at 544 eV and pre-edge windows placed at 503 and 519 eV, respectively. Exposure times of $\sim$ 5 s were adopted to obtain each energy filtered image. To detect the Si dots we used energy selected bright field imaging with an energy loss tuned to 17 eV, with a window of 4 eV. Due to the energy difference between the Si plasmon signal and the SiO_2 bulk plasmon signal (22.5 eV), Si dots and the oxide host are clearly distinguished in the TEM image.

3. Results and discussion

Figure 1(a) shows the cross sectional TEM micrograph of the SiO_x layer annealed at 1100 °C for half hour under N_2 flux. The contrast is not uniform and a brighter surface layer (10 nm thick) is detected on top of the SiO_x layer. The Si and O maps, determined by using the three windows background subtraction method [20], are shown in Figs. 1(b) and 1(c), respectively. Both maps exhibit a quite uniform contrast except in the surface region, already detected in Figure 1(a), where the Si signal intensity decreases, Figure 1(b), whilst the O one increases, Figure 1(c).

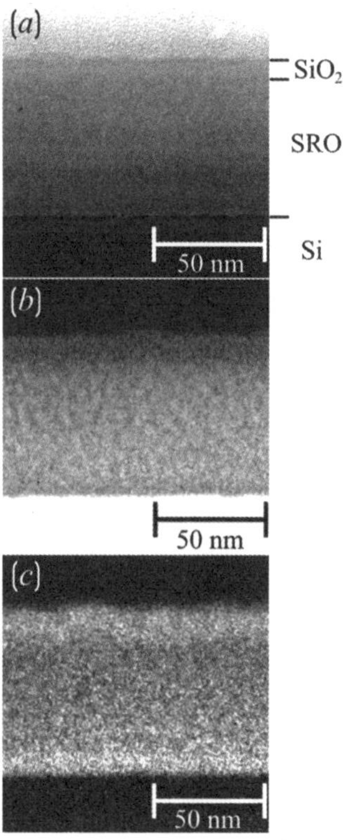

Figure 1. (a) TEM cross-sectional micrograph of a SiO_x layer annealed at 1100 °C for half hour under N_2 flux. (b) Silicon and (c) oxygen maps obtained by EFTEM of the same region.

Figure 2 shows the one-dimensional Si and O intensity profiles across the SiO_x layer obtained from the Si and O maps shown in Figs. 1(b) and 1(c), respectively. Both profiles are normalized to their maximum values and are plotted in arbitrary units. The presence of the O-rich surface layer is still evident. Convergent beam energy dispersion x-ray spectroscopy (EDS) indicates that this layer is a stoichiometric silicon oxide film, presumably formed during the high temperature anneal. The Si distribution is not uniform over all the SiO_x thickness and it smoothly decreases about 10 nm below the interface with the surface oxide film. In addition, the O profile decreases from the substrate up to the interface with the thin surface oxide film, with a total diminution of about 20%. Being the bulk Si concentration equal to 5×10^{22} atoms/cm^3, and the O one in the SiO_2 surface layer equal to 4.6×10^{22} atoms/cm^3, we can estimate an average Si and O concentration in the SiO_x film, equal to $C_{\mathrm{Si}} = 3.1 \times 10^{22}$ atoms/cm^3 and $C_{\mathrm{O}} = 3.7 \times 10^{22}$ atoms/cm^3, respectively. The silicon excess concentration, C_{excess}, defined as the concentration of Si atoms exceeding the value corresponding to the SiO_2 stoichiometry, amounts in our sample to $C_{\mathrm{excess}} = C_{\mathrm{Si}} - C_{\mathrm{O}}/2 = 1.25 \times 10^{22}$ atoms/cm^3, with a corresponding excess volume fraction equal to $f_{\mathrm{excess}} = 0.25$.

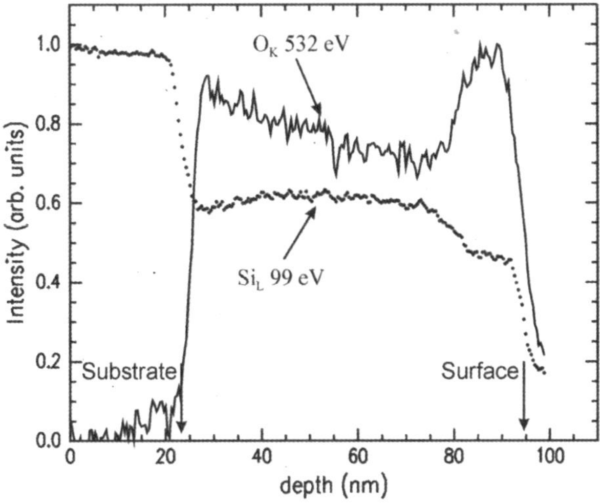

Figure 2. One dimensional Si (dotted curve) and O (full line) intensity profiles from the EFTEM Si and O maps shown in Figure 1(b) and 1(c), respectively.

The Si clusters formed as a consequence of the high temperature anneal, were detected by EFTEM images with the energy loss tuned to the value of the Si bulk plasmon (17 eV). The energy resolution of our spectrometer does not allow to take into account for the small energy shift of the plasmon resonance associated with the nanometer dimension of the clusters [21]. A typical micrograph is shown in Figure 3(a); the white spots are the Si clusters. They are distributed quite uniformly below the thin oxide surface layer and are characterized by a narrow size distribution with an average radius equal to 2 nm. This technique is particularly effective to detect Si dots since it is sensitive to the presence of silicon clusters independently of their particular phase, crystalline or amorphous, or orientation with respect to the direction of the incident electron beam. This is demonstrated by the micrographs shown in Figure 3 which refer to the same sample region analyzed by conventional high resolution TEM (Figure 3(b)) (HRTEM) or by Si plasmon EFTEM (Figure 3(c)), respectively. Although the high spatial resolution of the HRTEM technique detects the Si lattice plane spacing within an individual Si nanocrystal, the EFTEM analysis is able to reveal the presence of Si dots also in regions where they are not visible in the HRTEM micrograph (see the area marked as a circle in both images).

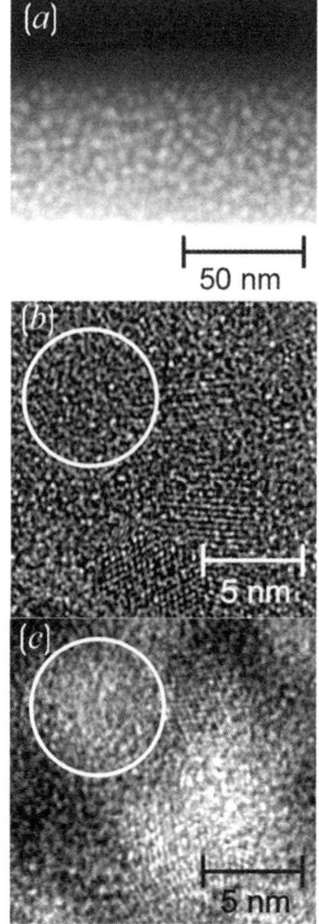

Figure 3. (a) EFTEM cross-sectional micrograph (energy loss tuned to 17 eV) of the same sample shown in Figure 1. (b) HRTEM image of a selected sample area. The (111) Si lattice planes are well visible in some dots. (c) EFTEM micrograph (energy loss tuned to 17 eV) of the same shown in (b). Note the presence of a bright area (marked by the circle) associated with a nanometer Si dot not revealed in the HRTEM micrograph.

Further information on the structure of our sample can be extracted from the analysis of the electron energy loss spectrum in the 0 – 35 eV range plotted in Figure 4. It was taken by converging the incident electron beam over a circular area, 40 nm in diameter, centered at a distance of about 30 nm from the interface with the substrate.

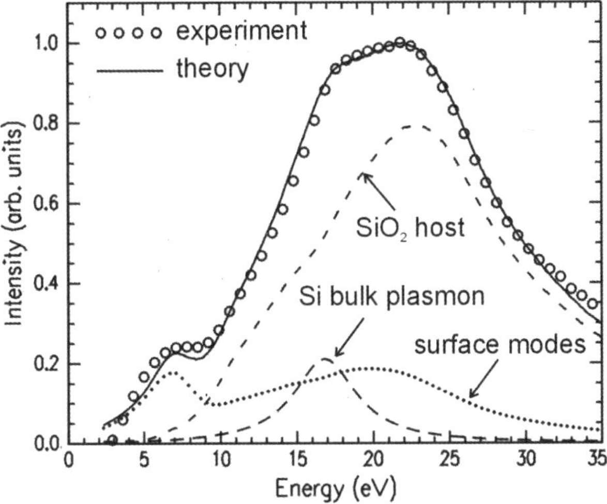

Figure 4. Electron energy loss spectrum of the SiO_x layer annealed at 1100 °C for half hour (open circles). The full line is a fit to the experimental spectrum using the model discussed in the text. Dotted-dashed, dotted, and dashed curves are the contribution to the total energy loss intensity coming from the SiO_2 host, from the surface or interface modes, and from the Si nanoclusters, respectively.

This spectrum is characterized by a very broad peak which is the result of the electron inelastic interaction with the Si nanoclusters, the surrounded host, and the Si/host interfaces. In order to describe these effects and to relate them to the clustered Si volume fraction and to the average dot size, we follow the theoretical approach developed by Barrera and Fuchs [18]. The model calculates the energy loss rate for electrons passing through a system of spherical particles of equal radii, located at random in a host material. In our case we assume that the presence of a high concentration of Si spherical clusters points the occurrence of a phase separation between silicon and SiO_2. The system is then describe by Si spheres of equal radii (2 nm) embedded in a SiO_2 host. Under these conditions, the spectral representation of the effective inverse dielectric function $\varepsilon^{-1}(k, \omega)$ of such a layer, for any finite wave vector k and frequency ω, is given by [18]:

$$\begin{aligned}\varepsilon^{-1}(k,\omega) = & \frac{1-f_c}{\varepsilon_{SiO_2}}\left[1-\frac{f_c}{1-f_c}\sum_s\left(\frac{1}{n_s}-1\right)C_s\right] \\ & +\frac{f_c}{\varepsilon_{Si}}\left[1-\sum_s C_s\right]+f_c\sum_s\frac{C_s/n_s}{n_s\varepsilon_{Si}+(1-n_s)\varepsilon_{SiO_2}}\end{aligned} \tag{3.1}$$

where f_c is the volume fraction of clustered Si, ε_{Si} and ε_{SiO_2} are the dielectric functions of Si and SiO_2, respectively, whilst n_s and C_s are the depolarization factors and the strengths of each surface mode (s^{th} order), respectively. In Eq. (3.1) the first two terms are the contributions coming from the host and the spheres, respectively. The presence of the negative coefficients within the square brackets takes into account for the depression of the bulk modes of both the host and the clusters due to the finite size of the spheres (*Begrenzung* effect) [22]. The third term in Eq. (3.1) corresponds to the contribution of the surface or interface modes of the system. The location of these modes in the energy loss spectrum is determined by the zeros of the denominators $n_s\varepsilon_{Si} + (1 - n)\varepsilon_{SiO_2}$. The properties of the surface modes (n_s and C_s) depend on interactions between spheres and are determined by the eigenvalues and eigenvectors of a real symmetric matrix $H_{ll'}$:

$$H_{ll'} = \frac{l}{2l+1}\delta_{ll'} + 3f_c\sqrt{ll'/(2l+1)(2l'+1)}\frac{(l+l')!}{l!(l')!}\left(\frac{1}{2}\right)^{l+l'-2}\frac{j_{l+l'-1}(2ka)}{2ka} \tag{3.2}$$

where $j_\nu(x)$ is the spherical Bessel function of order ν, and a is the sphere radius ($a = 2$ nm in our case). The depolarization factors n_s are the eigenvalues of this matrix, whilst the strength C_s of the surface modes depends on the correspondent eigenvectors U_{sl} by the following relationship:

$$C_s = 3\sum_{ll'}\sqrt{ll'(2l+1)(2l'+1)}(ka)^{-2}j_l(ka)j_{l'}(ka)U_{ls}U_{l's} \tag{3.3}$$

In our calculation we extended all the summations up to an upper cutoff value equal to $l_{\max} = 9$. Indeed, for a cluster radius of 2 nm the strength of the surface modes strongly reduces and becomes negligible as the index s approaches this cutoff value.

The probability $F(E)$ per unit path length, per unit energy, and for all the diffusion angles, of scattering with energy loss E will be proportional to the quantity $\Xi(E)$ given by:

$$\Xi(E) = \frac{1}{\pi}\int_0^{Q_c}\frac{QdQ}{k^2}\,\mathrm{Im}\left[-\varepsilon^{-1}(k,\omega)\right] \tag{3.4}$$

where $Q^2 = k^2 - (\omega/v_I)^2$, v_I being the electron incident velocity, whilst Q_c is related to the largest wave vector k_c for which the bulk plasmon is a well defined excitation [23] ($k_c \sim 0.25$ nm^{-1} in our experimental conditions). Actually, since we are interested only to the relative intensities of all the above mentioned contributions, the final results depend only very weakly on the particular choice of k_c. Eq. (3.4) was used to fit the experimental spectrum of Figure 4. For the host we adopted a dielectric function determined by *Kramers-Kronig* analysis [23] of the experimental energy loss spectrum of a reference PECVD SiO_2 layer, whilst for silicon we used a *Drude* free-electron dielectric function [23] given by $\varepsilon_{Si} = 1 - \omega_p^2 / [\omega(\omega + i\gamma)]$, where $\hbar\omega_p = 17$ eV and $\hbar\gamma = 5$ eV. The latter parameters were determined by a fit of the bulk silicon substrate electron energy loss spectrum. Then the only free parameter we needed to fit the SiO_x electron energy loss spectrum was the clustered Si filling fraction f_c. The agreement of the calculation (continuous line in Figure 4) with the experimental data (open circles) is excellent over all the explored energy range. The contributions to the energy loss coming from the host (dotted-dashed line), from the excitation of the interface modes (dotted curve), and from the Si dots (dashed line), are also plotted in Figure 4. From this fitting procedure we find $f_c = 0.12$ (corresponding to a concentration equal to 6×10^{21} Si/cm^3), a value which is significantly lower than the volume fraction $f_{excess} = 0.25$ of the Si excess (i.e. 1.25×10^{22} Si/cm^3). The difference is well above our experimental accuracy and, to our knowledge, this result represents the first clear evidence that Si agglomeration in a sub-stoichiometric silicon oxide film is not complete even after high temperature anneal (1100 °C) for a relatively long time (half hour). A comparable amount of Si is dissolved in the host as aggregates whose radius is presumably below 0.5 nm.

In order to evaluate the concentration profile of the clustered silicon we used the EFTEM micrograph of Figure 3(a). At any depth the signal intensity was taken by the integral of the EELS spectrum in the 4 eV energy window centered at 17 eV. According to the proposed approach this intensity is proportional to the clustered silicon volume fraction, f_c, (at least in the range 0 –0.15) and depends only very weakly on the average dot size. Under these conditions we can normalized the intensity measured at the center of the layer to the clustered silicon concentration (6×10^{21} Si/cm^3) determined from the analysis of the EELS spectrum in Figure 4. The result we obtain is the clustered silicon concentration as a function of the depth shown in Figure 5. It is quite uniform in the central region of the layer while it decreases approaching the interfaces with the silicon substrate and with the surface thin oxide layer. The decrease at the interface with the substrate could be attributed to the coalescence of a certain amount of the silicon clusters with the bulk silicon substrate which acts as an infinite grain boundary. Conversely, the reduction toward the surface is probably associated with the oxidation process induced by the high temperature anneal. A continuos thin surface oxide film is formed with a shrink of the silicon clusters close to this interface. Indeed, the reduction of the silicon cluster sizes at the interface with the surface oxide film is an effect well visible in the micrograph of Figure 3(a).

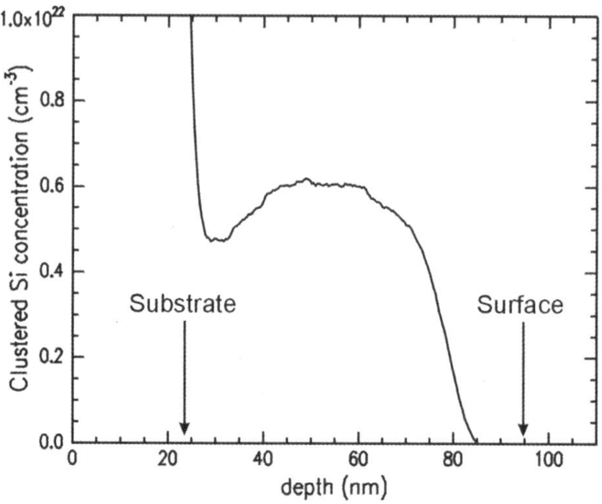

Figure 5. Clustered silicon concentration as a function of depth obtained from the one dimensional intensity profiles extracted from the micrograph shown in Figure 3(a).

4. Conclusions

In conclusion we have demonstrated that the combination of EFTEM imaging and EELS analysis provides an accurate and reliable determination of the chemical and structural properties of annealed SiO_x films. The analysis allows us to evaluate the one-dimensional clustered silicon concentration profile and to investigate the role of the surface oxidation during the high temperature anneal. The proposed methodology offers the opportunity to quantitatively investigate the kinetics of the Si agglomeration process and gives new insight for a full understanding of the thermodynamic properties of the Si/SiO_2 composite system.

References

[1] K. Yano, T. Ishii, T. Sano, T. Mine, F. Murai, T. Hashimoto, T. Kobayashi, T. Kure and K. Seki, Proc. IEEE **87** (1999), 633.

[2] L. Guo, E. Leobandung and S. Y. Chou, Science **275** (1997), 649.

[3] T. GEBEL, L. REBOHLE, J. ZHAO, D. BORCHERT, H. FRÖB, J. V. BORANY and W. SKORUPA, Mater. Res. Soc. Symp. Proc. **638** (2001), F18.1.

[4] C. GERARDI, G. AMMENDOLA, M. MELANOTTE S. LOMBARDO, I. CRUPI and E. RIMINI, Proceedings of European Solid-State Device Research Conference (2002), 475.

[5] S. LOMBARDO, C. GERARDI, D. CORSO, G. AMMENDOLA, I. CRUPI and M. MELANOTTE, Proceedings of Nonvolatile Semiconductor Memory Workshop (2003), 105.

[6] F. PRIOLO, G. FRANZÒ, D. PACIFICI, V. VINCIGUERRA, F. IACONA and A. IRRERA, J. Appl. Phys. **89** (2001), 264.

[7] F. IACONA, G. FRANZÒ and C. SPINELLA, J. Appl. Phys. **87** (2000), 1295.

[8] F. GOURBILLEAU, X. PORTIER, C. TERNON, P. VOIVENEL, R. MADELON and R. RIZK, Appl. Phys. Lett. **78** (2001), 3058.

[9] T. INOKUMA, Y. WAKAYAMA, T. MURAMOTO, R. AOKI, Y. KURATA and S. HASEGAWA, J. Appl. Phys. **83** (1998), 2228.

[10] J. G. ZHU, C.W. WHITE, J. D. BUDAI, S. P. WITHROW and Y. CHEN, J. Appl. Phys. **78** (1995), 4386.

[11] T. SHIMIZU-IWAYAMA, K. FUJITA, S. NAKAO, K. SAITOH, T. FUJITA and N. ITOH, J. Appl. Phys. **75** (1994), 7779.

[12] F. IACONA, C. BONGIORNO, C. SPINELLA, S. BONINELLI and F. PRIOLO, J. Appl. Phys. **95** (2004), 3723.

[13] R. H. RITCHIE, Phys. Rev. **106** (1957), 874.

[14] K. S. SINGWI and M. P. TOSI, "Solid State Physics", vol. 36, H. Ehrenreich, F. Seitz, D. Turnbull (eds.), Academic Press, 1981, p. 177.

[15] N. BARBERAN and J. BAUSSELS, Phys. Rev. B **31** (1985), 6354.

[16] P. M. ECHENIQUE, J. BAUSSELS and A. RIVACOBA, Phys. Rev. B **35** (1987), 1521.

[17] A. RIVACOBA, N. ZABALA and P. M. ECHENIQUE, Phys. Rev. Lett. **69** (1992), 3362.

[18] R. G. BARRERA and R. FUCHS, Phys. Rev. B **52** (1995), 3256.

[19] J. M. PITARKE, J. B. PENDRY and P. M. ECHENIQUE, Phys. Rev. B **55** (1997), 9550.

[20] O. L. KRIVANEK, M. KUNDMANN and X. BOURRAT, Mat. Res. Soc. Symp. Proc. vol. 332 (1994), 341.

[21] M. MITOME, Y. YAMAZAKI, H. TAKAGI and T. NAKAGIRI, J. Appl. Phys. **72** (1992), 812.

[22] M. SCHMEITZ, J. Phys. C **14** (1981), 1203.

[23] R.F. EGERTON, In: "Electron Energy-Loss Spectroscopy in the Electron Microscopy", second Edition, edited by Plenum Press, New York, 1996.

The color of intrinsically fluorescent proteins: a density functional study

R. Nifosì and V. Tozzini

Fluorescent Proteins (FPs) are a large class of proteins extracted from a variety of sea organisms (jellyfish, sea anemones and corals) that share the same peculiar fold and contain a chromophore, formed by the cyclization of three amino acids in the protein sequence. The first member of this family to be discovered, and the most famous one, is the Green Fluorescent Protein (GFP) from jellyfish *Aequorea Victoria*. The use of GFP and its mutants in fluorescence microscopy of living cells has lead to great advances in molecular biology, and it is expected that further and important advances from application of the newly discovered FPs are around the corner. FPs feature a wide range of spectral properties, covering, in absorption or emission, the whole visible spectrum plus the near UV and IR. The origin of this diversity can be associated with the different structure of the chromophore and of its close environment, although this issue is not completely clarified since only fragmentary studies on this family of chromophores are available.

This paper is aimed at clarifying the relationship between the chromophore structure and the spectroscopic properties within the FP family. Based on Density Functional Theory calculations, we report a systematic study of the whole series of chromophores of the FP family, including new mutants of GFP and recently discovered far-red absorbing homologues DsRed and Kaede. From the comparison of calculations with experimental data we deduce some rules on the relationship between structure and color, which explain the diversity within the FP family.

1. The FP family

The first member of the fluorescent protein (FP) family was extracted from jellyfish *Aequorea Victoria* in 1962 [1]. Thirty years later, with the cloning of its gene and the discovery that expression of the gene in heterologous (i.e. non-jellyfish) organisms led to fluorescence, it was

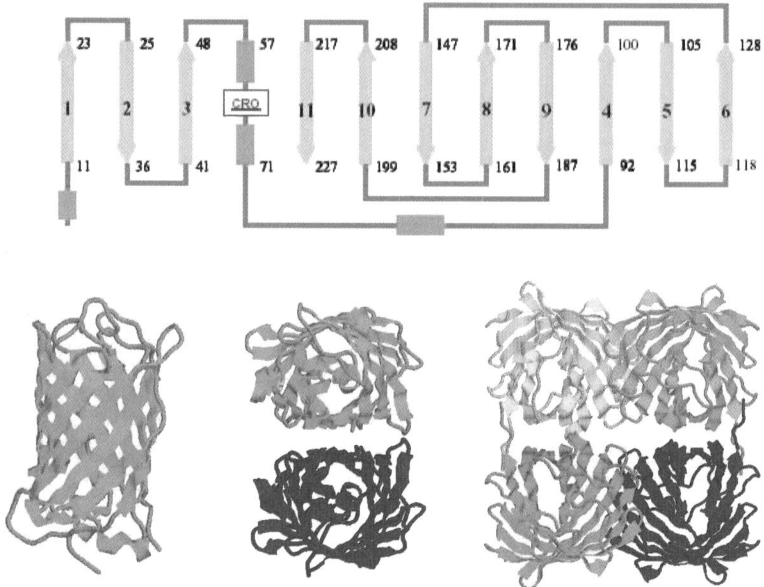

Figure 1. GFP structure and fold. Top: scheme of the secondary structure of GFP, constituted by eleven β strands (cyan) and four segments of α-helix (green). Bottom: Tertiary and quaternary structure. The eleven β strands form a β-can capped by the α-helix segments. The chromophore is buried in the middle of the can. The quaternary structure can be dimeric (such as in GFP) or tetrameric (such as in DsRed).

recognized that FPs were going to have a great impact in molecular biology. Their peculiarity stems from their fluorescence, which is bright and *intrinsical*, i.e. it does not need any external cofactor, from their stability with respect to environmental changes (temperature, pH, quenchers), and from the fact that they can be genetically fused to other proteins and expressed in living cells without significantly altering the cell physiology. For these reasons they are almost ideal fluorescent tags to monitor protein trafficking and gene expression [2, 3]. These circumstances have led in the last decade to the development of a new branch of biotechnologies based on the Green Fluorescent Protein from *Aequorea Victoria* (avGFP or simply GFP) and to the engineering of more than one hundred artificial mutants with enhanced fluorescence, different absorption and emission wavelengths and peculiar sensitivity to environmental conditions [4].

Only recently it was recognized that GFP was just the tip of the iceberg. A variety of homologues (about thirty, up to now) of GFP were extracted from different sea organisms (mainly corals or sea anemones from the *Anthozoa* group) and cloned [5, 6]. Unlike GFP, which is used by *Aequorea* jellyfish as phototransducer for the blue luminescence produced by another protein, *Aequorin*, their function in *Anthozoa* seems to be mainly photoprotection from UV radiation. Although the homology with GFP can be as low as about 30%, these proteins share the same "β-can" tertiary fold (see Figure 1). The quaternary structure can be tetrameric, as in most of the *Anthozoa*, or dimeric (see Figure 1), as in GFPs, which can assume also a monomeric form in diluted solutions and/or in presence of specific mutations.

The absorption-emission spectra of *Anthozoa* FPs lie in a much larger range than that covered by GFP mutants, extending especially in the far-red region. This circumstance, besides giving rise to the wonderful coloration of corals and sea anemones and pens, is particularly important for biotechnological applications, since the red and far-red region of the spectrum is not covered by GFP mutants.

In this paper we shall discuss the relationship between color and the structure of the chromophore and the active site. Density Functional Theory (DFT) and Time Dependent DFT studies on the chromophores are reported. We shall restrict to those proteins in the FP family whose chromophore structure is known. Table 1 reports the characteristics of the FPs and their chromophores. Most of them have been crystallized, so that the chromophore structure is known from direct observation. The chromophore structure of Kaede is based on the stoichiometry of the reaction involved in the protein maturation an on mass spectroscopy of the products [17] (see below for a description of the chromophore formation). Similar analyses are available also for asCP and zFP538 [12, 18, 19], but they lead to ambiguous conclusions on the chromophore structure. Thus they are not included in the our study. Some of the proteins included in Table 1 (asCP562 and Rtms5) are only very weakly fluorescent in their native form and should be more properly called chromo-proteins (CP) [13]. However, it was shown that they can be converted to a fluorescent form by irradiation at specific wavelengths or by specific mutations [20].

2. Chromophore formation and protein maturation

As it is apparent from Table 1, a great variety of colors spanning all the visible spectrum results from the diversity of the chromophore structures. In all cases the chromophore forms spontaneously in the post-translational phase, after the protein folding, through the cyclization of three subsequent amino acids located in the inner part of the β-can, result-

name	pdb code [ref.]	Organism, mutation	Backb. fragm.	chromophore structure	λ_{exc} (nm)	λ_{em} (nm) (QY)
BFP	[7]	*Æ. Victoria* Y66F	NO	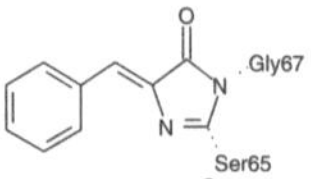	355	427 ($\sim$ 0.01)
BFP	1BFP [8]	*Æ. Victoria* Y66H	NO	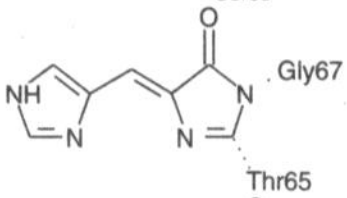	382	448 ($\sim$ 0.3)
CFP	1OXD [9]	*Æ. Victoria* Y66W	NO	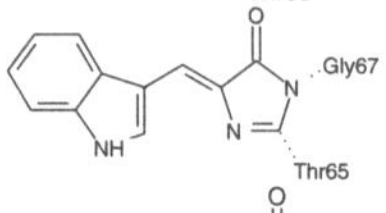	434 452	477 505 ($\sim$ 0.4)
GFP	1GFL [10]	*Æ. Victoria*	NO	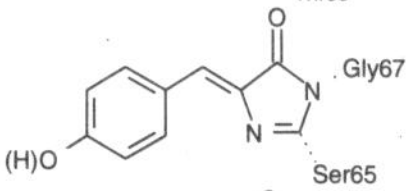	395 475	509 (0.75)
YFP	1YFP [11]	*Æ. Victoria* T203Y	NO	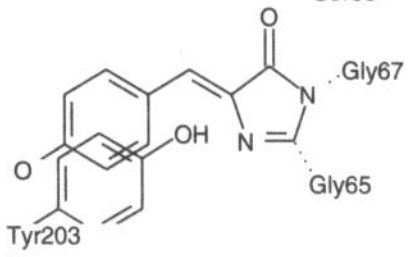	515	527 ($\sim$ 0.7)
zFP538	[12]	*Zoanthus sp.*	YES	?	494 528	538 (0.42)
GdFP	1OXF [9]	*Æ. Victoria* Y66W+NH_2	NO	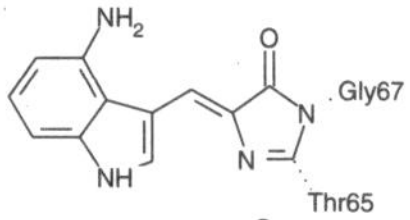	466	574
DsRed	1GGX [14]	*Discosoma sp.*	NO	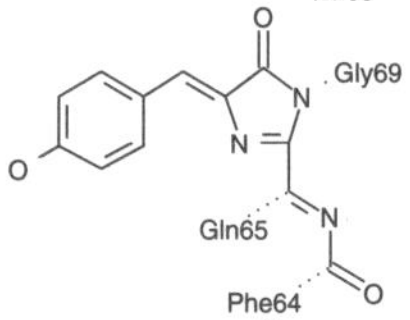	558	583 (0.23)
Kaede	[17]	*Trachyphyllia geoffroyi*	YES	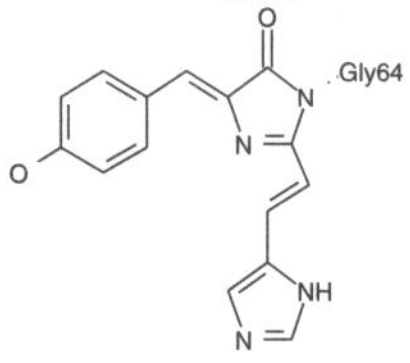	508 572	582 627
asCP562 asFP595	[18]	*Anemonia sulcata*	YES	?	562 572	– 595 (0.012)
eqFP611	1UIS [16]	*Entacmæa quadricolor*	NO	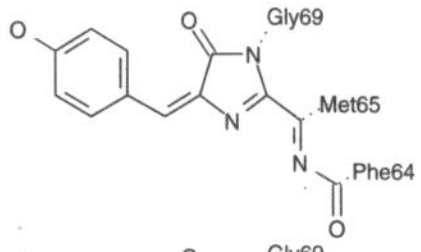	559	611 (0.45)
Rtms5	1MOU [15]	*Montipora efflorescens*	NO	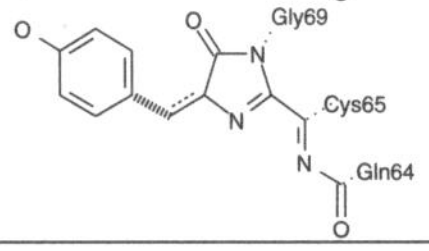	595	–

Table 1. Chromophore structures and optical properties of FPs. The conventional names, the parent organism and the characteristic mutation (where present) are reported, together with the Protein Data Bank (pdb) codes. The fourth column indicates whether the protein undergoes backbone fragmentation during maturation. The chromophore structure and the absorption/emission wavelengths (and quantum yields) are reported. In the case of BFP-Y66F we assumed that the cyclization process is the same as in the other GFP mutants.

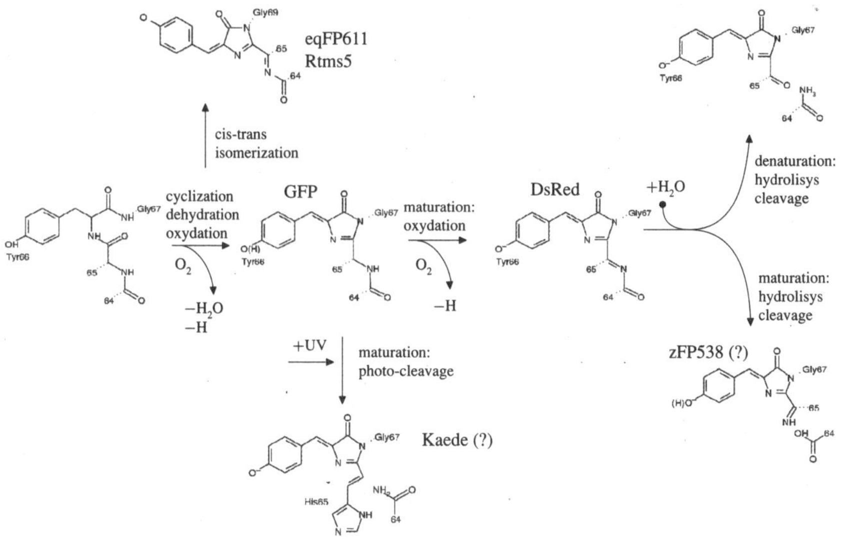

Figure 2. Schematic representation of the chromophore formation process.

ing in a perfectly shielded built-in photoactive bi-cyclic structure. Differences in the chromophore structures can arise either from a different cyclization-oxidation process or from mutations of Tyr66 (the central of the three amino acids involved in the cyclization), as in BFP, CFP, YFP and GdFP. In the last case Tyr66 was mutated to an artificial tryptophan variant substituted with an NH_3 [9].

The commonly accepted formation process for GFP and its mutants involves a cyclization occurring through a dehydration of the NH group of Gly67 and the carbonyl group of amino acid at position 65, followed by an oxydation of the C_α–Tyr66 bond, which is the bottleneck of the whole process [21]. The fluorescence mechanism in GFP involves an equilibrium between two forms of the chromophore, namely neutral (protonated on Tyr66 oxygen) and anionic (deprotonated on Tyr66), although the emission occurs prevalently through an anionic form [22].

In DsRed these early stages of the process lead to an immature green form, which is converted to the mature red form through a subsequent oxydation of the C_α–N bond of residue 65 [23]. Denaturation can occur through an hydrolisys of the same bond leading to the separation of the peptide chain in two parts [23]. An hydrolysis is also thought to occur during the final stage of maturation of asFP538, again leading to a backbone cleavage. However in this case there are indications that the cleaved bond is the peptide bond between amino acids 65 and 66 [12]. The formation of the trans chromophore of eqFP611 and Rtms5 occurs presumably through the same stages as those of DsRed, with an additional step where the chromophore assumes the trans configuration. No indications are available about the stage in which this transition occurs. A more complex reaction is believed to be the basis of the green-to-red photoconversion in Kaede. After the formation of a GFP-like chromophore, a cleavage of the C_α–N bond followed by an oxydation of the C_α–C bond of His65 should lead to an extension of the aromatic π system to histidine [17].

3. Models and Methods: Time Dependent Density Functional theory

The model structures for the chromophores are reported in Table 1. In the case of BFP-Y66H we considered two possible protonation forms, with the proton on N_δ or N_ϵ of histidine. For GFP, DsRed and Kaede we included both neutral and anionic forms. In all cases the dotted bonds are cut and saturated with a methyl group. Overall, we considered 11 structures, which are reported in Table 2.

The structures were optimized with Density Functional Theory, using the well established scheme described in ref. [28]: Troullier-Martins pseudopotentials, BLYP functional for exchange and correlation, plane-wave basis set with a 90 ryd cutoff [24] and supercells large enough to leave at least 5Å empty space between periodic images [25].

Excitation energies were evaluated within the frame of Time Dependent Density Functional Theory using the adiabatic local density approximation [37] and the Tamm-Dancoff approximation [38]. The accuracy of this computational scheme is known to be limited: the locality of the self consistent potential underestimates the effects due to charge transfer upon excitation [39]. This was shown to lead to systematic underestimations of the excitation energy of tenths of eV in systems where the excitation is mainly $\pi - \pi^*$ [40]. Conversely, TDDFT was shown to be quite robust and accurate in describing excited state geometries, dipoles and force constants [41], in addition to having a very low computational cost with respect to other first-principle methods for excited states. Furthermore, since the underestimation of the excitation energy is systematic in similar systems, the trends are fairly well reproduced. For all these reasons it results particularly proper to analyze excited-state properties of a large family of molecules. To our knowledge, this is the first study in which the electronic properties of ground and excited states of all the chromophores of the FP family are uniformly evaluated and compared.

All calculations are performed with the CPMD code [42].

4. Results: Electronic structure and excitation energies

Calculations of the electronic structure and excitation energies for the FP chromophores are available only for the GFP chromophore and its blue (BFP) and cyan (CFP) mutants, and for DsRed. For GFP, semiempirical calculations span the range $\sim$ 2.6–$\sim$ 2.9 eV for the anionic form and $\sim$ 3.2–$\sim$ 4.1 for the neutral [26–29]. Hartree-Fock yields much higher values [30], while SAC-CI [31] gives lower values. The CASSCF excitation energy (2.67eV) [32] of the anionic form is similar to the semiempirical values, and should be compared with the only one available experimental datum for the anionic chromophore in the gas phase, 2.58 eV [33]. Semiempirical calculations are available for the blue and cyan GFPs and DsRed, and give 4.19, 3.81, 3.65,3.67 eV [26] and 2.20 eV [36] for, respectively, GFP-Y66F, the two differently protonated forms of BFP, CFP and DsRed (anionic). In spite of the variety of values for the excitation energies, all previous calculations indicate that the transition is mainly HOMO-LUMO (in percentages depending on the calculation and on the system, but around 60-80%) and $\pi - \pi^*$ in character. Table 2 reports the results of our DFT and TDDFT calculations on the whole series of chromophores. For GdFP and Kaede these are the first electronic structure calculations to our knowledge.

According to our TDDFT calculations, the transitions with larger oscillator strength are mainly HOMO-LUMO in character for all the chromophores, in percentages varying from 60 to 90% depending on the case. This is in agreement with previously published calculations for GFP and BFPs, and turns out to be true also for DsRed, GdFP and Kaede. In all cases the transition is a $\pi - \pi^*$.

HOMO orbitals display a very similar shape in all cases. The π orbital on the double C=C bond on the bridge and on the C=N of imidazolinone are clearly visible in all neutral forms, while in the anionic forms the electronic charge is localized on the C=C of the imidazolinone. In all structures the aromatic pattern on the phenol (or on histidine or triptophane at position 66) ring is clearly visible. It is interesting to note that even when an extended aromatic system is present, as in Kaede and DsRed, the electronic charge of the HOMO orbital is mainly localized on the byciclic structure formed by residue 66 and the imidazolidinone, with only small charge transfer to the "tail" formed by the backbone or side chain of residue 65. Conversely, in the LUMO state a large charge transfer is observed to the tail of Kaede and DsRed. This is particularly evident comparing LUMO states of DsRed or Kaede with those of GFP. While in GFP the charge of LUMO is transferred to C=C–O of the imidazolinone, in DsRed and Kaede these regions are almost completely depleted

	HOMO	LUMO	HOMO-LUMO	H-L gap	exc en	exp
BFP_F	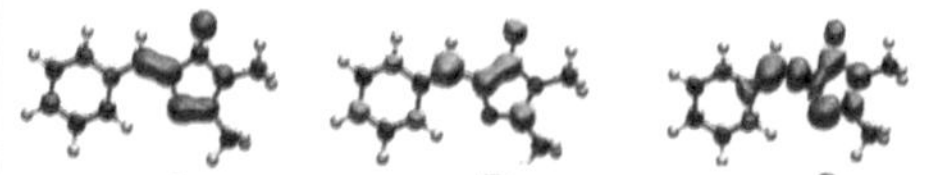			2.32	3.35	3.49
BFP_δ	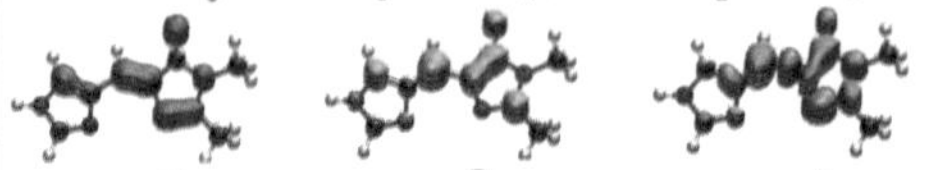			2.38	3.49	3.24
BFP_ϵ	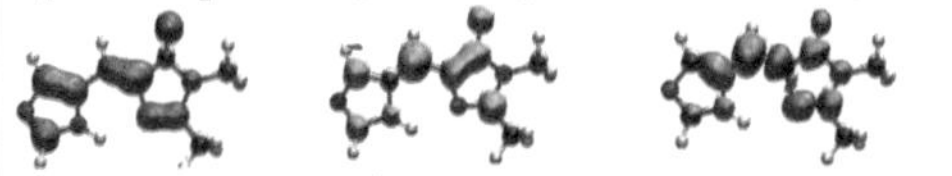			2.20	3.45	
CFP	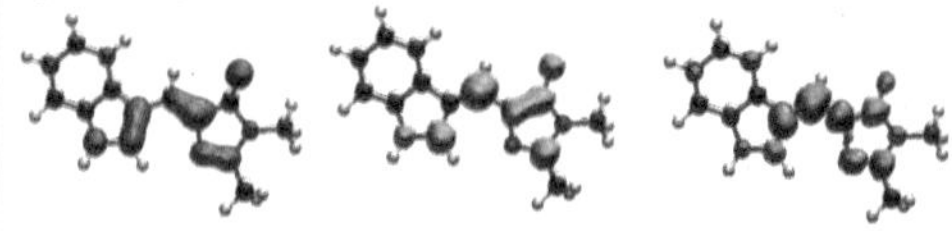			2.15	3.21	2.85
GFP_n	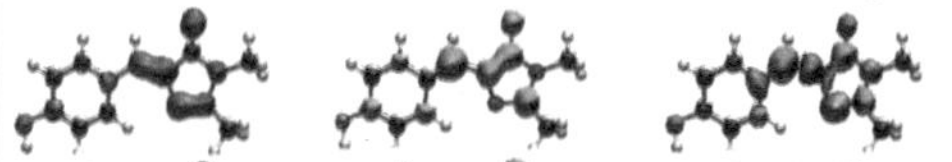			2.23	3.30	3.22
GFP_a	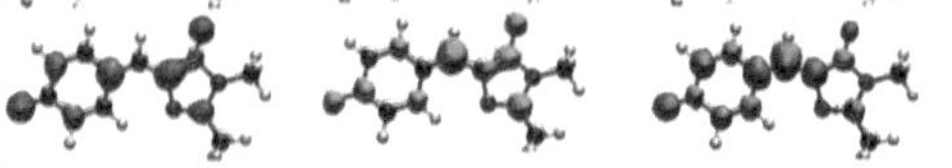			1.79	2.7	2.58
GdFP	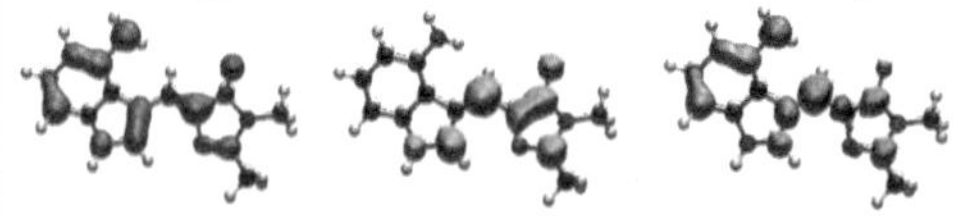			1.91	2.47	2.66
$DsRed_n$	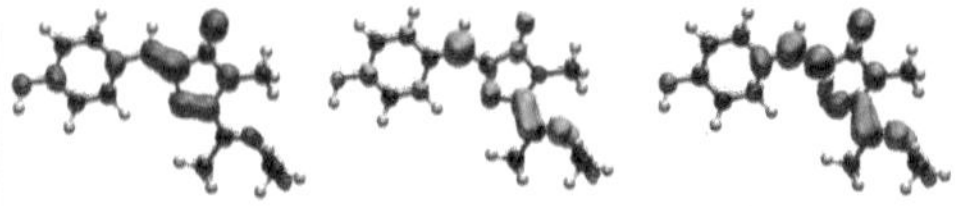			1.85	2.78	
$DsRed_a$	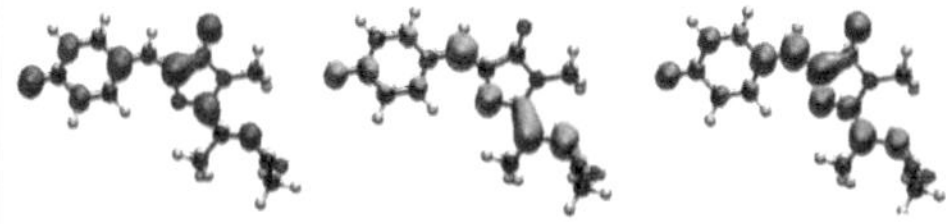			1.38	2.58	2.7
$Kaede_n$	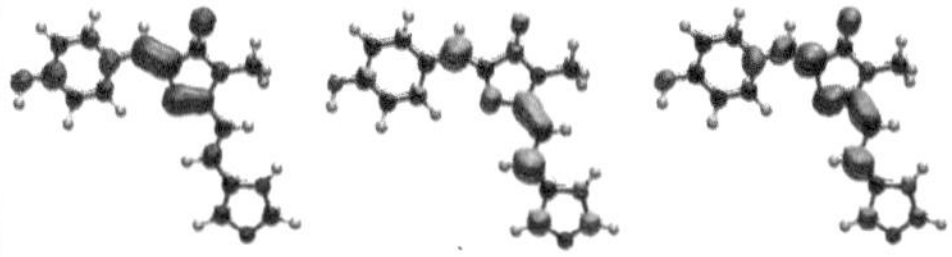			1.61	2.47	2.44
$Kaede_a$	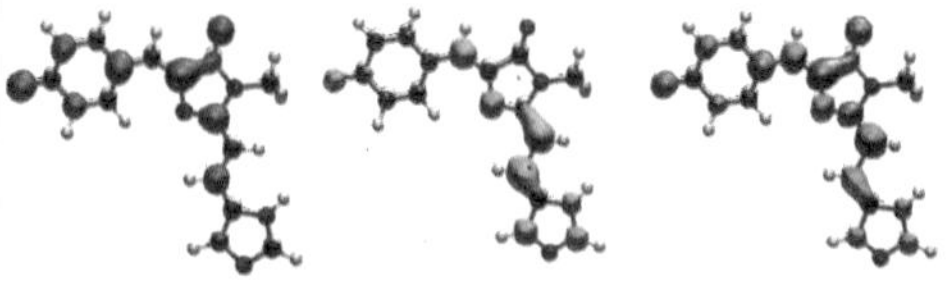			0.70	2.20	2.16

Table 2. Electronic structure and excitation energy. HOMO and LUMO electronic densities are reported in magenta and green. The HOMO–LUMO column reports the variation of the electronic density in a HOMO→LUMO transition: regions of charge accumulation and depletion are represented in pink and in cyan respectively. HOMO-LUMO energy gaps and excitation energies are in eV. Experimental absorption energies of the most popular mutant for each protein are reported in the "exp" column. "n" and "a" indicate the neutral and anionic form respectively. For BFP, δ and ϵ indicate two different protonation of the neutral histidine.

and the charge is transferred to the tail, "far" from the phenolic oxygen. This can be seen more clearly in the third column of Table 2, where the electronic density difference between HOMO and LUMO is represented. Cyan regions are those where the electronic charge is transferred upon HOMO-LUMO transition, while pink regions are those where charge is depleted. This effect is more evident in Kaede and DsRed, because the region where the charge is transferred extends to the whole aromatic tail. GdFP is also interesting. Although no tail is present in this case, the depleted region is larger because the tryptophan is substituted with an electronegative group, which attracts charge in the HOMO state. As a result, the net effect of charge transfer is again larger than in the "conventional" GFP mutants, included its parent CFP. In the ground state, the concentration of electronic charge on the extremum of residue 66 produces an unusual HOMO structure with respect to all other FPs and might be correlated with the particularly large Stokes shift.

The sixth column of Table 2, reporting the excitation energy evaluated with TDDFT, shows that the amount and distance of the charge transfer is strongly correlated with the excitation energy: the larger and more distant the charge transfer, the smaller the excitation energy. This a direct demonstration of the rule correlating the extension of the aromatic system with the absorption wavelength.

The comparison between TDDFT values for the excitation energy and the experimental values for several FPs is reported in Figure 3. Green labels report the experimental values of the excitation energies for the proteins versus the calculated TDDFT values of the correspondent chromophores. As it can be seen the trend is very well reproduced, with an average accuracy of about 5%. It should be noted that here we are comparing our *in vacuo* calculations with experimental values for the chromophore in the proteins. A comparison with other *in vacuo* calculations is possible only for the few available cases (inset in Figure 3, red labels) and shows that TDDFT values are on average 10% smaller than semiempirical values. This is in agreement with previously reported TDDFT performances (see "Methods" section). The agreement between TDDFT values and protein data can be attributed to the effect of interaction with the protein environment, which red shifts the excitation energy of the chromophore, at least for the neutral forms [28].

We showed that TDDFT calculations reproduce quantitatively the relationship between color and chromophore structure in FPs. In particular, the hypothetical chromophore structure of Kaede is compatible with the two absorption peaks observed in the protein absorption spectrum at 572 and 508 nm. Our calculations indicate that these might correspond to the anionic and neutral forms of the chromophore, respectively.

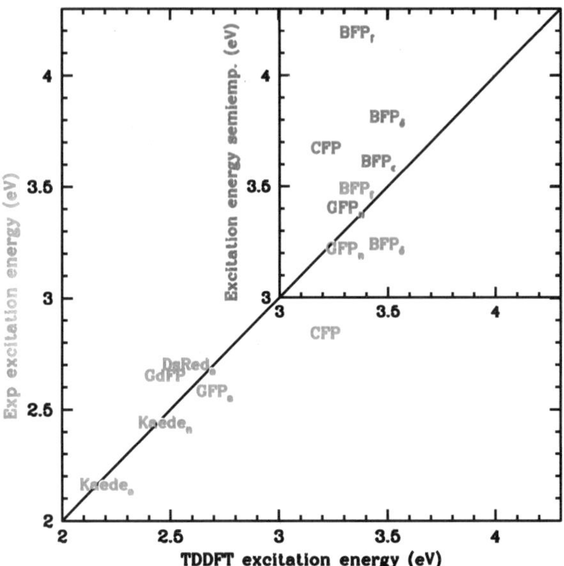

Figure 3. Plot of the experimental excitation energy versus the corresponding TDDFT value. Green labels report the name of the corresponding protein and are used in the place of dots. In the inset, the semiempirical calculations [26] versus TDDFT values are plotted in red with the same convention. The black line is to guide the eye.

5. Conclusions

In this paper we reported a DFT and TDDFT study of the properties of the ground and excited state of all the chromophores of the FP family whose structure is known. Our study provides a rationale to the commonly accepted empirical rule correlating the extension of the aromatic system to the resonance excitation wavelength: the more extended the aromatic system, the larger the excitation wavelength (and the smaller the excitation energy). Our study indicates that this is mainly due to differences in the excited states, rather than in the ground states: the extension of the aromatic system in DsRed and Kaede influences mainly the charge transfer of the LUMO state to the added aromatic tail, leaving the HOMO state almost unchanged with respect to the GFP chromophore. An exception to this behavior is GoldFP, where the LUMO is very similar to that of its parent structure, the CFP chromophore, while the HOMO is quite different. This could explain the very large red shift of GoldFP with respect to CFP, detected especially in emission. This work also demonstrates that, in spite of its known limitations, TDDFT is a powerful method in the study of excited state properties when general trends in a family of similar molecules are analyzed.

ACKNOWLEDGMENTS. We wish to thank Prof. Fabio Beltram and Pietro Amat for useful discussions.

References

[1] O. SHIMOMURA, F. H. JOHNSON and Y. SAIGA, J. Cell. Comp. **59** (1962), 223-239.
H. MORIS, O. SHIMOMURA, F. H. JOHNSON and J. WINANT, Biochemistry **13** (1974), 2656-2662.

[2] D. C PRASHER, V. K. ECKENRODE, W. W. WARD and F. G. PRENDERGAST, Gene **111** (1992), 229-33.

[3] M. CHALFIE, Y. TU, G. EUSKIRCHEN, W. W. WARD and D. C. PRASHER, Science **263** (1994), 802-805.

[4] In the last years many reviews and books on GFP have been published. See for instance: "Green Fluoresecent Proteins: Properties, Applications and Protocols", Chalfie, M., Chain S. (eds.); Wiley-Liss: New York, 1998.
"Green Fluoresecent Protein", Vol. 302, Conn P. M. Ed., Academic press: San Diego, 1999;
"Green Fluorescent Protein: Methods in Cell Biology", Vol. 58, Sullivan K. F., Kay S. A. (eds.), Academic Press: San Diego, 1998.
M. ZIMMER, Chem. Rev. **102** (2002), 759-782.
R. Y. TSIAN (1998) *The Green Fluorescent Protein*, Annu. Rev. Biochem. **67**, 509-44.
A. ZUMBUSCH and G. JUNG, Single Mol. **1** (2000), 261-270.

[5] Y. A. LABAS, N. G. GURSKAYA, Y. G. YANUSHEVICH, A. F. FRADKOV, K. A. LUKYANOV, S. A. LUKYANOV and M. W. MATZ, Proc. Natl. Acad. Sci. USA **99** (2002), 4256-4261.

[6] V. V. VERKHYSHA and K. A. LUKYANOV, Nat. Biotechnol. **22** (2002), 289-296.

[7] A. D. KUMMER,J. WIEHLER, T. A. SCHÜTTRIGKEIT, B. W. BERGER, B. STEIPE and M. E. MICHEL-BEYERLE, Chem. Bio. Chem. **3** (2002), 659-663.

[8] R. M. WACHTER, B. A. KING, R. HEIM, K. KALLIO, R. Y. TSIEN, S. G. BOXER and S. J. REMINGTON, Biochemistry **36** (1997), 9759-9765.

[9] J. H. BAE, M. RUBINI, G. JUNG, G. WIEGAND, M. H. J. SEIFERT, M. K. AZIM, J. S. KIM, A. ZUMBUSH, T. A. HOLAK, L. MORODER, R. HUBER and N. BUDISA, J. Mol. Biol. **328** (2003), 1071-1081.

[10] F. YANG, L. G. MOSS and G. N. JR PHILLIPS, Nat. Biotechnol. **14** (1996), 1246-1251.

[11] R. M. WACHTER, M.-A. ELSLIGER, K. KALLIO, G. T. HANSON and S. J. REMINGTON, Structure **6** (1998), 1267-1277.

[12] V. E. ZAGRANICHNY, N. V. RUDENKO, A. Y. GOROKHOVATSKY, M. V. ZAKHAROV, Z. O. SHENKAREV, T. A. BALASHOVA and A. S. ARSEIEV, Biochemistry **43** (2004), 4764-4772;
M. V. MATZ, A. F. FRANDKOV, Y. A. LABAS, A. P. SAVITSKY, A. G. ZARAISKY, M. L. MARKELOV and S. A. LUKYANOV, Nat. Biotechnol. **17** (1999), 969-973.

[13] The rational nomenclature for these proteins consists in: initials of the name of the organism in lower case, followed by FP or CP depending on if it is a fluorescent or non fluorescent form, followed by the emission wavelength for fluorescent or absorption wavelength for non fluorescent forms.

[14] M. A. WALL, M. SOCOLICH and R. RANGANATHAN, Nat. Struct. Biol. **12** (2000), 1133-1138, The structural basis for red fluorescence in the tetrameric GFP homolog DsRed.
D. YARBROUGH, R. M. WACHTER, D. KALLIO, M. V. MATZ and S. J. REMINGTON, Proc. Natl. Acad. Sci. USA **98** (2000), 462-467.

[15] M. PRESCOT, M. LING, T. BEDDOE, A. J. OAKLEY, S. DOVE, O. HOEGH-GULDBERG, R. J. DEVENISH and J. ROSSJOHN, Structure **11** (2003), 275-284.

[16] J. PETERSEN, P. G. WILMANN, T. BEDDOE, A. J. OAKLEY, R. J. DEVENISH, M. PRESCOTT and J. J. ROSSJOHN, Biol. Chem. **278** (2003), 44646-44631.

[17] H. MIZUNO, T. K. MAL, K. I. TONG, R. ANDO, T. FURUTA, M. IKURA and A. MIYAKAWAKI, Mol. Cell. **12** (2003), 1051-1058.

[18] V. I. MARTYNOV, A. P. SAVITSCKY, N. Y. MARTYNOVA, P. A. SAVITSKY, K. A. LUKYANOV and S. A. LUKYANOV, J. Biol. Chem. **24** (2001), 21012-21016.

[19] V. E. ZAGRANICHNY, N. V. RUDENKO, A. Y. GORONOKHOVATSKY, M. V. ZAKHAROV, T. A. BALASHOVA and A. S. ARSENIEV, Biochem. **43** (2004), 13598-13603.

[20] D. M. CHUDAKOV, A. V. FEOFANOV, N. N. MUDRIK, S. LUKYANOV and K. A. LUKYANOV, J. Biol. Chem. **278** (2003), 7215-7219.

[21] B. G. REID and G. C. FLYNN, Biochemistry **36** (1997), 6789-6791.

[22] K. BREJC, T. K. SIXMA, P. A. KITTS, S. R. KAIN, R. TSIEN, M. ORMÖ and S. G. REMINGTON, Proc. Natl. Acad. Sci. USA **94** (1997), 2306-2311.

[23] L. A. GROSS, G. S. BAIRD, R. C. HOFFMAN, K. K. BALDRIDGE and R. Y. TSIEN, Proc. Natl. Acad. Sci. USA **97** (2000), 11990-11995.

[24] Unlike ref. [28] in this case we used sligthly harder pseudopotential in order to have a better accuracy. This pseudopotential converges to a cutoff which is larger than the commonly used one (70 ryd).

[25] The Hockney technique is used to remove interaction between periodic images.

[26] A. A. VOITYUK, M.-E. MICHEL-BEYERLE and N. RÖSCH, Chem. Phys. Lett. **272** (1997), 162-167;
A. A. VOITYUK, M.-E. MICHEL-BEYERLE and N. RÖSCH, Chem Phys **231** (1998), 13-25;
A. A. VOITYUK, M.-E. MICHEL-BEYERLE and N. RÖSCH, Chem Phys Lett. **296** (1998), 269-276;
A. A. VOITYUK, M.-E. MICHEL-BEYERLE, A. D. KUMMER and N. RÖSCH, **269** (2001), 83-91.

[27] W. WEBER, V. HELMS, J. A. MCCAMMON and P. W. LANGHOFF, Proc. Natl. Acad. Sci. USA **96** (1999), 6177-6182.

[28] T. LAINO, R. NIFOSÌ and V. TOZZINI, Chem. Phys. **298** (2004), 17-28.

[29] A. TONIOLO, M. BEN-NUM and T. J. MARTINEZ, J. Phys. Chem. A **106** (2002), 4679-4689;
A. TONIOLO, G. GRANUCCI and T. J. MARTINEZ, J. Phys. Chem. A **107** (2003), 3822-3830.

[30] V. HELMS, C. WINSTEAD and P. W. LANGHOFF, J. Mol. Struct. (Theochem) **506** (2000), 179-189.

[31] A. K. DAS, J. Y. HASEGAWA, T. MIYAHARA, M. EHARA and H. NAKATSUJI, H. J. Comput. Chem. **24** (2003) 1421-1431.

[32] M. E. MARTIN, F. NEGRI and M. OLIVUCCI, J. Am. Chem. Soc. **126** (2004), 5452-5464.

[33] S. B. NIELSEN, A. LAPIERRE, J. U. ANDERSEN, U. V. PEDERSEN, S. TOMITA and L. H. ANDERSEN, Phys. Rev. Lett. **87** (2001), 228102-5.

[34] L. J. ANDERSEN, A. LAPIERRE, S. B. NIELSEN, S. U. PEDERSEN, U. V. PEDERSEN and S. TOMITA, Eur. Phys. J. F **20** (2000), 597-600.

[35] X. HE, A. F. BELL and P. J. TONGE, J. Phys. Chem. B **106** (2002), 6056-6066.

[36] L. A. GROSS, G. S. BAIRD, R. C. HOFFMAN, K. K BALDRIDGE and R. Y. TSIEN, Proc. Natl. Acad. Sci. USA **97** (2000), 11990-1995.

[37] CASIDA, In: " Recent Advances in Density Functional Methods", Part I, Singapore, World Scientific, 1995, p. 155.
[38] J. HÜTTER, J. Chem. Phys. **118** (2003), 3928-3934.
[39] A. DREUW, J. L. WEISMAN and M. HEAD-GORDON, J. Chem. Phys. **119** (2003), 2943-2946.
[40] M. WANKO, M. GARAVELLI, F. BERNARDI, T. A. NIEHAUS, T. FRAUENHEIM and M. ELSTNER, J. Chem. Phys. **120** (2004), 1674-1692.
[41] F. FURCHE and R. AHLRICHS, J. Chem. Phys. **117** (2002), 7433-7447.
[42] CPMD, Copyright IBM Corp 1990-2004, Copyright MPI für Festkörperforschung Stuttgart 1997-2001.

On the bathochromic shift in chromophore-protein interaction

F. Buda

The absorption properties of chromophores are strongly modified when the chromophore is embedded in a protein environment. To resolve the molecular basis of the color shift due to the formation of chromophore-protein complexes, we used time-dependent density functional theory (TDDFT) calculations. We investigated the bathochromic shift in the photoactive yellow protein and in α-crustacyanin, the blue carotenoprotein in the lobster shell. TDDFT appears to be quite accurate in predicting absorption energies and allows to separately analyse the relative importance of conformational changes, polarization effects, and aggregation effects.

1. Introduction

The relevance of photochemical reactions in biological systems and for life in general is well known, most noticeable examples being the photosynthesis in plants and the photochemical reactions in our visual photoreceptors. The basic photochemical processes occur in the so called chromophores, that are organic molecules bound to the protein and able to detect light in the visible region. The chromophore imparts some decided color to the compound of which it is an ingredient. The shift of a spectral band to lower frequencies (longer wavelengths) owing to the influence of substitution or a change in environment (e.g., solvent), is called bathochromic shift, also informally referred to as a red shift. A bathochromic shift is generally observed upon formation of a chromophore-protein complex, the extent of which depends on the specific interactions between the chromophore and the protein environment.

One example is provided by the green fluorescent protein (GFP) found in jellyfish that acts as a wavelength shifter. The GFP and its mutants have become very popular tools in molecular and cell biology for visualizing gene expression in cells and for monitoring protein-protein interactions [1]. Another remarkable example is color vision due to the bathochromic shift of the retinal chromophore in cone pigments. The binding sites of these pigments change from red to green to blue via systematic shifts in the position/delocalization of the primary counterion relative to the protonated Schiff base retinal [2].

A detailed understanding of the molecular origin of the bathochromic shift in photoactive proteins is a current theoretical challenge. This goal is of technological relevance as well as it is important from the point of view of fundamental science. In fact the understanding of the operational principles optimized by evolution may be used to engineering new optical devices. Possible mechanisms that can contribute to the bathochromic shift include: (i) protein induced conformational changes, (ii) polarisation effect due to bonding and/or long-range interaction with charged residues, (iii) protonation/deprotonation of the chromophore, and (iv) aggregation effects when more than one chromophore is present in the protein complex.

In this paper we will show that time-dependent density functional theory (TDDFT) calculations are a powerful tool to investigate the bathochromic shift mechanisms. We will focus here on two proteins, the photoactive yellow protein (PYP) [3,4] and the α-crustacyanin, the blue carotenoprotein in the lobster shell [5].

2. Photoactive Yellow Protein

The Photoactive Yellow Protein (PYP) belongs to a new class of photoreceptors found in several eubacteria. PYP, first isolated from *Ectothiorhodospira halophila* [6], is the most studied photoreceptor of this family, which displays negative phototaxis upon absorption of blue light. The PYP exhibits a photocycle involving several intermediates and recovers the initial state on a subsecond time scale [7]. The chromophoric group of PYP is a *p*-coumaric acid (pCA) (4-hydroxycinnamyl group) covalently bound to the unique cysteine in the apoprotein through a thiol ester bond.

The interest in PYP has increased since a high-resolution crystal structure has been obtained, providing the first example of a detailed atomic characterization of a protein with a photocycle [8]. In fact, the knowledge of the atomic structure of the chromophore and of its protein binding pocket constitutes a prerequisite to a better understanding of the functionality of these biologically relevant photoreceptors.

The 4-hydroxycinnamyl chromophore undergoes a *trans*-to-*cis* isomerization upon absorption of a blue photon. More specifically, the pCA in the PYP ground state (usually denoted as pG) is found in a *trans* configuration with a net negative charge, and an absorption peak at 446 nm; a red-shifted intermediate (denoted as pR) develops on a short time scale of few ps after photo excitation and the chromophore is found to be in a *cis* configuration and still deprotonated. Before recovering the ground state pG, the system is converted to at least one more long lived, blue shifted intermediate (denoted as pB), where the chromophore is in the *cis* configuration and protonated. It is noticeable the analogy with the rhodopsins family, where the first photochemical event involves also a photoisomerization of the chromophore followed by a proton transfer step.

2.1. Models and computational methods

We have investigated the structural and electronic properties of the pCA chromophore both in the vacuum and in the protein environment represented by the surrounding residues, taken according to the crystallographic structure of PYP [8]. We have considered the neutral pCA (Figure 1a) and the anionic pCA (Figure 1b) with the phenolic oxygen deprotonated. We have also considered the anionic 4-hydroxycinnamyl thiomethyl (TMpCA) model compound (Figure 1c) to check the effects induced by the thiol ester bond on the geometry and electronic excitations of the chromophore. Finally we have studied an extended model of the pCA chromophore in the binding pocket of PYP both in the *trans* ground state (Figure 1d) and in the *cis* first photo-intermediate configuration. The comparison between the more realistic PYP active site model with the isolated pCA and TMpCA models allows to investigate how the protein environment induces the bathochromic shift.

The structure of all the models have been optimized using the *ab-initio* molecular dynamics Car-Parrinello method [9] where the electronic ground state is obtained within the DFT [10]. We used the Becke-Perdew exchange-correlation functional. Only valence electrons are treated explicitly in the calculation: The interaction with the frozen core is described by soft first-principles pseudopotentials [11] for C, N, and O, and by norm-conserving pseudopotentials for H [12] and S [13]. The Kohn-Sham orbitals are expanded on a plane wave basis set up to an energy cutoff of 20 Ry. Since we use periodic boundary conditions, the size of the box has to be chosen large enough to minimize the spurious effects due to interaction with the periodic images. For the isolated pCA we took a box of $13.76 \times 8.47 \times 8.47$ Å^3, while for the extended model of the PYP active site the size was increased to $20.0 \times 17.5 \times 15.0$ Å^3. The initial coordinates for our extended model have been taken from the 1.4 Å resolution PYP crystalline structure [8] available in the Brookhaven Protein Data Bank. For the various model compounds considered here, we have computed the excitation energies and corresponding oscillator strengths in the energy range relevant for the photochemistry by means

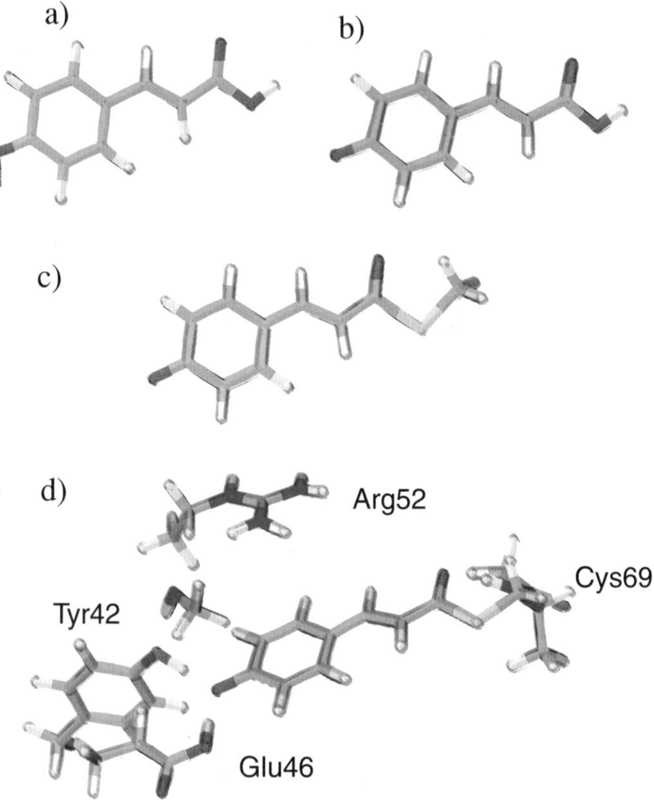

Figure 1. Models of the PYP chromophore used in this study: a) neutral pCA, b) anionic pCA, c) anionic TMpCA, d) extended model of the PYP active site including the residues surrounding the pCA chromophore, specifically, Cys69, Glu46, Tyr42, Thr50, and the counterion Arg52.

of time-dependent DFT [10]. For the calculation of these properties we used the Amsterdam Density Functional (ADF) code [14]. In the ADF code the electronic orbitals are written in terms of Slater Type Orbitals (STO). We used a triple-ζ basis set with one polarization function.

2.2. Results

The absorption peak in the PYP ground state (446 nm) is strongly red-shifted compared to the absorption of free pCA chromophore in aqueous solvents (310 nm at pH=2, 284 nm at pH=7). Several chromophore-protein complexation effects can contribute to this bathochromic shift: (i) formation of the thioester bond between pCA and Cys69, (ii) deproto-

Model compound	Excitation En. (nm)	Oscillator Strength	Exp. λ_{max} (nm)
pCA-Neutral	329.4	0.577	310
pCA-Anionic	383.2	0.745	
TMpCA-Anionic	402.0	0.863	384
PYP (pG)	417.6	0.462	446
PYP (pR)	422.5	0.626	465

Table 1. Excitation Energies and Oscillator Strengths computed with TDDFT. Neutral *trans* p-coumaric acid (pCA-Neutral), Anionic *trans* p-coumaric acid (pCA-Anionic), Anionic thiomethyl pCA (TMpCA-Anionic), and extended model of the PYP active site in the ground state (pG) and in the *cis* intermediate (pR). For comparison we show also the experimental value of the main absorption peak (λ_{max}) in the UV spectra for pCA-Neutral (ref. [15]), TMpCA-Anionic (ref. [4]), and for PYP (ref. [6, 16]).

nation of the phenolic oxygen of the chromophore, and (iii) hydrogen bond network and electrostatic interactions. In Table 1 we show the computed excitation energies and the corresponding oscillator strengths for the model compounds considered here. From these calculations we can analyze the importance of the various contributions to the absorption shift, and ascertain which states are responsible for the photochemistry of PYP.

For the neutral *trans*-pCA we find two dominant singlet-singlet excitations in the range up to 5.5 eV, at 329 and 238 nm, respectively, with a 238:329 nm oscillator strength ratio of about 0.2. The oscillator strength of the other excitations in this energy range is at least one order of magnitude lower than the largest oscillator strength at 329 nm. This result compares reasonably well to the experimental UV absorption spectrum of 4-hydroxycinnamic acid [15], where two broad peaks are observed with maxima at 310 and 228 nm and with a 228:310 nm absorbance ratio of 0.5. The 329 nm excitation with the largest oscillator strength is predominantly related to the HOMO-LUMO transition, while the 238 nm excitation is dominated by the transition from a lower orbital (HOMO-4) to the LUMO. The analysis of these molecular orbitals shows that the HOMO-4 has a π-bonding character on the C7-C8 double bond, and the HOMO has a π-bonding character on the C7-C8 bond and on the phenolic ring. The LUMO instead has clearly a $\pi^\star$-antibonding character on the C7-C8 isomerizable bond and a π bonding character on the adjacent single bonds. Therefore both excitations can be characterized as π to $\pi^\star$ transitions mostly localized on the C7-C8 double bond. The HOMO-LUMO transition, which turns out to be the one relevant in the photochemistry, results in displacing the π-bonding character from the double bond to the adjacent single bonds.

Similarly, in the anionic *trans*-pCA compound, we find that the singlet-singlet excitation with the largest oscillator strength - in the energy range relevant for the photochemistry - is predominantly the HOMO-LUMO transition. The excitation energy shows a large red-shift of about 54 nm (see Table 1) in going from the neutral (protonated) to the anionic pCA in the vacuum.

A further red-shift of about 20 nm is found in the excitation energy of the anionic TMpCA compound when compared to the anionic pCA. This result shows that the thiol ester bond contributes significantly in the total red-shift measured experimentally in PYP.

One relevant conclusion of this analysis is that the photo-excitation is related to the HOMO-LUMO transition and results in a displacement of electronic charge from the C7-C8 isomerizable double bond to the adjacent single bonds in the conjugated chain of the chromophore. Therefore, the electronic excitation weakens the double bond, thus lowering the barrier to isomerization.

In Figure 1d we show the optimized ground state structure of the extended model with the pCA chromophore in the *trans* configuration. The chromophore is covalently bound to the Cys69 via a thioester linkage. We observe a network of hydrogen bonds in the active-site as suggested by the crystallographic data [8]. Specifically, the phenolic oxygen of the pCA forms two strong hydrogen bonds with Glu46 and Tyr42. A third weaker hydrogen bond is formed between Thr50 and Tyr42. On the opposite side of the chromophore an hydrogen bond is formed between the thioester oxygen and the backbone amyde of Cys69: The computed O-N distance of 2.76 Å is slightly longer than the experimental value of 2.69 Å [8]. In agreement with the experiment we find that the negative charge on the chromophore is strongly delocalized and that the bond length between the phenolic oxygen O1 and the aromatic ring shows a considerable double bond character [15]. In Table 1 we show the computed excitation energies and oscillator strengths for this model of the PYP chromophore binding pocket in the *trans* and *cis* configuration. We find that the first excitation energy with a large oscillator strength is dominated by the transition between two electronic orbitals localized on the chromophore and whose charge density distribution is very similar to that of the HOMO and LUMO of the isolated anionic pCA chromophore. We note that the excitation energy computed for PYP (pG) (see Table 1) is further red-shifted compared to the TMpCA-anionic compound ($\Delta\lambda = 15$ nm). Therefore, according to our model, the counterion Arg52 and the hydrogen bonding network, contribute to only a small amount of the total bathochromic shift of PYP when compared to the larger contributions due to the deprotonation of the chromophore and to the formation of the thiol ester bond.

In Table 1, we can see that the excitation energy shows a small red-shift of 5 nm in going from the *trans* (pG) to the *cis* configuration (pR). This result is in qualitative agreement with the experiment (19 nm measured at room temperature [6, 16], 8 nm in the intermediate PYP_L formed at -80°C [17]).

3. α-crustacyanin: the coloration mechanism in the lobster shell

Carotenoids are the most widespread class of pigments in both plants and animals. They are responsible for many natural yellow, orange or red colors. In addition, when carotenoids are associated with proteins, these colors can be modified to green, purple or blue by the formation of carotenoid-protein complexes. A well-known example of this phenomenon is provided by the lobster Homarus gammarus, which is deep blue (λ_{max} = 632 nm, aq. PO4-buffer). Heat causes the lobster to change color from blue to red. The high temperature denatures the protein under liberation of the "red" astaxanthin (λ_{max} = 478 nm in acetone) The bathochromic shift of the astaxanthin chromophore in α-crustacyanin has intrigued scientists for many years.

In a seminal paper, Buchwald and Jencks [18] reported that α-crustacyanin can be irreversibly dissociated in eight β-crustacyanin units. β-crustacyanin consists of two apoprotein units and two bound astaxanthin units. They also reported the visible, optical rotary dispersion and circular dichroism (CD) spectra of α- and β-crustacyanin. In both systems, the main CD band exhibits a change of sign at the absorption maximum with a negative and a positive peak at wavelengths above and below the absorption maximum, respectively. The negative chirality demonstrated by the CD spectra indicates an exciton coupling of the astaxanthins similar to that predicted by exciton theory. Resonance Raman studies [19] have suggested that the bathochromic shift is caused by a charge polarization mechanism possibly induced by charged groups and/or by hydrogen bonds in the binding site. More recently, ^{13}C NMR and resonance Raman spectroscopy combined with semi-empirical theoretical calculations that fitted the electronic absorption properties, led to the hypothesis that there is a strong electrostatic polarization originating from the keto groups, most likely a double protonation [20]. Recently the X-ray structure of β-crustacyannin at 3.2 Å resolution has been published [21]. The X-ray study showed that in β-crustacyanin two astaxanthins are present in parallel planes that cross at a distance of 7 Å a little outside the centers of the chromophores, as shown in figure 2. The vectors of the central axis are oriented at 120°. The conformation around the 6-7 and 6'-7' bonds is planar s-*trans*, in contrast to free astaxanthin, which has a s-*cis* conformation. The oxygen at the 4'-keto group is close to His-90 and His-92 for astaxanthin 1 and 2, respectively. The two other keto groups have interactions with Asp, two Tyr residues, a Ser and a bound water molecule. Although the increased conjugation due to the end rings coplanarization should lead to a bathochromic shift compared to free astaxanthin, it has been stated that this effect is not sufficient to explain the full color shift.

In spite of the considerable progress made in structural and spectroscopic studies, a clear converging picture on the dominant mechanism for the bathochromic shift in crustacyanin is still missing. There are two main hypotheses that are still debated: (i) The "on-site" (intramolecular) mechanism involving protein induced conformational changes and/or charge-polarization effects modifying the electronic ground state of the chromophore; (ii) the aggregation (intermolecular) mechanism due to the interaction of the chromophores in the subunits of crustacyanin inducing an exciton coupling of the transition dipole moments in the excited state.

In order to resolve the precise molecular basis of the coloration mechanism of α-crustacyanin we have used time-dependent DFT calculations for several models based on the structural information provided by X-ray data on β-crustacyanin [21] combined with ^{13}C NMR data on α-crustacyanin [5]. The theoretical investigation is crucial in identifying the relative importance of different mechanisms that may contribute to the total bathochromic shift.

3.1. Models and computational methods

In our theoretical analysis we have considered the following models: (I) free astaxanthin, (II) 6-s-*trans* astaxanthin with the end rings coplanar to the polyene chain, (III) 6-s-*trans* astaxanthin including a hydrogen bonded histidine analogue (methylimidazole) and a water molecule according to the X-ray data on β-crustacyanin. Comparison of model I and model II gives a measure of the shift induced by the increased conjugation due to the ring coplanarization. Model III shows the importance of the polarization effects due to the hydrogen bonding with the keto oxygens.

The calculation of excitation energies and transition dipole moments to the optically allowed ^1B_u excited state were performed using TDDFT. We used the BLYP Generalized Gradient Approximations for the exchange and correlation functional. The calculations were done with a

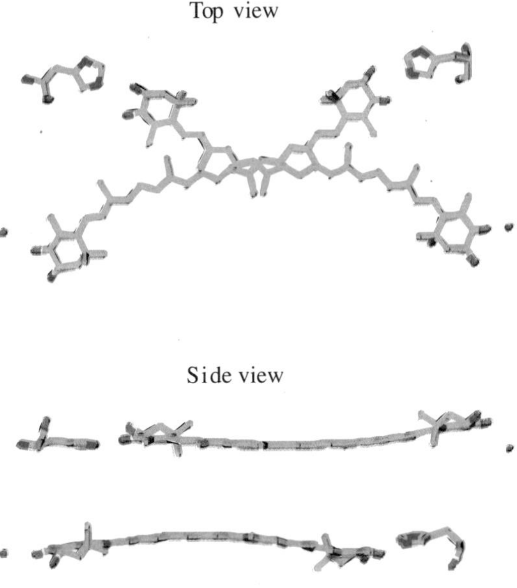

Figure 2. Top view and side view of two astaxanthin molecules in β-crustacyanin. The atomic coordinates have been extracted from the Protein Data Bank file (PDB code 1GKA).

triple-ζ Slater-type orbital (STO) basis set with a single set of polarization functions using the Amsterdam Density Functional (ADF) package. It is known that TDDFT shows a systematic underestimation of the excitation to the optically allowed ${}^1B_u$ state in linear polyenes by about 0.5-0.7 eV. However, the general trend of decreasing excitation energy with chain length is correctly reproduced. Since here we are interested in the bathochromic shift of the absorption energy rather than in the absolute values, TDDFT represents a good compromise between accuracy and computational efficiency for our models.

The aggregation effects are estimated using a simple dipole-dipole approximation of the interaction between the two astaxanthins in the geometry of the β-crustacyanin dimer. Within this approximation the interaction, that gives an estimate of the exciton splitting, can be written as

$$V_{12} = \frac{1}{4\pi\epsilon_0} \frac{\mu_1 \cdot \mu_2 - 3(\mu_1 \cdot \mathbf{R})(\mu_2 \cdot \mathbf{R})}{R^3}$$

Here μ is the transition dipole moment, $\mathbf{R}$ is the unit vector connecting the two molecules and R is the distance between the two molecules. The transition dipole moment is obtained from the TDDFT calculation and the orientation factor is estimated using the β-crustacyanin structure. We used the dielectric constant in the vacuum, though inside the protein it may generally be larger.

3.2. Results

^{13}C NMR spectra for isotopically labeled astaxanthin reconstituted in the protein have been used to explore the hypothesis of a strong polarization effect on the ground state electron charge in the chromophore [5]. We found that only minor changes in the ground state electron density of the C-atoms 8-11 of astaxanthin occur upon binding. The rather small NMR chemical shift differences are indicative of small changes in the ground state electronic distribution due to the rotation of the end-rings and to polarization effects due possibly to hydrogen bonding with the protein. The small effect for the C8 effectively rules out the hypothesis of a polarization originating from the keto-groups of the chromophore. NMR studies on protonated canthaxanthin show that a strong electrostatic effect, like a single or double protonation, should give increasingly larger changes in the chemical shift values of the even numbered atoms closer to the ring, in contrast with our data. We can definitely discard now the earlier hypothesis that a large charge effect, such as protonation, near the keto-groups of the chromophore is responsible for the bathochromic shift. These conclusions are consistent with the X-ray analysis of β-crustacyanin. Although the resolution of the X-ray data is not sufficient to determine protonation states, no charged amino acids were observed in the proximity of the astaxanthin-ring.

Model compound	ΔE (eV)	ΔE (eV)+0.7 eV (λ_{max})
I (s-*cis* astaxanthin)	1.89	2.59 (478 nm)
II (s-*trans* astaxanthin)	1.77	2.47 (502 nm)
III (s-*trans* astaxanthin + water + His)	1.70	2.40 (517 nm)
III + exciton splitting		1.91 (650 nm)

Table 2. Excitation energies for carotenoid models I, II and III obtained with TDDFT. In the second column we have shifted the TDDFT results for a comparison with the experimental data (see text). In parenthesis we give the corresponding λ_{max} in nm. The excitation energy in the last row is obtained adding the estimated exciton splitting to the TDDFT result of model III.

In light of this NMR data, we concentrate our theoretical analysis only to neutral models. In table 2 we report the first excitation energy with a considerable oscillator strength obtained with TDDFT for the three carotenoid models described above. This excitation corresponds to the allowed ^1B_u-like excited state and can be described to a large extent as a HOMO to LUMO transition (contributing for more than 80% of the excitation) in all cases. In order to compare the shift in nm with the experimental values we have also reported in the second column a set of values shifted by 0.7 eV. With this offset the excitation energy of the free astaxanthin model corresponds closely to the experimental value for astaxanthin. As discussed above, the underestimation of the excitation energy within TDDFT is in line with previous theoretical investigations for long polyenes. Nevertheless, the shift due to a change in conformation or to the presence of hydrogen bonds is reliable.

In the optimized s-*cis*-free astaxanthin model the end rings form a dihedral angle of about 46° with the polyene chain. This is in close agreement with the 43° angle observed in X-ray studies of canthaxanthin. According to the X-ray structure of β-crustacyanin [21], the astaxanthin chromophores bound to the protein are in an all-E conformation with the end rings essentially coplanar to the polyene chain. Thus, the main conformational change from free astaxanthin to the chromophore in β-crustacyanin consists of the coplanarization of the end-rings, thereby inducing a more extended conjugated chain. Comparing model I and II, we find that this protein induced conformational change produces a shift of 0.12 eV ($\Delta\lambda_{max} = 24$ nm). This is by far not sufficient to account for the full bathochromic shift that occurs upon binding. The next step is the inclusion of polarization effects induced by hydrogen bonding interaction. The X-ray structure of β-crustacyanin has revealed an asymmetric environment of the chromophore with a histidine and a water molecule hydrogen bonded to the keto-oxygens. In model III we include these interactions explicitly and we obtain a further shift in the excitation energy, though relatively small, giving a total shift of 0.19 eV ($\Delta\lambda_{max} = 39$ nm) compared to the s-*cis* astaxanthin model.

We have also calculated Mulliken partial atomic charge differences between the model I and model III. We find that the hydrogen bonds induce a small positive charge on the chromophore, which is consistent with the small positive charge difference estimated on the basis of ^{13}C chemical shift data for α-crustacyanin, thus further supporting the conclusion that keto groups are not protonated and no strong polarisation occurs upon binding. An analysis of the HOMO and LUMO orbitals for models I and III shows that a charge displacement from the centre of the chromophore towards the rings occurs upon excitation in the model including the hydrogen bonding with the protein environment, while little charge displacement is observed for the free chromophore.

We turn then our analysis on the importance of aggregation effects for the bathochromic shift mechanism. A first order estimate of the exciton splitting induced by the proximity of the chromophores in the protein can be obtained by using the dipole-dipole approximation. We use the X-ray structural information available for β-crustacyanin to evaluate the geometrical factor and we compute the transition dipole moment within the TDDFT for the model III. The two bound astaxanthins in β-crustacyanin approach each other within 7 Å at a position close to the C11-C12 bond. As can be seen in figure 2, the long axes of the molecules form an angle of about 120°. The computed transition dipole moment is very intense ($\approx$ 19 Debye) and oriented parallel to the long axis of the molecule pointing towards the ring hydrogen bonded to the His residue. By using this information together with the orientation factor in the dipole-dipole approximation, we obtain an exciton splitting of 0.49 eV. If we add the exciton splitting to the previously computed "on-site" shift for model III, we obtain a $\lambda_{max} = 650$ nm (see table 2).

Therefore, based on our TDDFT calculations, we can conclude that the total protein induced "on site" effects on the bathochromic shift are not very large and can account for only about 30% of the total observed bathochromic shift in α-crustacyanin and 40% of the bathochromic shift in β-crustacyanin. Our study strongly supports the hypothesis of exciton coupling as the main source of the bathochromic shift. The exciton splitting and the angle of about 120° between the dipole moments are consistent with the negative chirality observed in CD spectra of crustacyanin [18]. The calculated total shift ($\lambda_{max} = 650$ nm) is even larger than the observed bathochromic shift. In fact our estimate of the exciton splitting is based on the structural information on the dimer in β-crustacyanin, where $\lambda_{max} = 586$ nm. It is very likely that the simple approximation of the dipole-dipole interaction in the excited state used here overestimates the exciton splitting. Further refinement in the modeling is needed to obtain a more quantitative estimate of the exciton splitting. Nevertheless, we can clearly conclude that aggregation effects give the predominant contribution to the observed color shift.

4. Conclusions

We have explored the bathochromic shift in the photoactive yellow protein and in α-crustacyanin by using quantum chemical calculations based on time-dependent DFT. We have shown that in PYP the deprotonation of the chromophore gives the largest absorption shift, though also the thioester bond and the protein induced polarization contribute considerably to the total shift. TDDFT calculations for models of the crustacyanin active site show that the protein induced conformational changes and polarization effects contribute for about 30% of the total bathochromic shift. The exciton coupling due to the proximity of the astaxanthin chromophores, estimated using the β-crustacyanin dimer structure, is found to be large and can account for the additional absorption shift. In general TDDFT calculations appear to be a powerful tool to analyse with good accuracy the molecular mechanisms that are responsible for the absorption shift in chromophore-protein complexes.

ACKNOWLEDGMENTS. I would like to acknowledge the coworkers that have contributed to the results presented here. In particular, A. Sergi, M. Ferrario, M. Grüning and L. Premvardhan for the work on PYP, and A. C. van Wijk, A. Spaans, N. Uzunbajakava, C. Otto, H. J. M. de Groot and J. Lugtenburg for the work on crustacyanin.

References

[1] V. HELMS, Current Opinion in Structural Biology, **12** (2002), 169.

[2] R. E. STENKAMP, S. FILIPEK, C. A. DRIESSEN, D. C. TELLER and K. PALCZEWSKI, Biochim. et Biophys. Acta - Biomembranes, **1565** (2002), 168.

[3] A. SERGI, M. GRÜNING, M. FERRARIO and F. BUDA, J. Phys. Chem. B **105** (2001), 4386.

[4] L. PREMVARDHAN, F. BUDA, M. A. VAN DER HORST, D.C. LÜHRS, K. J. HELLINGWERF and R. VAN GRONDELLE, J. Phys. Chem. B **108** (2004), 5138.

[5] A. C. VAN WIJK, A. SPAANS, N. UZUNBAJAKAVA, C. OTTO, H. J. M. DE GROOT, J. LUGTENBURG and F. BUDA, J. Am. Chem. Soc. **127** (2005), 1438.

[6] T.E. MEYER, Biochim. Biophys. Acta **806** (1985), 175.

[7] K. J. HELLINGWERF, J. HENDRIKS and T. GENSCH, J. Phys. Chem. A **107** (2003), 1082.

[8] G. E. O. BORGSTAHL, D. R. WILLIAMS and E. D. GETZOFF, Biochemistry, **34** (1995), 6278.

[9] R. CAR and M. PARRINELLO, Phys. Rev. Lett. **55** (1985), 2471.

[10] R. M. DREIZLER and E. K. U. GROSS, "Density Functional Theory. An approach to the Quantum Many-Body problem", Springer-Verlag, Berlin, 1990.

[11] D. VANDERBILT, Phys. Rev. B **41** (1990), 7892.

[12] L. PAVESI, P. GIANNOZZI and F. K. REINHART, Phys. Rev. B **42** (1990), 1864.

[13] G. B. BACHELET, D. R. HAMANN and M. SCHLÜTER, Phys. Rev. B **26** (1982), 4199.

[14] (a) C. FONSECA GUERRA, J. G. SNIJDERS, G. TE VELDE and E. J. BAERENDS, Theor. Chem. Acc. **99** (1998), 391; (b) S. J. A. VAN GISBERGEN, J. G. SNIJDERS and E. J. BAERENDS, Comp. Phys. Commun. **118** (1999), 119.
[15] M. BACA, G. E. O. BORGSTAHL, M. BOISSINOT, P. M. BURKE, D. R. WILLIAMS, K. A. SLATER and E. D. GETZOFF, Biochemistry **33** (1994),14369.
[16] W. D. HOFF *et al.* Biophys. J. **67** (1994), 1691.
[17] Y. IMAMOTO, M. KATAOKA and F. TOKUNAGA, Biochemistry **35** (1996), 14047.
[18] M. BUCHWALD and W. P. JENCKS, Biochemistry **7** (1968), 844.
[19] V. R. SALARES, N. M. YOUNG, H. J. BEUKERS and P. R. CAREY, Biochim. Biophys. Acta **576** (1979), 176.
[20] R. J. WEESIE, J. C. MERLIN, H. J. M. DE GROOT, G. BRITTON, J. LUGTENBURG, F. J. H. M. JANSEN and J. P. CORNARD, Biospectroscopy **5** (1999), 358.
[21] M. CIANCI, P. J. RIZKALLAH, A. OLCZAK, J. RAFERTY, N. E. CHAYEN, P. F. ZAGALSKY and J. R. HELLIWEL, Proc. Natl. Acad. Sci. USA **99** (2002), 9795.

Coherent control of resonant light-matter interaction

G. C. La Rocca, F. Bassani, and M. Artoni

Quantum coherence and interference have recently enabled one to control with success the resonant interaction of light with atoms while effects of photon confinement in semiconductor microstructures have long been investigated. We here examine the possibility of combining quantum coherence and interference with photon confinement to control the resonant interaction of light waves in media with low dimensionality. Specifically we first examine the case of a semiconductor based microcavity in which all-optical coherent control is achieved via a coupling beam resonant with the exciton-biexciton transition and then the case of a one-dimensional periodic lattice built from resonant absorbing ultracold atoms. Tayloring of the photon modes dispersion is in this case achieved directly through coherent all-optical resonant nonlinearities.

The possibility of modifying the photon density of states through optical confinement, and thus engineering the light-matter interaction, has attracted great interest in the field of microcavities and photonic band gap systems. Planar microcavities based on inorganic as well as organic semiconductors allow to couple strongly a Fabry-Perot photon mode with an electronic excitation, typically a quantum well Wannier-Mott exciton or a Frenkel exciton. In this regime, which has been extensively studied both theoretically and experimentally [1], the light-matter coupling determined by the exciton oscillator strength is sufficiently strong to overcome broadening mechanisms due to the finite quality factor of the optical microcavity or to the exciton absorption. As a consequence, the new system eigenstates are a coherent superposition of excitons and cavity photons, *i.e.*, the upper and lower cavity polaritons separated by the Rabi splitting. The photon confinement in microcavities is only along the growth direction, while the cavity polaritons exhibit a two-dimensional dispersion along the microcavity plane. More generally, photonic crystals are inhomogeneous material media exhibiting periodic variations in space in one or more directions of their refractive index on length scales comparable to optical wavelengths. The periodic variation in their optical response leads to Bragg scattering of light and electromagnetic wave propagation becomes best described in terms of a photonic band structure, akin to the electronic band structure in a crystalline solid. In particular, when the light wavevector is close to the Brillouin zone boundary the propagation of light is strongly affected, leading to the existence of a range of frequencies, known as a photonic band gap, for which light does not propagate. In recent years, photonic band gap systems of different dimensionalities have been much investigated from the scientific as well as technological viewpoint [2]. We note that the photon modes of a microcavity can be simply considered as the defect states in a one dimensional photonic crystal. The microcavity layer inserted in the photonic band gap system of the distributed Bragg mirrors supports a photon mode localized along the growth direction analogous to an electronic defect state within the gap of a semiconductor. The high transmission of a Fabry-Perot mode is just due to tunneling via this localized photon mode. In the case of photonic crystals, the interest has been mainly focussed on molding the flow of light, rather than controlling a resonant interaction with a material excitation.

Quite a distinct approach to the control of light-matter interaction is via the use of all-optical resonant nonlinearities in multilevel atomic systems where quantum coherence and interference play a major role. Over the past decade and long after Gozzini's pioneering work [3], coherent population trapping in atoms has certainly revived the interest of the scientific community as it lays at the heart of topical phenomena such as electromagnetically induced transparency, subrecoil laser cooling, lasing without inversion, adiabatic transfer and slow light, just to mention a few [4]. For many potential applications solid state media are preferred. In such media, however, decoherence takes place rather quickly which makes effects associated with quantum coherence and interference rather difficult to observe. Electromagnetically induced transparency in solids has been so far attained in a class of materials exhibiting defect states such as, e.g., rare-earth impurities [5] or color center states [6]. In both cases, inhomogeneous broadening plays a very significant role. Conversely, coherent control schemes could be implemented in bulk semiconductors exhibiting narrow intrinsic resonances associated with delocalized free exciton and biexciton levels [7]. In particular, it has been recently anticipated that forbidden yellow exciton states of Cu_2O [8] can be used to obtain effects analogous to those in atomic systems. It is the purpose of this work to discuss the possibility of combining photon confinement and quantum coherence and interference effects in order to achieve novel and efficient all-optical control schemes. We will first consider the case of a semiconductor microcavity and then the case of a one-dimensional photonic band gap structure built from ultracold atoms.

Recently, we have studied the effects of quantum coherence between exciton and biexciton levels in CuCl [9]. CuCl is a prototype example of a semiconductor having an allowed interband transition, quite pronounced exciton and biexciton resonances [10] and exhibiting, in particular, a fully developed polaritonic stop-band. We have shown that in bulk CuCl a pump driven exciton-biexciton transition allows for a well developed transparency within the stop-band where a probe pulse may propagate with very low group velocity. More specifically, the transparency of a probe beam within the Z_3-exciton polariton stop-band can be controlled via a pump light beam resonant with the transition from the Z_3-exciton to the Γ_1-biexciton. The large oscillator strength of the exciton-biexciton transition and the very narrow linewidth and long coherence time of the biexciton state in the small wave-vector region appear to favor quite appreciable degrees of transparency. The phenomenon is reminiscent of quantum coherence and interference effects occurring in three-level atomic systems [4, 11], except that delocalized electronic excitations in a crystalline structure are instead involved in CuCl [12]. Besides, the physics underlying induced transparency within an otherwise reflecting stop-band relies on a frequency and wave-vector selective polaritonic mechanism. Transparency can be induced in our case through a ladder scheme in which a circularly polarized probe beam is nearly resonant with the transition from the crystal ground state to the Z_3 exciton with dispersion $\omega_x(k) = \omega_T + \hbar k^2/(2m_x)$ and linewidth γ_x, while a pump beam having opposite circular polarization couples the Z_3 exciton to the Γ_1 biexciton with dispersion $\omega_m(k) = \omega_M + \hbar k^2/(2m_m)$ and linewidth γ_m. The CuCl response to a weak probe beam of frequency ω and wavevector $\vec{k}$, in the presence of the strong coupling beam of frequency ω_c and wave-vector $\vec{k}_c$ opposite to $\vec{k}$, can be described by the following dressed dielectric constant [13]

$$\varepsilon(k,\omega) = \varepsilon_b + \frac{\varepsilon_b \Delta_{LT}}{\hbar\omega_x(k) - \hbar\omega - i\gamma_x + \Sigma} \quad , \tag{1}$$

$$\text{with} \quad \Sigma = \frac{\beta}{\hbar\omega + \hbar\omega_c - \hbar\omega_m(k-k_c) + i\gamma_m}.$$

Here Σ describes the nonlinearity due to the coherent pump and β is proportional to the pump intensity and the oscillator strength of the exciton-biexciton transition with $\beta \simeq 10^{-8}\,\mathrm{eV}^2$ at a pump power of 10 kW/cm^2. When Maxwell's equations are solved with such an $\varepsilon(k,\omega)$ in the absence of the pump ($\beta \to 0$) one obtains the usual upper and lower polariton dispersion branches and a polaritonic stop-band within which the probe is nearly completely reflected [14]. The exciton-biexciton coherent coupling induced by the pump leads to three solutions of the probe dispersion equation $(c^2k^2/\omega^2) = \varepsilon(k, \omega)$ [15] with a third polariton branch appearing in the frequency region of the exciton resonance. A pronounced transparency window in correspondence of the pump-induced dispersion branch opens up within the polaritonic stop-band around a frequency $\omega \simeq \omega_m(k-k_c)-\omega_c$. The transparency frequency can be coherently controlled over a rather wide spectral range of several meV corresponding to the entire polaritonic stop-band. However, we remark that even within the transparency window the absorption is here still quite significant unless sub-micron slab thicknesses are used. Unlike the exciton linewidth, the biexciton linewidth at a small wave-vector ($\gamma_m \simeq 0.05$ meV) affects critically the appearance of a transparency window [9].

Effects associated with biexciton resonances in strongly coupled semiconductor microcavities are an area of current interest both theoretically and experimentally [16]. Even in microcavities, the exciton-biexciton coherent coupling is expected to originate a third cavity polariton branch besides the usual upper and lower ones separated by the Rabi splitting. Using a material with well developed and strongly bound exciton and biexciton resonances such as CuCl would make the coherent coupling effects quite visible. No experimental results, to the best of our knowledge, have been reported on CuCl based microcavities. In the following, we use model parameters extrapolated from the bulk values to examine the behaviour of a microcavity containing a CuCl quantum well. Such an extrapolation procedure should be sufficient at least from a qualitative point of view. Specifically, we take a Rabi splitting Δ_R of about 50 meV, the two-dimensional biexciton binding energy of about 100 meV and the two dimensional exciton and biexciton broadenings of the order of 1 meV, allowing for interface roughness scattering. The coupling $\beta \approx 10^{-4}\,\mathrm{eV}^2$ is taken to be about 100 times larger than that used for the free standing slab [9]; such a choice takes into account the enhancement of the oscillator strength of the exciton-biexciton transition due to the 2D confinement and, most importantly, the resonant effect of the cavity on the pump electric field. Thus, the actual incident pump power need not be higher than in the bulk case. The microcavity probe transmission and reflection spectra are obtained by using a transfer matrix method whereby the CuCl quantum well is included via a local dressed dielectric function of the same form as above. The cavity is tuned below the exciton resonance with a detuning close to the biexciton binding energy. The exciton-biexciton quantum coherence can then be driven by a resonant pump entering the cavity at near normal incidence.

The microcavity reflectivity dips are shown in Figure 1a as a function of probe frequency and parallel wavevector q with the inclusion of the third cavity polariton branch induced by the pump. The corresponding section for q close to the bare exciton-cavity mode anticrossing is shown in Figure 1b, while Figure 1c shows the corresponding absorption spectrum. When compared with the case of the bulk, it has to be noticed that much larger biexciton (and exciton) broadenings do not prevent the observation of the pump induced transparency associated with the third cavity polariton branch. It is worthwhile to notice, in addition, that the exciton-biexciton coherent coupling effects here considered are of the same nature as those described by the "average polarization model" of Ref. [17], yet quite distinct from effects related to the presence of an incoherent exciton population that would transfer oscillation strength to the biexciton transition [18].

In particular, a very good approximation for the dispersion $\omega(k)$ of the three cavity polariton modes of the coherently dressed system can be obtained from the following determinantal equation of a coupled oscillator model:

$$\det\begin{pmatrix} \hbar(\omega_{MC}-\omega) & \Delta_R/2 & 0 \\ \Delta_R/2 & \hbar(\omega_x-\omega) & \sqrt{\beta} \\ 0 & \sqrt{\beta} & \hbar(\omega_m-\omega_c-\omega) \end{pmatrix} = 0 \quad ; \qquad (2)$$

where $\omega_{MC}(k)$ is the empty cavity photon mode and the three states corresponding to the rows of the above matrix from top to bottom are: i) one cavity photon, no exciton, no biexciton, N photons in the pump beam ($N \gg 1$), ii) no cavity photon, one exciton, no biexciton, N photons in the pump beam, iii) no cavity photon, no exciton, one biexciton, $(N-1)$ photons in the pump beam.

In the above example, photonic confinement is provided by the microcavity mirrors while the coherent control by the coupling beam resonant with the exciton-biexciton transition. Similarly, the spatial dependence of the optical response in familiar photonic band gap materials and the corresponding photonic band structure are determined once and for all by the way the photonic crystal is grown. Yet, it would be of great interest the possibility to tune directly the photonic confinement by optical means. Schemes to modify the optical properties of photonic crystals have relied on slow electro and thermo optics effects in infiltrated liquid crystals [19], or on fast resonant and nonresonant optical nonlinearities in semiconductors [20].

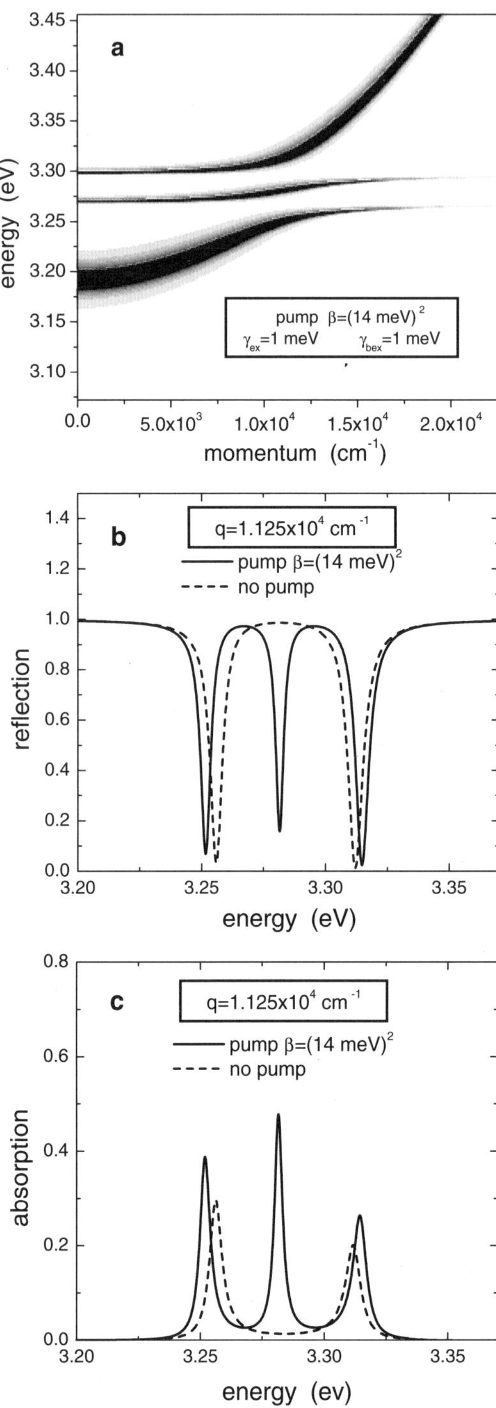

Figure 1. Microcavity reflectivity dips as a function of probe frequency and parallel wave-vector (a). Microcavity reflectivity (b) and absorption (c) spectra near the anticrossing

In such cases, the medium periodic structure determines the Brillouin zone, the gross features of the photonic bands and, in particular, the band-edge position which can easily be tuned. A completely different approach to the realization of tunable photonic crystals is that of directly creating the periodic modulation of the optical response via all-optical nonlinearities. In this case, the underlying material medium is homogeneous and the photonic band structure is fully determined, rather than modified, by one or more control beams. A suitable mechanism to achieve very large modulations of the optical properties relies on quantum coherence and interference in multilevel systems, the simplest of which is a three level atom in a Lambda configuration commonly used to observe electromagnetically induced transparency with a monochromatic control beam. When travelling waves are used as control beams, the dressed dielectric function experienced by a weak probe beam does not vary in space, yet when a standing wave configuration for the pump [21] is employed, the probe optical response is modulated periodically in space realizing a photonic band gap system [22–24]. We show in the following how to realize an all-optical tunable and fully developed photonic band gap in a three level system employing for the pump a modified standing wave configuration without nodes, in which the resonant absorption of the probe at the pump nodes no longer prevents the full development of a photonic stop band [25].

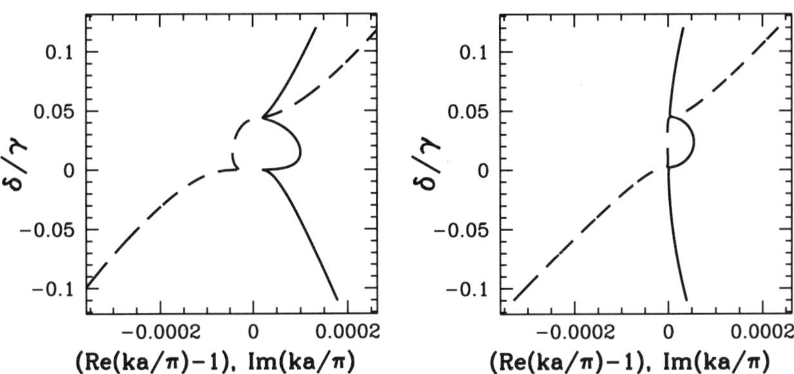

Figure 2. Probe photonic band structure near the Brillouin zone boundary: to the left a perfect ($\eta = 0$) and to the right a modified ($\eta = 0.05$) standing wave pump configuration. Both real (dashed lines) and imaginary (solid lines) parts of the Bloch wave vector are shown.

Let us consider a sample consisting of three level stationary atoms having a ground level g, an excited level e and a metastable level m, being the transitions $e-g$ and $e-m$ electric dipole allowed and the level g the only one populated. To be definite, numerical results will be calculated using parameters appropriate to the D_2 line of cold rubidium atoms, in which case g and m are long-lived hyperfine sublevels of the electronic ground state $S_{1/2}$, while e corresponds to the electronically excited $P_{3/2}$ state. A weak probe beam of frequency ω and propagating in the x-direction is nearly resonant with the $e - g$ transition. A strong control beam of frequency ω_c and Rabi frequency Ω_o is used to resonantly drive the $e-m$ transition as in a Lambda configuration for electromagnetically induced transparency. The dressed dielectric function experienced by the probe can be written as

$$\epsilon(\omega, x) = 1 - \frac{3\pi \mathcal{N} \gamma (\delta + i\gamma_m)}{(\delta + i\gamma)(\delta + i\gamma_m) - \Omega_c^2(x)}\,, \tag{3}$$

with

$$\frac{\Omega_c^2(x)}{\Omega_o^2} = \left(1 + \sqrt{R_m}\right)^2 \cos^2\left(\frac{\omega_c}{c}x\right) + \left(1 - \sqrt{R_m}\right)^2 \sin^2\left(\frac{\omega_c}{c}x\right)\,.$$

Here $\mathcal{N}$ is the sample scaled average atomic density $\lambda\!\!\!^{-}{}_o^3 n$ where λ_o is the wavelength of the $e - g$ transition. We denote by $\delta = \omega - \omega_{eg}$ the probe detuning while the coupling beam frequency is exactly resonant with the $e-m$ transition. The overall dephasings of the excited and metastable levels are γ and γ_m, respectively [27]. The Rabi frequency of the pump beam $\Omega_c(x)$ corresponds to a standing wave configuration obtained retroreflecting a beam of Rabi frequency Ω_o on a mirror of reflectivity R_m, and its spatial periodicity a is half the control beam wavelength $a = \lambda_c/2$ [26]. For a perfect standing wave $R_m = 1$ and $\Omega_c^2(x) \propto \cos^2(\omega_c x/c)$ varies periodically from 0 to $4\,\Omega_o^2$. In particular, the pump intensity vanishes at the nodal positions around which the atoms remain normally absorbing. Yet, it is sufficient to slightly reduce the reflectivity of the mirror to make the pump intensity nowhere vanishing with nodes becoming quasi-nodes. In terms of the tunable parameter $\eta = (1 - \sqrt{R_m})/(1 + \sqrt{R_m})$ with $0 \le \eta \ll 1$, the minimum value of $\Omega_c^2(x)$ at a quasi-node is approximately $4\,\eta^2\,\Omega_o^2$.

The probe progates in the coherently dressed system described by Eq. (3) as in a one-dimensional photonic crystal. Because $\omega \simeq \omega_c$, the probe wavevector is close to the corresponding Brillouin zone boundary (π/a) where a photonic stop band is expected to open [28]. As a first step we calculated numerically the 2×2 unimodular transfer matrix [29] $M(\omega)$ for a probe propagating through a single period of length a. Then, the translational invariance of the periodic medium is fulfilled by imposing the Bloch condition on the photonic eigenstates:

$$\begin{pmatrix} E^+(x+a) \\ E^-(x+a) \end{pmatrix} = M(\omega) \begin{pmatrix} E^+(x) \\ E^-(x) \end{pmatrix} = \begin{pmatrix} e^{ika}\,E^+(x) \\ e^{ika}\,E^-(x) \end{pmatrix}, \tag{4}$$

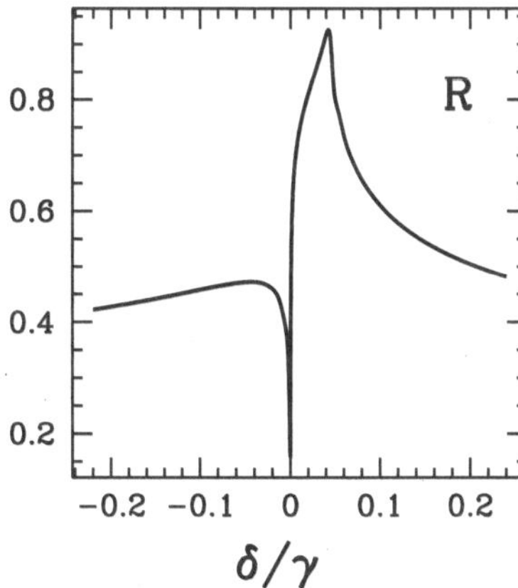

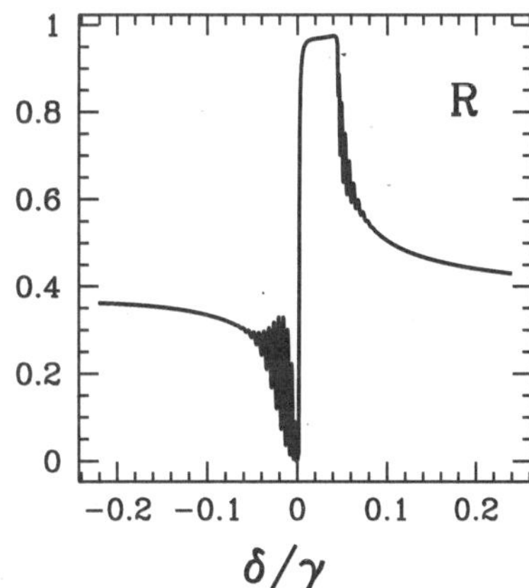

Figure 3. Reflection spectra corresponding to the photonic bands of Figure 2 for samples with $N = 2^{16}$.

where E^+ and E^- are the electric field amplitudes respectively for the forward and backward (Bragg reflected) propagating probe, and $k = k' + ik''$ is the probe complex Bloch wavevector. The one-dimensional photonic band structure is obtained from the solution of the corresponding determinantal equation $e^{2ika} - \text{Tr}(M(\omega))e^{ika} + 1 = 0$ ($det\ M$=1). By noting that if k is a solution also $-k$ is a solution, one directly has [31] $ka = \pm \cos^{-1}[\frac{\text{Tr}(M(\omega))}{2}]$.

We show in Figure 2 the photonic band structure around the lowest photonic gap for the two cases of $\eta = 0$ (perfect standing way) and $\eta = 0.05$ (the atomic density being $n = 2 \cdot 10^{12}$ cm^{-3}). In the latter case, dissipative effects are nearly negligible and the gap is characterized as the frequency range in which $k' = \pi/a$ and $k'' \neq 0$, while states in the bands have $k'' \simeq 0$. In the former case, the residual absorption at the pump nodal positions is appreciable and the gap is less clearly defined [25]. In particular, k' deviates significantly from π/a and, in the bands, k'' is larger.

The photonic band structure just discussed refers to the probe Bloch modes in an infinite periodic stack. Typical experimental investigations [2, 23] focus on the transmission and reflection of electromagnetic waves through samples of finite length. The transfer matrix approach is ideally suited for calculating such spectra. The total transfer matrix $M_{(N)}$ of a sample of thickness $L = Na$ ($N \gg 1$) is simply given in terms of the single period transfer matrix M as $M_{(N)} = M^N$. Because M is unimodular, $M_{(N)}$ can be expressed in closed form in terms of the elements of

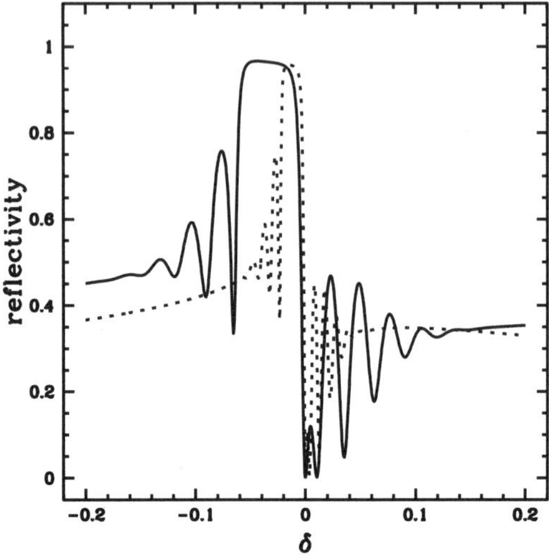

Figure 4. Reflection spectra for a longer spatial periodicity ($\theta \simeq 10^{-2}$) in a denser sample ($n = 10^{13}$ cm^{-3}) with $N = 2^{13}$, for two different control beam intensities: $\Omega_o = 5\gamma_e$ (solid line) and $\Omega_o = 3\gamma_e$ (dotted line)

M and of the complex Bloch wavevector k. Finally, the reflection ($\rho(\omega)$) and transmission ($\tau(\omega)$) amplitudes for a sample of N periods are

$$\rho = \frac{M_{(N)12}}{M_{(N)22}} = \frac{M_{12}\ \sin(kaN)}{M_{22}\ \sin(kaN) - \sin(ka(N-1))}, \tag{5}$$

$$\tau = \frac{1}{M_{(N)22}} = \frac{\sin(ka)}{M_{22}\ \sin(kaN) - \sin(ka(N-1))}, \tag{6}$$

from which, in turn, the directly measurable reflection ($R = |\rho|^2$), transmission ($T = |\tau|^2$), and absorption ($A = 1 - R - T$) spectra follow. We show in Figure 3 the probe reflection spectra corresponding to the photonic band structures of Figure 2 for thick samples [32]. It is clear how the use of a modified standing wave configuration ($\eta = 0.05$) leads to a fully developed stop band with $R \simeq 1$, contrary to the case of a perfect standing wave [33] in which absorption is significant in the spectral region of the gap. The photonic band gap is very narrow and is contained within the transparency window at the quasi-nodes where the pump intensity is minimum. Finally, in Figure 4 we show that the position and the width of the stop band can be tuned by changing the standing wave period [26, 28] and the control beam Rabi frequency Ω_o ($\eta = 0.05$ as above), while a denser sample leads to a well developed stop band at smaller sample thicknesses.

The present results demonstrate that photon confinement combined with coherent nonlinearities may be very efficient toward all-optical control of the resonant response of both condensed matter and atomic samples. The possibility to tune via a modified standing wave pump a fully developed one-dimensional photonic band gap can be extended to two and three dimensions [34]. Most importantly, time-dependent modulations of light signals propagating through such photonic band-gap materials may be quite relevant for applications to optical switching, light storage and quantum nonlinear optics.

We are honoured to present this work to the Festschrift celebrating Mario Tosi's 72nd birthday. Mario has been for many years, and continues to be, a clear example of scientist and teacher. We wish him continued excellent health and we look forward to his contributions to our common field of condensed matter physics for a long time to come.

References

[1] G. Khitrova, H. M. Gibbs, F. Jahnke, M. Kira and S. W. Koch, Rev. Mod. Phys. **71** (1999), 1591; A. Kavokin and G. Malpuech, *Cavity Polaritons*, In: "Thin Films and Nanostructures", Vol. 32, Elsevier, 2003; *Electronic excitations in organic based nanostructures*, V. M. Agranovich and F. Bassani (eds.), In: "Thin Films and Nanostructures", Vol. 31, Elsevier, 2003.

[2] S. Johnson, J. Joannopoulos, R. Meade , 1591 J. Winn, "Photonic Crystals: The Road from Theory to Practice", Kluver, Norwell, 2002; R. Slusher and B. Eggleton, "Nonlinear Photonic Crystals", Springer, Berlin, 2003.

[3] G. Alzetta, A. Gozzini, L. Moi and G. Orriols, Nuovo Cimento **36** (1976), 5.

[4] S. Harris, Physics Today **50** (1997), 36; E. Arimondo, In: "Progress in Optics XXXV", E. Wolf (ed.), Elsevier Science, Amsterdam, 1996, p. 257; A. B. Matsko, O. Kocharovskaya, Y. Rostovtsev, G. R. Welch, A. S. Zibrov and M. A. Scully, Adv. At. Mol. Opt. Phys. **46** (2001), 191.

[5] B. S. Ham, M. S. Shahriar and R. P. Hemmer, Opt. Lett.s **22** (1997). 1138; K. Ichimura, K. Yamamoto and N. Gemma, Phys Rev. A **58** (1998), 4116; A. V. Turukhin, V. S. Sudarshanam, M. S. Shahriar, J. A. Musser, B. S. Ham and P. R. Hemmer, Phys. Rev. Lett. **88** (2002), 23602.

[6] C. Wei and N. B. Manson, Phys. Rev. A **60** (1999), 2540.

[7] Quantum coherence and interference effects have also been studied in semiconductor quantum wells: G. B. Serapiglia, E. Parpa-

LAKIS, C. SIRTORI, K. L. VODOPYANOV and C. G. PHILLIPS, Phys. Rev. Lett. **84** (2000), 1019; L. SILVESTRI, F. BASSANI, G. CZAJKOWSKI and B. DAVOUDI, Eur. Phys. J. B **27** (2002), 89; M. C. PHILLIPS, H. WANG, I. RUMYANTSEV, N. H. KWONG, R. TAKAYAMA and R. BINDER, Phys. Rev. Lett. **91** (2003), 183602.

[8] M. ARTONI, G. C. LA ROCCA and F. BASSANI, EuroPhys. Lett.s **49** (2000), 445; M. ARTONI, G. C. LA ROCCA, I. CARUSOTTO and F. BASSANI, Phys. Rev. B **65** (2002), 235422.

[9] S. CHESI, M. ARTONI, G.C. LA ROCCA, F. BASSANI and A. MYSYROWICZ, Phys. Rev. Lett. **91** (2003), 57402; F. BASSANI, G. C. LA ROCCA and M. ARTONI, Jour. of Luminesc. **110** (2004), 174.

[10] J. B. GRUN *et al.*, Chapter 11 in: "Excitons", E. I. Rashba and M.D. Sturge (eds.), North Holland, Amsterdam, 1982; A. L. IVANOV, H. HAUG and L. V. KELDYSH, Phys. Rep. **296** (1998), 237.

[11] T. W. HÄNSCH and P. E. TOSCHEK, Z. Phys. **236** (1970), 213; G. S. AGARWAL and R. W. BOYD, Phys. Rev. A **60** (1999), R2681.

[12] R. SHIMANO and M. KUWATA-GONOKAMI, Phys. Rev. Lett. **72** (1994), 530.

[13] V. MAY, K. HENNEBERGER and F. HENNEBERGER, Phys. Stat. Sol. (b) **94** (1979), 611; R. MÄRZ, S. SCHMITT-RINK and H. HAUG, Z. Phys. B **40** (1980), 9; K. CHO, J. Phys. Soc. Japan **54** (1985), 4444; E. HANAMURA, Phys. Rev. B **44** (1991), 8514; N. MATZUURA and K. CHO, J. Phys. Soc. Japan **64** (1995), 651.

[14] V. M. AGRANOVICH and V. L. GINZBURG, "Crystal Optics with Spatial Dispersion and Excitons", Springer, Berlin, 1984.

[15] As the pump light has a frequency sufficiently below that of the Z_3 exciton, its dispersion relation is simply approximated by $k_c = 2.55\,\omega_c/c$.

[16] T. BAARS, G. DASBACH, M. BAYER and A. FORCHEL, Phys. Rev. B **63** (2001), 165311, and references therein.

[17] U. NEUKIRCH, S. R. BOLTON, N. A. FROMER, L. J. SHAM and D. S. CHEMLA, Phys. Rev. Lett. **84** (2000), 2215.

[18] M.SABA, F. QUOCHI, C. CIUTI, U. OESTERLE, J. L. STAEHLI, B. DEVEAUD, G. BONGIOVANNI and A. MURA, Phys. Rev. Lett. **85** (2000), 385.

[19] K. BUSH *et al.*, Phys. Rev. Lett. **83** (1999), 967; D. KANG *et al.*, ibidem **86** (2001), 4052.

[20] P.M. JOHNSON *et al.*, Phys. Rev. B **66** (2002), 81102(R); H. W. TAN *et al.*, Phys. Rev. B **70** (2004), 205110.

[21] R. CORBALÁN *et al.*, Opt. Commun. **133** (1997), 225; H. Y. LING *et al.*, Phys. Rev. A **57** (1998), 1338; M. MITSUNAGA and N.

IMOTO, Phys. Rev. A **59** (1999), 4773; F. SILVA *et al.*, Phys. Rev. A **64** (2001), 33802.

[22] A. ANDRE *et al.*, Phys. Rev. Lett. **89** (2002), 143602.

[23] M. BAJCSY *et al.*, Nature **426** (2003), 638.

[24] H. KANG *et al.*, Phys. Rev. Lett. **93** (2004), 73601.

[25] Absorption in general blurs the distinction between photonic bands and gaps and may even supress it (I. DEUTSCH *et al.*, Phys. Rev. A **52** (1995), 1394; A. TIP *et al.*, J. Phys. A **33** (2000), 6223). A detailed treatment of resonantly absorbing photonic band gap systems including a discussion of the group velocity around the stop band will be published elsewhere.

[26] It is also of interest a configuration in which the forward and backward control beams are slightly misaligned by an angle θ, in which case the spatial periodicity given by $(\lambda_c/2)/\cos(\theta/2)$ can be varied. Such an instance is shown in Figure 4 below where $\theta \simeq 10^{-2}$ rad.

[27] In our case, $\gamma/(2\pi) \simeq 6$ MHz and $\gamma_m/(2\pi) \simeq 1$ KHz.

[28] For the case of Figures 2 and 3, the $e - g$ transition wavevector $(2\pi/\lambda_o)$ is slightly smaller than the Brillouin zone boundary $(\pi/a = 2\pi/\lambda_c)$, while for the case of Figure 4 it is slightly larger $(\pi/a = 2\pi\cos(\theta/2)/\lambda_c)$.

[29] M. BORN and E. WOLF, "Principles of optics", 6th Edition, Cambridge University Press, Cambridge, 1980.

[30] F. BASSANI and G. PASTORI PARRAVICINI, "Electronic states and optical transitions in solids", Pergamon Press, Oxford, 1975.

[31] The solutions of this equation can be folded into the first Brillouin zone expressing k', the real part of k, modulus $2\pi/a$. In Figure 2, however, we plot the photonic bands near the Brillouin zone boundary without folding them.

[32] The oscillations at the gap edges are Fabry-Perot fringes due to the assumption of a sample with sharp boundaries.

[33] The data presented in Ref. [23] for a perfect standing wave are very similar to the present calculations, but do not show the very sharp reflectivity dip at $\delta = 0$. However, that experiment is done at $T = 90\ ^{o}$C and special considerations for the Doppler effect would be required.

[34] C. S. ADAMS and E. RIIS, Progr. Quant. Electr. **21** (1997), 1.

Advances in optical writing on nematic azobenzene polymers: relaxation processes and length scales

L. Andreozzi, M. Faetti, F. Zulli, M. Giordano, and G. Galli

The possibility of achieving devices for rewritable optical storage is provided by polymers with azobenzene side groups, because they undergo photochemically induced trans-cis reversible isomerization. Main crucial parameters in data storage are bit stability, homogeneity at molecular level, and working temperature range. In this work we overview our investigations into liquid crystalline, nematic PMA4 homopolymers and copolymers of a side group azobenzene methacrylate, as candidate materials for optical nanowriting. Relaxation processes over different time-length scales are discussed with particular reference to understanding the interplay between homogeneity at the molecular level, bit stability and working temperature range. We emphasize the significance of conformational rearrangements of the polymer chain backbone in stabilizing structurally homogeneous substrates. Suitability of topography and birefringence writings is illustrated with examples of imprinted information at the nanometer scale.

1. Introduction

Side group azobenzene polymers are suitable materials as media for optical information storage. While quite a few examples of such polymers are investigated by several research groups (see for recent examples [1–4]), we are especially interested in nematic side group azobenzene polymethacrylates (PMA4 homopolymers and copolymers) that possess great potential for application (Figure 1).

The basic process in optical writing is the photoisomerization of the mesogenic trans azobenzene molecule (Figure 2a), and alignment can be effectively photoinduced by the combination of photoorientation and thermotropic self-organization. UV radiation ($h\nu$) excites the $\pi - \pi^*$ trans-cis isomer transition, while blue radiation ($h\nu'$), with an energy density below a certain threshold produces trans-cis-trans cycles leading to

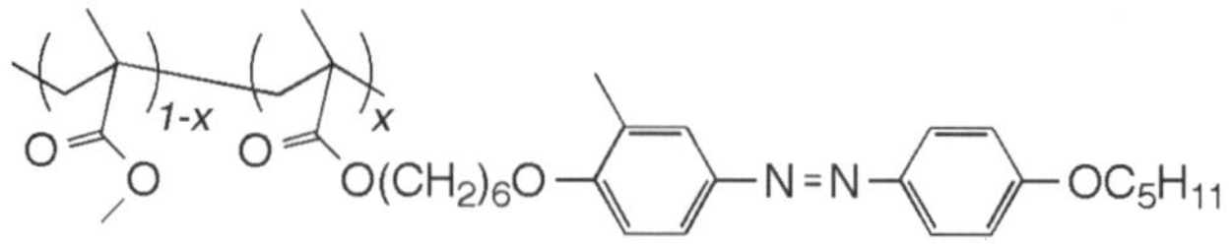

Figure 1. Chemical structure of the PMA4 homopolymer ($x = 1$) or copolymer ($x = 0.9$, 0.8, or 0.7) samples.

the reorientation of the azobenzene electric dipole (Figure 2b). Above threshold, the fraction of non-mesogenic cis isomer increases, resulting in a frustration of the nematic phase, and a cooperative rearrangement of the polymer chain takes place. Therefore, birefringence and/or topographic reliefs can be locally induced in dependence of the optical irradiation. The trans isomer form is energetically favored through a thermal activation process (Δ) leading to spontaneous relaxation.

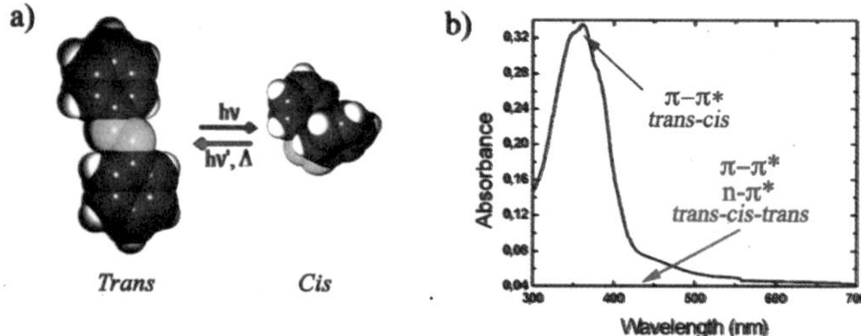

Figure 2. (a) Trans-cis photoisomerization of the azobenzene moiety. (b) Absorption spectrum of a PMA4 thin film [5].

The ultimate size of the obtainable optically induced bit is not determined yet. A guess of such dimension is of the order of the cooperativity or entanglement lengths [6, 7], that is about $3 - 10$ nm, respectively. In terms of data storage density that would correspond to about 1 Tbit/cm^2.

Main crucial parameters in data storage are bit stability, homogeneity at molecular level, and working temperature range. Bit stability and working temperature range depend on both glass temperature and conformational temperature values, the latter being associated with conformational spontaneous rearrangements of the polymer main chain. These can be modulated by varying the molar mass and molar mass distribution of the polymers, and in fact the temperature range of stability of the optical bit can be enlarged in polymers possessing a greater molar mass and a larger content of high molar mass components. Moreover, heterogeneities at the molecular level may substantially affect the bit stability, thus seriously limiting the effectiveness of the azobenzene-based polymer matrices as erasable storage devices at the nanometer length scale.

Homogeneity of the polymeric matrix strongly depends on its thermal history and specific annealing procedures.

On the other hand relaxation processes, effective at different time and length scales, can affect the bit stability, even if backbone conformational rearrangements can stabilize stored information over a larger temperature range. Therefore, spectroscopic techniques covering different time and length scales must be employed in order to fully characterize the matrices of interest.

Here we overview our recent results of studies on polymers and copolymers of nematic polymethacrylates PMA4 and discuss the relevance of various relaxation processes over different time-length scales. We also provide examples of optical writing from millimeter to nanometer resolution.

2. Materials

The PMA4 polymers under investigation (both homopolymers and copolymers with methyl methacrylate (MMA)) are nematic polymethacrylates containing the (3-methyl-4'-pentyloxy)azobenzene mesogenic unit connected at the 4-position by an hexamethylene spacer to the main chain [7]. The methacrylate monomer MA4 was synthesized following a standard procedure [8,9].

The different PMA4 homopolymer samples S1-S5 were prepared in five polymerization batches in which the experimental conditions (concentration of monomer, initiator, and chain transfer agent, polymerization time) were adjusted to yield samples with different molar masses and molar mass distributions (Table 1). Additionally, 10/90, 20/80, and 30/70 copolymers incorporating different contents of non-mesogenic MMA counits (10, 20, or 30 mol%, respectively) were prepared from comonomer feed mixtures of varied MMA/MA4 composition.

Homopolymer	M_n	M_w	M_z	M_4	M_w/M_n
S1	18600	59000	171200	310400	3.17
S2	15300	35500	64400	93900	2.32
S3	14300	42300	85700	127900	2.96
S4	15700	45900	91800	134600	2.92
S5	29900	72600	130500	186800	2.43

Table 1. Average molar masses and first polydispersity index of the homopolymer samples as determined by size exclusion cromatography (SEC) with polystyrene reference.

All of the PMA4 polymers formed a nematic phase (Tables 2 and 3) between the glass transition temperature (T_g) and the nematic-isotropic transition temperature (T_{NI}), as detected by differential scanning calorimetry (DSC) (Figure 3a). Beside these transitions, the homopolymer samples exhibited one further event pointing to the occurrence of a conformational transition of the polymer backbone at an intermediate temperature (T_c) (Figure 3b). By contrast, such a transition was not detected in any of the copolymers investigated (Figure 3c), probably because of the spreading of the azobenzene units by the non-mesogenic MMA units. Copolymerization also affected both T_{NI}, that was weakly depressed, and T_g, that was slightly increased, relative to the respective values of the homopolymers.

Homopolymer	S1	S2	S3	S4	S5
T_g (K)	294	296	302	303	305
T_c (K)	320	333	330	329	337
T_{NI} (K)	353	351	353	354	357

Table 2. Transition temperatures of the homopolymer samples.

Sample	S1	10/90	20/80	30/70
T_g (K)	294	306	308	314
T_c (K)	320	—	—	—
T_{NI} (K)	353	352	348	346

Table 3. Transition temperatures of the copolymers and S1 homopolymer sample.

The values of the first four moments of the molar mass distribution and the first polydispersity index, as evaluated by SEC for the homopolymers, were rather scattered and no clear correlation appeared to be possible with the DSC transition temperatures. A better understanding was achieved in [10] where the experimental SEC curves were reproduced by two logarithmic normal distribution functions [11].

3. Thermo-rheological behavior

The structural relaxation mechanisms of the azobenzene polymers and copolymers was investigated by using DSC and rheological techniques.

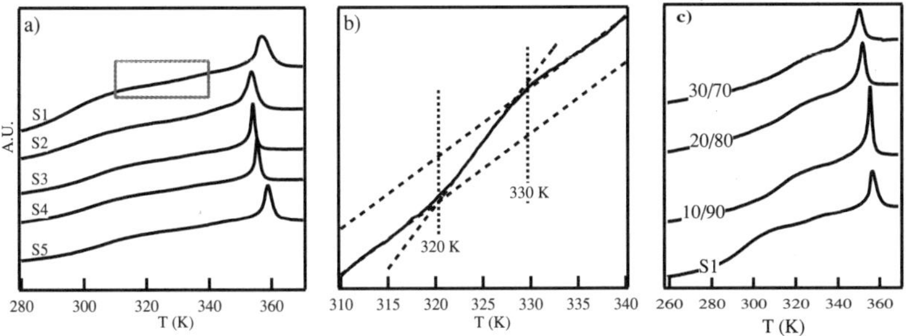

Figure 3. DSC heating curves for the S1-S5 homopolymers (a) with an enlarged region for the S1 homopolymer (b). DSC curves of the 30/70, 20/80, 10/90 copolymers compared to the S1 homopolymer (c).

In particular, long time relaxations were detected by carrying out physical aging experiments via DSC.

DSC is a widely used experimental technique in polymer science that allows one to record the temperature dependence of the heat capacity of a sample during a heating scan through the glass transition temperature. Before the measuring scan, the sample is subjected to a more or less complex thermal history usually involving an annealing stage of the glass. Due to the off-equilibrium character of the glassy state, during the annealing procedure the enthalpy of the system decreases, and what is measured by DSC in the following heating scan is the recovery of equilibrium near T_g evidenced in the experimental thermograms by an endothermic peak overshooting at the glass transition [12]. The position and shape of this peak strongly depends on the adopted annealing conditions (Figure 4).

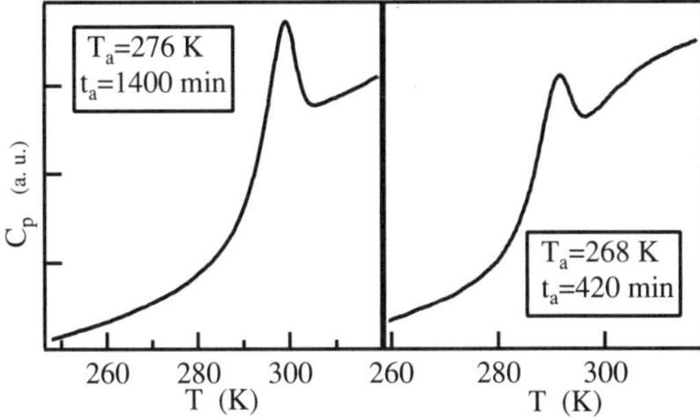

Figure 4. DSC heating curves for S1 homopolymer after two different thermal annealings in the glass. The annealing temperatures and times are shown in the insets.

The analysis of these experiments is quite complex because the typical techniques of linear response theory can not be directly used since the structural relaxation mechanism is strongly non linear. However, by adopting specific theoretical models [12, 13], it is possible to obtain from the DSC experiments the equilibrium temperature dependence of enthalpy relaxation time with good confidence. As an example, in Figure 5 a simultaneous fit of six different experimental DSC curves recorded in the 20/80 copolymer are reported. The configurational entropy approach [13] was used to reproduce the DSC experiments [14]. With the parameters obtained by the fit the enthalpy relaxation mechanism of the sample can be completely characterized. In particular, in Figure 6 the temperature dependence of the equilibrium relaxation time is reported for the 20/80 copolymer, the S1 homopolymer, and a commercial PMMA [15] for comparison.

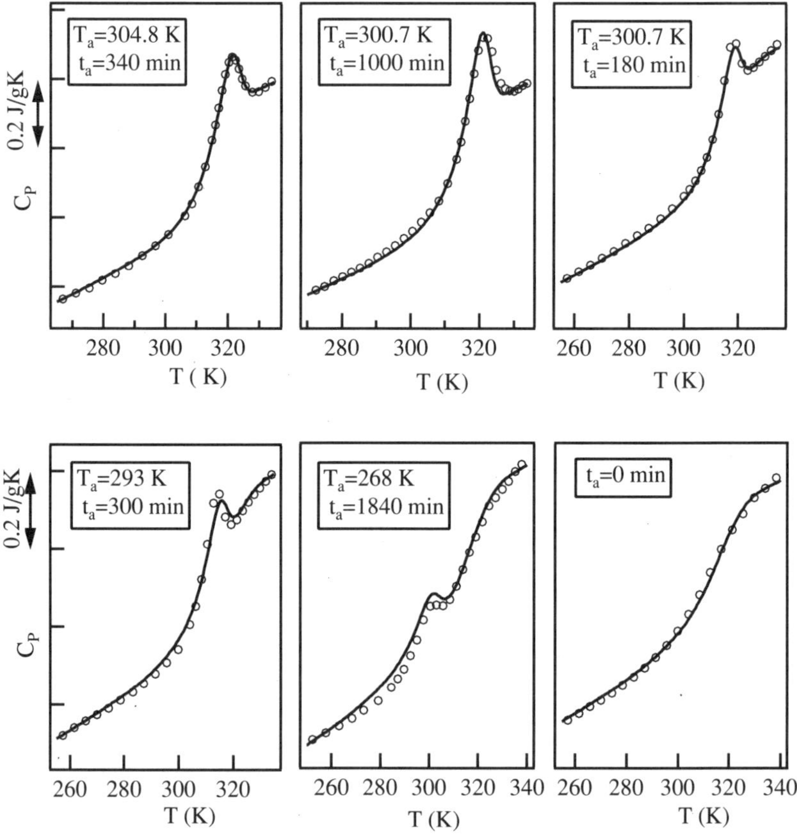

Figure 5. Simultaneous fitting procedure of six different DSC heating curves for the 20/80 copolymer by adopting the configurational entropy model.

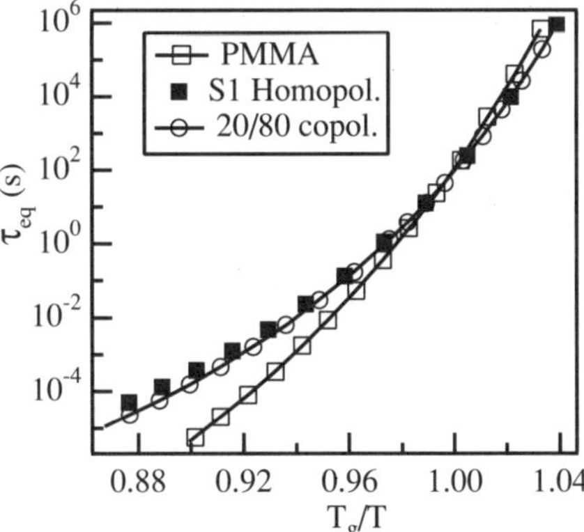

Figure 6. Equilibrium temperature dependence of the structural relaxation time of S1 homopolymer, 20/80 copolymer and commercial PMMA as obtained by the analysis of DSC experiments.

It can be noted that the S1 homopolymer and the 20/80 copolymer show very similar behaviors. On the contrary, the PMMA sample behaves in a quite different manner. By evaluating the steepness index at the glass transition:

$$m = \left.\frac{d \log \tau}{d(T_g/T)}\right|_{T=T_g} \tag{3.1}$$

representing a measure of the kinetic fragility of a glass former, we conclude [14] that the presence of the azobenzene side groups shifts the response of the polymer toward the strong character in the strong/fragile classification proposed by Angell [16].

The rheological measurements were performed in the temperature ranges of 303 – 403 K and 323 – 394 K for the S1 homopolymer and the 30/70 copolymer, respectively.

According to the time-temperature superposition (TTS) principle [17], the dependence of the shear elastic complex modulus on ω and T can be written as a function of shear elastic complex modulus measured at the reference temperature T_r:

$$G^*(\omega, T) = b_{T_r}(T)G^*(a_{T_r}(T)\omega, T_r) \tag{3.2}$$

The vertical and horizontal shift parameters, b_{T_r} and a_{T_r}, are two real temperature dependent functions.

Reference temperatures for the S1 homopolymer and the 30/70 copolymer were chosen at the same T_r/T_g ratio ($T_r = 363$ K for the homopolymer and $T_r = 388$ K for the copolymer) [18]. Single master curves were found for the storage G' and loss G'' moduli in both the nematic and isotropic phases for the two polymer samples [18]. This finding indicates that the rheological properties are dominated by the polymer backbone [19]. The structural times follow the Vogel-Fulcher (VF) law of the equation:

$$\tau^{st} = \tau_0^{st} \exp\left(\frac{T_b}{T - T_0}\right) \tag{3.3}$$

T_0 and T_b being the Vogel temperature and the activation pseudo-energy (in K), respectively. The relevant parameters in Eq. (3.3) were $T_0 = 259 \pm 5$ K, $T_b = 1300 \pm 50$ K and $\tau_0^{st} = (7.0 \pm 0.8) \times 10^{-13}$ s for the homopolymer, and $T_0 = 266 \pm 5$ K, $T_b = 1570 \pm 50$ K and $\tau_0^{st} = (1.7 \pm 0.2) \times 10^{-12}$ s for the copolymer.

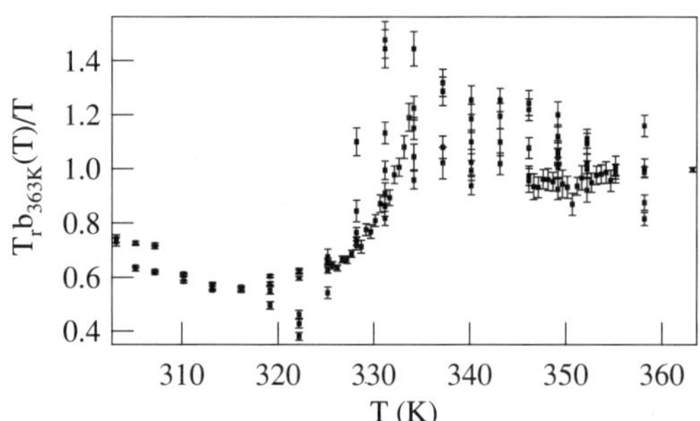

Figure 7. Temperature dependence of the vertical shift parameter $b_{363\ \mathrm{K}}(T)$ for the S1 homopolymer.

A peculiar phenomenon observed in the vertical shift parameter $b_{T_r}(T)$ for the homopolymer requires comment. Usually the $b_{T_r}(T)$ plots are very unstructured with varying temperature and consist of slightly scattered points around the reference line T/T_r. It is well known that this vertical correction factor accounts for the changes with temperature of the unrelaxed and/or relaxed shear moduli [17,20], which usually depend only on the small density variations with temperature. However, this behavior is not followed by PMA4 homopolymer, for which the $b_{363\ \mathrm{K}}(T)$ mean value doubles within a narrow range of 6 K when crossing the temperature of 328 K (Figure 7). It is important to note that previous electron spin resonance (ESR) and longitudinally detected ESR (LODESR) studies of the homopolymer dynamics firstly suggested [10, 21] that, on lowering temperature, a conformational transition of the polymer backbone took place in this same range of temperature, driven by the increasing nematic order in the side groups. This finding is also consistent with the

DSC thermograms shown in Figure 3, where a conformational change in the main chain polymer appeared to occur in the 320 – 330 K temperature range. This additional phenomenon possibly alters the usual trend of $b_{T_r}(T)$ of amorphous linear polymers, revealing itself as a not trivial temperature dependence of the limiting modulus.

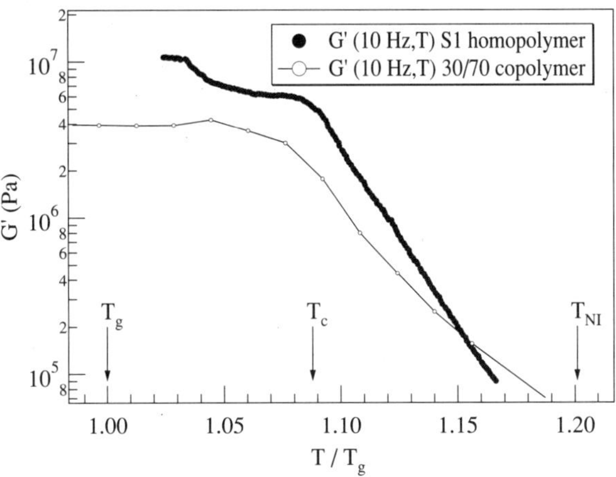

Figure 8. Temperature scan of storage modulus at the constant frequency of 10 Hz for the S1 homopolymer and 30/70 copolymer. Temperatures T_c and T_{NI} refer to S1 homopolymer.

As a consequence of the conformational transition, a structured increase in the polymer main chain rigidity is expected at lower temperatures below T_c. Evidence that the polymer is more rigid below T_c is provided by the data in Figure 8, which illustrates an oscillatory temperature scan. The trend of G' with temperature, at the fixed frequency of 10 Hz, clearly shows a plateau at $\approx 10^7$ Pa up to T_g and a second plateau at $\approx 6 \times 10^6$ Pa down to approximately T_c. Interestingly enough, the G' value in the plateau regions approximately doubles, the same 2-factor having been observed in the trend of $b_{T_r}(T)$. A pronounced drop in modulus is then detected as the temperature is increased in the nematic phase. It is worth noting that 10 Hz measurements are on a fast time scale with respect to the polymer dynamics around T_g, and the hardest plateau lasts for several degrees above the DSC glass transition temperature. However, in the vicinity of T_c the polymer dynamics is fast enough with respect to 10 Hz oscillatory measurements and the conformational crossover is signaled more precisely. Such a peculiar behavior does not occur in the 30/70 copolymer in which the presence of the non-mesogenic methyl methacrylate counits as side groups would inhibit the conformational transition of the polymer main backbone.

4. Heterogeneities at molecular level and thermal procedure

The inhomogeneities induced in the azobenzene homopolymers and copolymers by isothermal annealing in the isotropic phase were investigated by ESR spectroscopy [21–23]. The cholestane spin probe is an excellent molecular tracer, especially in the study of reorientation processes of liquid crystal polymers [24]. As an example, the temperature dependence of the spinning correlation time in the fast and slow molecular sites of cholestane dissolved in the 30/70 copolymer or homopolymer, after annealing at 383 K or 358 K, is shown in Figure 9 for a large temperature range. The study of the rotational dynamics of the cholestane spin probe dissolved in the polymer samples evidenced the presence of slow and fast sites available for reorientation. Both slow and fast components, whenever the latter was present, typically exhibited an Arrhenius behavior in the temperature region below T_g. The activation energy ranged approximately from 31 kJ/mol to 33 kJ/mol. Above T_g, two different dynamic regions, namely high temperature and intermediate temperature regions, were detected (Figure 9). The temperature dependences of the spinning correlation time for both fast and slow components could individually be described by the VF law:

$$\tau_{\|} = \tau_{\|0} \exp\left(\frac{T_b}{T - T_0}\right), \tag{4.4}$$

where $\tau_{\|}$ and T_b, the activation pseudo-energy in K, are constants depending also on the spin probe, and T_0 is the Vogel temperature. The values of the fit parameters are compared in Table 4. Since T_0 resulted in all cases, in both the high and intermediate temperature regions, of the same value as the corresponding Vogel temperature from rheological measurements, $\tau_{\|}$ can be expressed by a fractionary law of the structural relaxation time τ^{st}:

$$\tau_{\|} \propto \left[\tau^{st}(T)\right]^{\xi}. \tag{4.5}$$

In Eq. (4.5), the fractional exponent ξ may vary between 0 and 1, with $\xi = 1$ corresponding to a complete coupling of the probe dynamics to the structural relaxation of the host matrix. The fractional exponent resorts to be the ratio of the activation pseudo-energy, relevant to the ESR dynamics, to the T_b value of the structural relaxation time τ^{st}. In [8] the decoupling of the dynamics of a paramagnetic tracer dissolved in a liquid crystalline polymer was taken as a measure of the decoupling degree of the probe dynamics from the α relaxation of the polymer sample.

The stability with temperature of the molecular sites in both the homopolymer and copolymer was found to be very different (Figure 9).

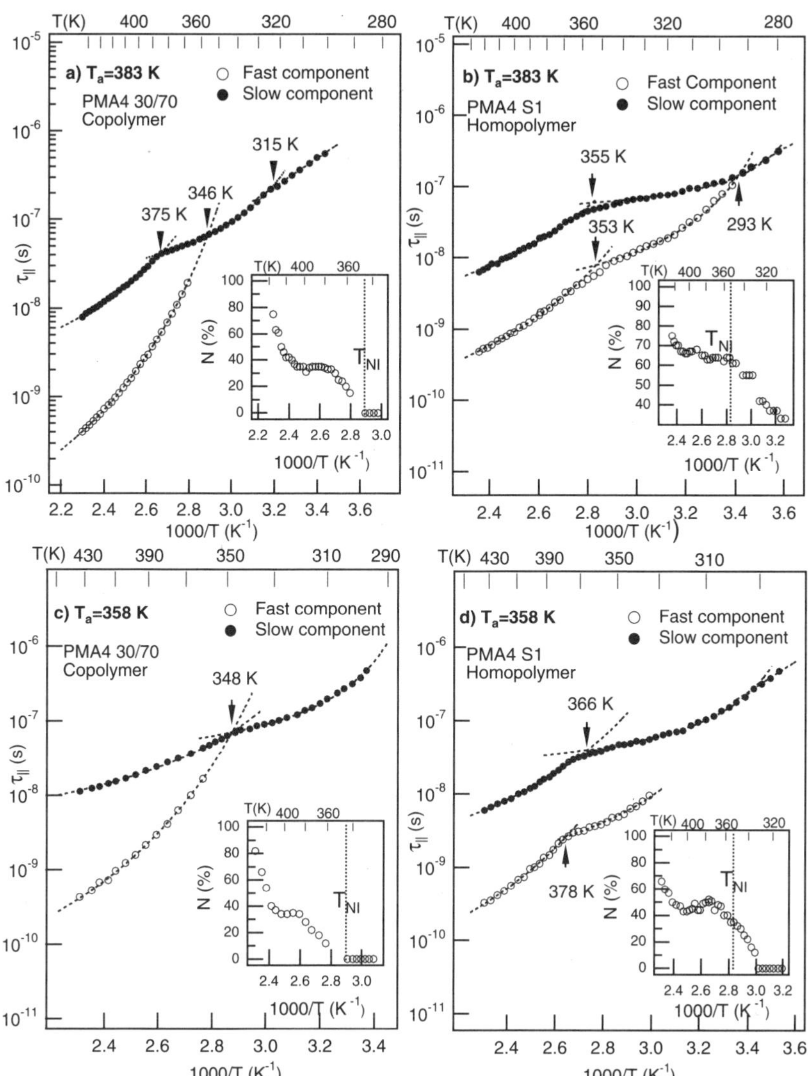

Figure 9. Temperature dependences of the spinning correlation time in the fast and slow sites of the cholestane spin probe after annealing at $T_a = 383$ K of the 30/70 copolymer (a) and the S1 homopolymer (from [10]) (b), or at $T_a = 358$ K of the copolymer (from [25]) (c) and the homopolymer (from [22]) (d). The insets show the percent population of the fast sites as a function of temperature.

The comparison between the probe dynamics and the polymer structural relaxation allowed to locate the reorientational sites at different positions of the polymer structure, that in turn allows to evaluate the inhomogeneity degree induced by the adopted annealing treatment, and therefore the suitability of the polymeric substrate for optical writing and the convenient temperature range.

	T region	30/70 Copolymer $\tau_{\|0}$ (s)	T_0 (K)	T_b (K)
$T_a = 383$ K	HT (F)	$(4.8 \pm 0.3) \times 10^{-12}$	266 ± 4	778 ± 62
	IT (F)	—	—	—
	HT (S)	$(4.4 \pm 0.3) \times 10^{-10}$	266 ± 5	493 ± 39
	IT (S)	$(1.1 \pm 0.2) \times 10^{-8}$	266 ± 5	147 ± 12
$T_a = 358$ K	HT (F)	$(2.9 \pm 0.2) \times 10^{-12}$	266 ± 7	820 ± 70
	IT (F)	—	—	—
	HT (S)	$(2.0 \pm 0.1) \times 10^{-9}$	266 ± 6	290 ± 20
	IT (S)	$(2.2 \pm 0.1) \times 10^{-8}$	266 ± 8	94 ± 6
	T region	S1 Homopolymer $\tau_{\|0}$ (s)	T_0 (K)	T_b (K)
$T_a = 383$ K	HT (F)	$(1.2 \pm 0.1) \times 10^{-11}$	259 ± 3	608 ± 27
	IT (F)	$(1.4 \pm 0.1) \times 10^{-9}$	258 ± 8	164 ± 9
	HT (S)	$(3.3 \pm 0.3) \times 10^{-10}$	259 ± 3	497 ± 4
	IT (S)	$(3.6 \pm 0.3) \times 10^{-8}$	259 ± 4	47 ± 2
$T_a = 358$ K	HT (F)	$(3.3 \pm 0.2) \times 10^{-12}$	258 ± 9	799 ± 60
	IT (F)	$(3.0 \pm 0.2) \times 10^{-10}$	258 ± 7	258 ± 15
	HT (S)	$(3.0 \pm 0.2) \times 10^{-10}$	258 ± 7	527 ± 20
	IT (S)	$(1.6 \pm 0.1) \times 10^{-8}$	258 ± 8	97 ± 7

Table 4. Fit parameters of the VF temperature dependence of $\tau_{\|}$ (fast and slow) in different temperature ranges after annealing at T_a temperatures.

Of particular interest appears to note that the cholestane dynamics in the S1 homopolymer becomes homogeneous at temperatures at which the conformational transition was signaled by the thermo-rheological analyses. Moreover, the annealing adopted at various temperatures in the isotropic phase of the copolymer resulted in very similar homogeneities in the nematic phase. In fact the fast component appears to be quite unstable and disappears at about 348 K. This finding suggests that an optically induced bit in the copolymer could be stable well above the glass transition temperature at higher temperatures than in the corresponding homopolymer, for which homogeneity of the substrate as detected by ESR is expected to onset below the conformational transition at 320 K.

5. Optical writing: length-scale in topography and birefringence

In order to test the effectiveness of the polymer structure and thermal history on optical bit stability, optical experiments were performed by inducing trans-cis-trans isomerization cycles of the azobenzene side groups.

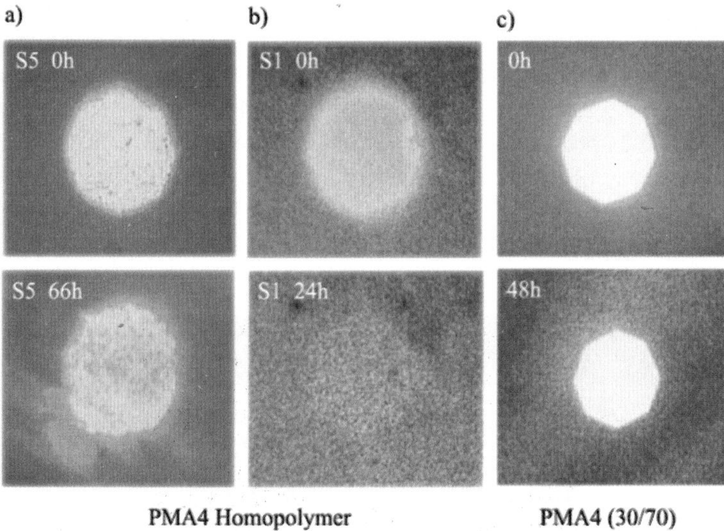

Figure 10. Optical writing performed at 338 K; samples observed with the polarizing microscope: S5 (a) and S1 (b) homopolymers, and 30/70 copolymer (c) [10, 25].

Figure 10 shows the optical writing in birefringence, at the mm scale, obtained in various originally isotropic samples. The experiment was performed at a temperature of 338 K which is above T_g and in the temperature region of homogeneity of the 30/70 copolymer (see Figure 9) and approximately coincident with the conformational transition temperature for S5 ($T_c = 337$ K, see Table 2). In the 30/70 copolymer, no appreciable relaxation effect took place during the time of 48 hours after irradiation, and the optically induced bit remained quite stable. A different behavior was observed in the optical bit in the homopolymer samples with the same experimental procedure, for which the optical bit rapidly became blurred and vanished. It can be envisaged that a main chain conformation modified by the optically induced orientation of the side chains was stabilized in the copolymer. It is apparent, however, that the temperature range of stability could be widened when selecting polymers with a higher molar mass, compare S5 to S1. This finding may address one crucial question of bit stability and suggests prescription of azobenzene polymethacrylates with optimized transition temperatures for information storage.

Experiments carried out at micrometer length scale [26] provided an estimation of the optical writing rate on 10 μm thick films of S1. The time necessary to induce a 5×5 μm^2 dot in birefringence was evaluated to be about 0.5 ms. The sample was linearly translated while illuminated by a focused blue spot, at various speeds in order to precisely adjust fluence. The writing effect obtained thereby consists of micron-sized lines, with molecular orientation perpendicular to the writing spot polarization direction, and with order parameter proportional to the absorbed energy. Lines produced at different fluence values show up as coloured in dependence of the local order parameter (Figure 11). The pattern was unchanged for 20 h at 318 K, that is slightly below the conformational transition temperature of S1 (Table 2).

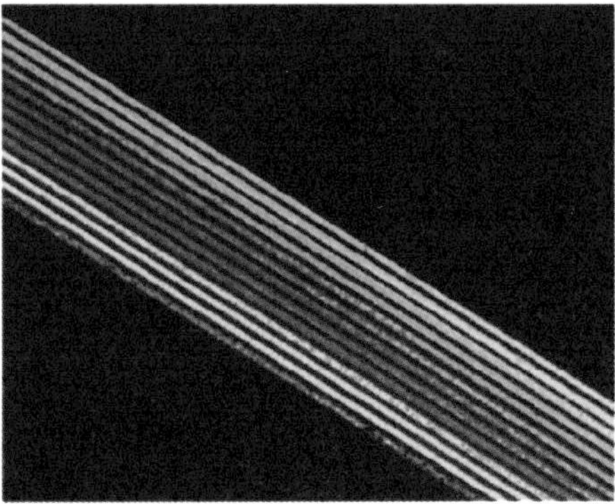

Figure 11. Polarizing optical microscopy image (1000×850 μm^2) of lines written by polarized light on a film of S1 homopolymer, at varying fluence [26].

Similar measurements were performed on films of the 10/90, 20/80, 30/70 copolymers of equal thickness, at room temperature and submitted to the same energy densities with the aim to evaluate the relative optical writing efficiencies at the micrometer length scale. Taking into account the spot dimensions (5 μm diameter) equal fluences were used (from 0.65 to 3.15 J/cm^2) [27]. The writing efficiency on the homopolymer turned out to be about 20 times greater than on the copolymers, Figure 12. Moreover, the writing threshold resulted to be much greater in the copolymers in which only lines written at a 400 μm/s rate can be detected.

Laser radiation through the aperture of a scanning near-field optical microscope (SNOM) was used to induce isomeric (cis-trans or trans-cis) transitions in the side units, obtaining their orientation, as well as mass

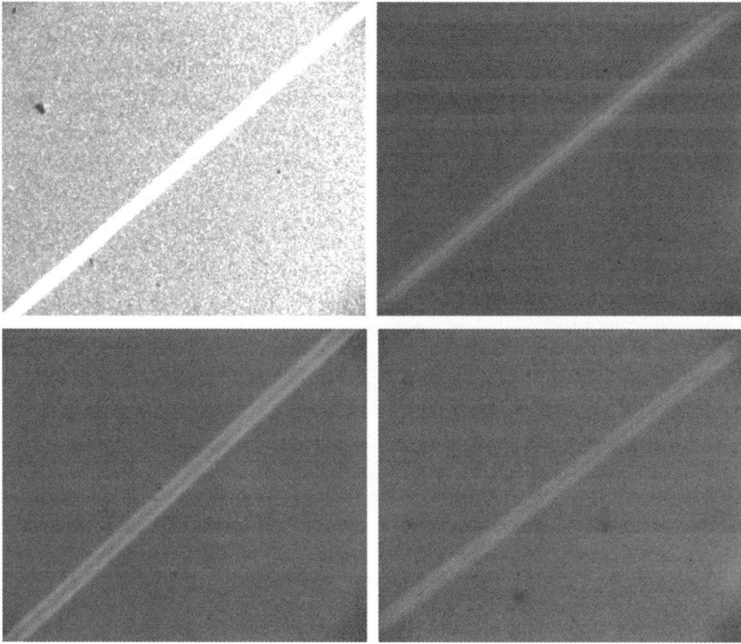

Figure 12. Examples of imprinted images at the μm scale by a energy flow of 0.65 J/cm^2 on the homopolymer (top left), 30/70 copolymer (top right), 10/90 copolymer (bottom left), and 20/80 copolymer (bottom right). The image on the homopolymer was obtained with a 10 times shorter exposition time.

migration due to induced displacement of the main chain. Both mechanisms can be used to imprint patterns on the sample surface. Optical writing and subsequent reading of sub-micron features were obtained on spin-coated thin films of S1 homopolymer and 30/70 copolymer samples. Topographic reliefs were detected by means of the SNOM distance control feedback loop. Optical modifications, induced by 458 or 488 nm light with fixed polarization state, were detected by SNOM by means of 690 nm laser light, that has shown not to perturb appreciably the material. Polarization-modulation SNOM provides additional information about the sample dichroism and birefringence properties [28, and references therein].

As a first application of SNOM, the check of the sample surface after the thermal treatment at 358 K is presented here. Topography and optical appearance as revealed by SNOM of S1 homopolymer thin film after thermal annealing at 358 K for 24 h and subsequent fast cooling to room temperature is shown in Figure 13 [29]. Flat topography, with occasional micron-size reliefs (100 nm high) is observed. We recall here that ESR spectroscopy provided us this recipe to obtain homogeneous substrates at the nanometer length scale of the polymers. Optical reading was accomplished by polarization modulation SNOM with 690 nm laser light (estimated out-put power $I = 500$ nW, tip diameter $a = 250$ nm). Randomly distributed optical features of micrometer size are present in both polarization modulation amplitude and phase images. The spontaneous optical domains look stable at least within a few hours, as demonstrated by repeated scans.

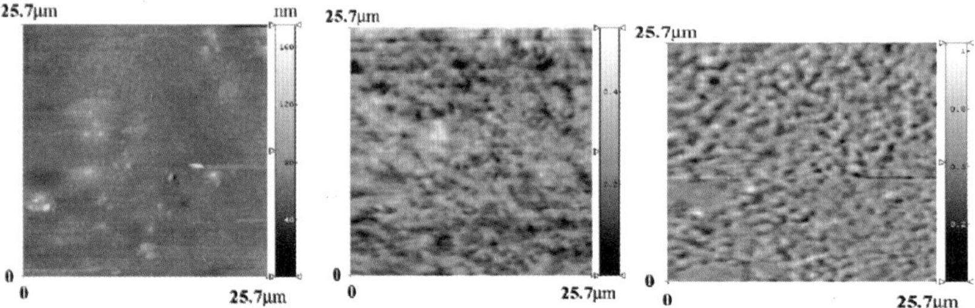

Figure 13. From the left: topography, polarization modulation amplitude (refractive index), polarization modulation phase (birefringence) of spontaneous optical domains in the S1 homopolymer thin film.

A sequence of optical images after writing vertical lines with 458 nm light is reported in the S1 homopolymer (Figure 14). The lines were imprinted by moving the SNOM tip at the constant speed of 30 nm/s light (estimated out-put power $I = 50$ nW, tip diameter $a = 250$ nm, estimated tip-sample distance $d = 5$ nm).

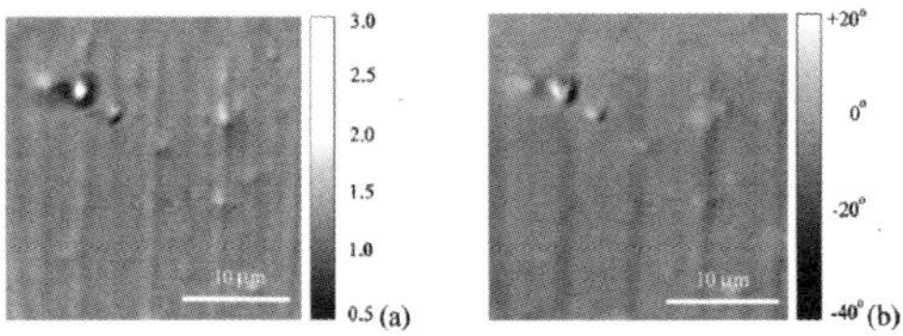

Figure 14. (a) Polarization modulation SNOM amplitude in arbitrary units, (b) polarization modulation phase of written optical lines in the S1 homopolymer spin-coated thin film [29]. The fluence in the experiment was enough to produce discontinuous topographic relief.

The lines in the amplitude image show a full width at half maximum (FWHM) of 600 nm, the same lines in the phase image present a FWHM of 1 μm. The imprinted optical information was shown to be stable at room temperature at least for one day. The different appearance and width of the lines provide evidence of the different character of the images. In particular, the optical orientation seems to spread over a wider region while the orientation degree (order parameter) is more pronounced in the center of the lines. Reliefs in the topography exhibit smaller width (400 nm) and are mostly uncorrelated with the optical features [29]. Therefore the information bit turns out to be blurred, even if to a less extent, also in topography.

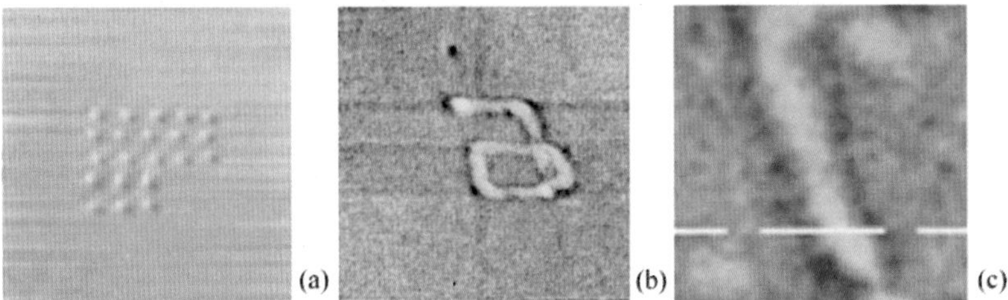

Figure 15. SNOM topography images of the optically nanostructured PMA4 thin films of (a) S1 homopolymer (exposure time = 1 s/dot, image size = 5 × 5 μm^2), (b) 30/70 copolymer (sensor speed = 20 nm/s, size = 2 × 2 μm^2) and (c) 30/70 copolymer (speed = 50 nm/s, size = 300 × 300 nm^2) [5,28].

The blurring of information bit is a serious drawback in data storage. So experiments in different materials were carried out to overcome this problem. Figure 15 shows examples of nanostructures created by SNOM with 325 nm laser light, enabling exclusively the trans-cis transition, thus eventually deforming the surface. The topographic structures were obtained at $T = 297$ K and have 180, 75 and 45 nm width, respectively. The aperture dimension of the SNOM probe was 100 nm (a), 80 nm (b), 50 nm (c). Thus, only in the 30/70 copolymer, the information bit reaches the intrinsic limit of the probe aperture, which corresponds here in the most favourable case to an information density of 40 Gbit/cm^2.

The comparison of the topographic behavior of the homopolymer and the 30/70 copolymer signals that structure and interactions at nanoscale appear to be relevant in order to obtain correspondence between the information bit and instrumental limit; more specifically at the nanoscale the greater the mesogenic interaction (homopolymer), the more blurred the information bit. This is at variance with the findings at the micrometric scale where instead the stronger mesogenic character stabilizes and mantains the light imprinted information in the homopolymer.

Birefringence writing experiments carried out on the 30/70 copolymer confirmed these features. In fact, optical patterns were written by tracing vertical lines spaced by 1000 and 500 nm, respectively, with the SNOM tip emitting about 5 nW of laser light at 488 nm for the used SNOM probe with $a \simeq 150$ nm (Figure 16). The tracing speed was 1 μm/s, providing a writing fluence of about 5 J/cm^2. Finally, Figure 17 shows a

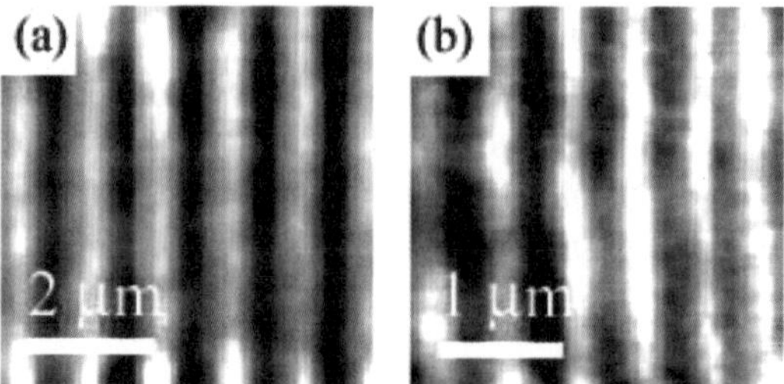

Figure 16. Birefringence SNOM images ($\lambda = 690$ nm) of optical gratings realized by SNOM on the 30/70 copolymer. The lines spacing is (a) 1000 nm, and (b) 500 nm [28]. The corresponding topographic relief along the inscribed lines are the order of 3 and 2 nm, respectively.

narrow vertical line obtained with a speed of 30 nm/s, corresponding to a fluence of about 150 J/cm^2. The imaging quality obtained in this case is limited by the intrinsic resolution of aperture SNOM due to the tip diameter ($a \simeq 100$ nm), so that the thin line appears with poor contrast and irregular cross-section. The line profile of Figure 17b demonstrates a purely optical writing/read-back resolution of the order of 100 nm.

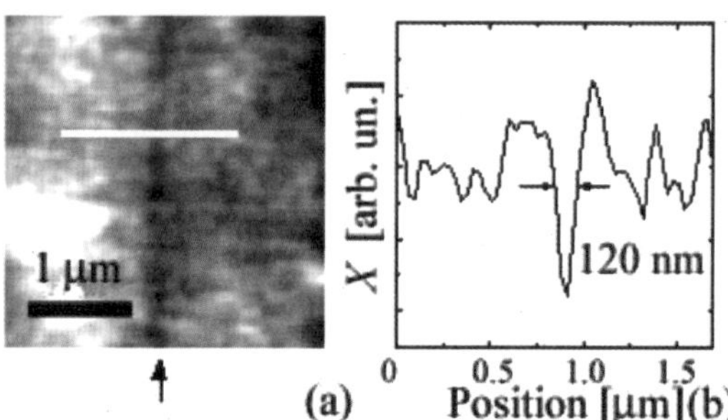

Figure 17. (a) Birefringence SNOM image of a line produced by previous writing with the SNOM probe at the arrow position. (b) Line profile at the horizontal stroke in (a) [28].

6. Conclusions

The photoinduced motion at molecular level in azobenzene polymers provides rewritable optical data storage as important possible application. The imprinted information is in principle limited only by the optics. However, when the writing process is pushed down to nanoscale, the chemical structure of the azobenzene polymers as well as their physical properties play a relevant role in the achievement of the limiting size of the imprinted information. In this paper, an overview on the physical parameters relevant to optical data storage in newly synthesized PMA4 azobenzene polymers is presented. ESR line-shape analysis permits both to single out a suitable thermal procedure to obtain homogeneous substrates convenient for optical writing and to understand the significance of the polymeric conformational rearrangements. We have shown that bit stability and working temperature range above the T_g can be modulated either by molar mass and molar mass distribution in homopolymers or by designing an adequate copolymer composition. Optical investigations, carried on millimeter and micrometer length scales, confirm the ESR findings. SNOM measurements proved the suitability of the adopted thermal procedure in order to obtain homogeneous substrates at the nanometer length scale. Due to the reduced influence of the nematic potential, films of 30/70 copolymer provide the best substrates among those taken into account so far. It has been shown that topographic patterning is obtained by SNOM with feature size as small as 50 nm, while birefringence experiments show imprinted information bit of 120 nm, each of them being determined by the SNOM aperture size. Further investigations of the ultimate bit size require optimization of the tip geometry, possibly including use of apertureless SNOM. This would allow systematic studies devoted to gain faster writing rates at the nanometer scale to be implemented in technological devices. On the other hand, we speculate that shorter response times over enlarged temperature ranges may alternatively be reached by chemical manipulation of the architecture of nematic azobenzene polymers that assemble in nanoscale morphologies.

ACKNOWLEDGEMENTS. We thank Italian MIUR for financial support (CIPE project cluster 26 P5BW5).

References

[1] M. NAKANO, Y. YU, A. SHISHIDO, O. TSUTSUMI, A. KANAZAWA, T. SHIONO and T. IKEDA, Mol. Cryst. Liq. Cryst. **398** (2003), 1.

[2] J. G. MEIER, R. RUHMANN and J. STUMPE, Macromolecules **33** (2000), 843.

[3] D. BUBLIZ, M. HELGERT, B. FLECK, L. WENKE, S. HVILSTED and P. S. RAMANUJAM, Appl. Phys. B **70** (2000), 803.

[4] O. YAROSHCHUK, T. BIDNA, M. DUMONTand J. LINDAU, Mol. Cryst. Liq. Cryst. **409** (2004), 229.

[5] S. PATANÈ, A. ARENA, M. ALLEGRINI, L. ANDREOZZI, M. FAETTIand M. GIORDANO, Opt. Commun. **210** (2002), 37.

[6] A. L. KHOLODENKO and T. A. VILGIS, Physics Reports **298** (1998), 251.

[7] E. DONTH, J. Polym. Sci. B **34** (1996), 2881.
[8] L. ANDREOZZI, M. FAETTI, M. GIORDANO, D. PALAZZUOLI and G. GALLI, Macromolecules **34** (2001), 7325.
[9] A. S. ANGELONI, D. CARETTI, M. LAUS, E. CHIELLINI and G. GALLI, J. Polym. Sci. Polym. Chem. Ed. **29** (1991), 1865.
[10] L. ANDREOZZI, M. FAETTI, M. GIORDANO, D. PALAZZUOLI, M. LAUS and G. GALLI, Mol. Cryst. Liq. Cryst. **398**, (2003), 97.
[11] J. BRANDRUP and E. H. IMMERGUT (eds.), "Polymer handbook", John Wiley & Sons, New York, 1989, 3rd ed.
[12] I. M. HODGE, J. Non-Cryst. Solids **169** (1994), 211.
[13] J. L. GÓMEZ RIBELLES and M. MONLEÓN PRADAS, Macromolecules **28** (1995), 5867.
[14] L. ANDREOZZI, M. FAETTI, M. GIORDANO and D. PALAZZUOLI, Macromolecules **35** (2002), 9049.
[15] J. L. GÓMEZ RIBELLES, M. MONLEÓN PRADAS, A. VIDAURRE GARAYO, F. ROMERO COLOMER, J. MAS ESTELLÉS and J. M. MESEGUER DUEÑAS, Macromolecules **28** (1995), 5878.
[16] R. BÖHMER and C. A. ANGELL, Phys. Rev. B **45** (1992), 10091.
[17] J. D. FERRY, "Viscoelastic properties of polymers" New York, 1980, 3rd ed.
[18] L. ANDREOZZI, M. FAETTI, M. GIORDANO, F. ZULLI, G. GALLI and M. LAUS, Mol. Cryst. Liq. Cryst. **429** (2005), 301.
[19] J. BERGHAUSEN, J. FUCHS and W. RICHTERING, Macromolecules **30** (1997), 7574.
[20] N. G. MCCRUM, B. E. READ and G. WILLIAMS, "Anelastic and dielectric effects in polymeric solids", Dover Publications, Inc., New York, 1991.
[21] L. ANDREOZZI, M. BAGNOLI, M. FAETTI and M. GIORDANO, Mol. Cryst. Liq. Cryst. **372** (2001), 1.
[22] L. ANDREOZZI, M. FAETTI, M. GIORDANO, D. PALAZZUOLI, M. LAUS and G. GALLI, Macromol. Chem. Phys. **203** (2002), 1636.
[23] L. ANDREOZZI, M. FAETTI, G. GALLI, M. GIORDANO, D. PALAZZUOLI and F. ZULLI, Mol. Cryst. Liq. Cryst. **429** (2005), 21.
[24] L. ANDREOZZI, M. GIORDANO and D. LEPORINI, In: "Structure and transport properties in organized polymeric materials", E. Chiellini, M. Giordano and D. Leporini (eds.), World Scientific, Singapore, 1997, Vol. 11 of "Contemporary Chemical Physics", chap. 7.
[25] L. ANDREOZZI, M. FAETTI, D. PALAZZUOLI, G. GALLI and M. GIORDANO, Mol. Cryst. Liq. Cryst. **398** (2003), 87.

[26] L. ANDREOZZI, P. CAMORANI, M. FAETTI and D. PALAZZUOLI, Mol. Cryst. Liq. Cryst. **375** (2002), 129.
[27] Internal report CIPE project cluster 26 P5BW5 (1998/2003).
[28] V. LIKODIMOS, M. LABARDI, L. PARDI, M. ALLEGRINI, M. GIORDANO, A. ARENA and S. PATANÈ, Appl. Phys. Lett. **82** (2003), 3313.
[29] M. LABARDI, N. COPPEDÈ, L. PARDI, M. ALLEGRINI, M. GIORDANO, S. PATANÈ, A. ARENA and E. CEFALÌ, Mol. Cryst. Liq. Cryst. **398** (2003), 33.

Investigation of the electronic core levels of $MnPS_3$ intercalated with the potassium ion

V. Grasso and L. Silipigni

With the aid of the X-ray photoelectron spectroscopy technique, we have investigated the electronic core levels of $K_{2x}Mn_{1-x}PS_3$, an intercalation compound of the manganese thiophosphate ($MnPS_3$). The attention is focused on the most intense XPS core-level peaks of the constituent atoms. All the core levels analyzed show a single-peak structure. On going from $MnPS_3$ to $K_{2x}Mn_{1-x}PS_3$, the lack of remarkable shifts on the binding energy positions of the investigated Mn, P and *S* core levels confirms that no electron transfer occurs from the guest species (K^+) to the host ($MnPS_3$). Moreover, the K $3p$ levels are discrete and well localized in good agreement with the presence of a weak link between potassium and the cluster $(P_2S_6)^{4-}$ of $MnPS_3$. The Mn $2p$ and $3p$ core levels show a satellite feature on the high binding energy side of each peak suggesting that $K_{2x}Mn_{1-x}PS_3$ is a large-gap insulating Mn compound, like $MnPS_3$.

1. Introduction

The incorporation of guest species (ion, atom or molecule) into the two-dimensional host material interlayer region (*intercalation*) can lead to advanced materials with interesting chemical, electronic, optical or mechanical properties. This class of materials are referred to as *nanocomposite* since the host matrix and the guest molecules are mixed on the molecular level [1]. The properties of nanocomposite materials depend not only on the properties of their individual parents but also on their morphology and interfacial characteristics. There is also the possibility of new properties which are unknown in the parent constituent materials. Therefore nanocomposites show themselves as promising candidates in a variety of technological applications, e.g. in energy storage, electrochromic and ferroelectric devices, reversible batteries, nonlinear optical and quantum size devices, and catalysis, and this their potentiality makes them particularly appealing.

In this framework, the $MnPS_3$ thiophosphate is a more and more interesting layered host lattice since in it, like in $CdPS_3$ and $FePS_3$ compounds, it is possible to substitute, by means of an ion-exchange intercalation reaction, a fraction of intra-layer manganese cations with alkali-metal ions such as K^+, Na^+, Li^+, and so on [2, 3]. This type of exchange reaction, that is unique to some MPS_3, creates therefore immobile and randomly distributed cationic vacancies in the layer. Moreover a significant increase in the interlayer spacing is observed indicating that the alkali metal ions are incorporated within the interlayer regions. This chemistry is probably a consequence of the fact that the bonding between the divalent metal ion and the thiophosphate group is fairly ionic [4,5] and allows one to have an extraordinary variety of ways for modifying (*tailoring*) at ambient temperature the chemical composition of the interlayer space of the host lattice while retaining its structural units. It is thus possible to obtain a variety of intercalation compounds starting from the ion-exchange intercalation reaction, whose physical properties of new materials will depend on the chemical nature of the guest species.

The ability of $MnPS_3$ to this type of exchange, which is described by the following reaction:

$$\mathrm{MnPS_3} + 2x\mathrm{M^ICl} \overset{\mathrm{H_2O}}{\rightarrow} \mathrm{M^I}_{2x}\mathrm{Mn}_{1-x}\mathrm{PS_3} + x\mathrm{MnCl_2} \tag{1.1}$$

where M^I is an alkali metal such as K^+, Na^+, Li^+, Cs^+, ..., is due to the presence of empty and accessible sites in the van der Waals gaps which separate each $S[Mn_{2/3}(P_2)_{1/3}]S$ basic building block from the next one, where the single sandwich $S[Mn_{2/3}(P_2)_{1/3}]S$ consists of two honeycomb sulphur atom sheets in which there are octahedral interstitial sites a third of which is occupied by the P-P pairs while the left over two thirds are filled with the manganese cations.

Among the alkali metals, potassium is particularly charming because its intercalation into the $MnPS_3$ host lattice allows one to insert subsequently also much bigger guest species unable to be inserted in pure $MnPS_3$ directly. In fact by means of a further ion-exchange process it is possible to replace some solvated K^+ ions with the desired species. Moreover it is reported in the literature that if one wants to intercalate $MnPS_3$ with polymer cationic species, it is more expedient to intercalate at first the K^+ ions since they are more favoured in the cation exchange than other alkali metal ions, such as for example Na^+ [6]. Consequently, since the structural and physical properties of $MnPS_3$ have been extensively studied and reviewed [2], it is necessary to know in detail the electronic properties of the $K_{2x}Mn_{1-x}PS_3$ compound.

In this context the x-ray photoemission spectroscopy (XPS) is a powerful tool because it, giving information about both the bonding state of the constituent atoms and the presence of non-equivalent atoms, can shed light upon the effects caused on the $MnPS_3$ electronic properties by the potassium intercalation. We are not aware of a detailed core level XPS study carried out on $K_{2x}Mn_{1-x}PS_3$. Therefore in this article we present the results of our x-ray induced electron emission study performed at room temperature on the $K_{2x}Mn_{1-x}PS_3$ compound at the Mn, P, S and K most intense core levels.

2. Materials and experimental setup

Preparation of the $K_{2x}Mn_{1-x}PS_3$ material is a two step process as reported by Lagadic *et al.* [7]:

1) *Synthesis of neat* $MnPS_3$: $MnPS_3$ powders were prepared via a high-temperature process by heating a stoichiometric mixture of the elements of high purity at 750 °C for a week in a sealed evacuated quartz tube.

2) *Potassium ions intercalation* ($K_{2x}Mn_{1-x}PS_3$): polycrystalline powdered $MnPS_3$ samples (1g) were first soaked at room temperature in 2 N KCl aqueous solution (50 ml) and stirred at room temperature for 1h to yield the intercalation $K_{2x}Mn_{1-x}PS_3$ compound. Then this compound was washed several times with distilled and de-ionized water and finally dried in an oven at 60 °C for 3 hours.

X-ray photoemission spectra, performed at room temperature, were collected with a VG Scientific spectrometer equipped with a standard twin-anode Mg/Al Kα X-ray source and a 150° concentric hemispherical analyzer CLAM 100 operating in the constant pass-energy (CAE) mode at 20 eV. The analyzer was fitted with a channeltron, the voltage of which was fixed at 2.5 kV. Pressed $K_{2x}Mn_{1-x}PS_3$ powder pellets were attached on the sample holder with silver paint. No traces of foreign surface contaminants were detected except for small amounts of carbon and oxygen. During each measurement the sample was irradiated with MgKα radiation ($h\nu = 1253.6$ eV) and the analysis chamber pressure was in the 10^{-9} mbar range. The observed charging effects, due to the high resistivity of the sample, were removed by referring the core peak binding energies to the carbon C 1s line at 285.0 eV. A Shirley-type inelastic background was subtracted to each analyzed XPS spectrum. Subsequently a least-square fitting procedure was carried out on each spectrum adopting Gaussian-Lorenztian cross-product profiles.

3. Results

XPS spectra of the Mn $2p$ and $3p$, S $2p$, P $2p$, K $2p$ and $3p$ core levels in $K_{2x}Mn_{1-x}PS_3$ are shown in Figures 1-6 respectively. The open circles represent the experimental data, the solid line the best fit while the dashed while the dashed lines the Gaussian-Lorentzian subbands. In Table 1, the fit-deduced binding energies of the investigated core levels, which are referred to the C 1*s* line at 285.0 eV, are listed together with those of the same levels measured in $MnPS_3$ [10] (in the round brackets) for comparison.

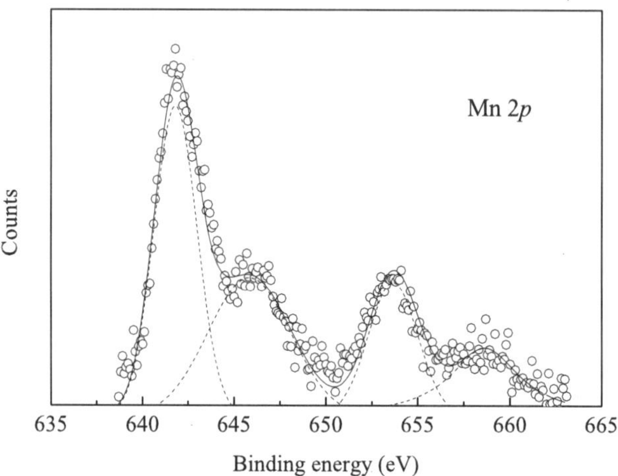

Figure 1. Mn 2*p* core level XPS spectrum in $K_{2x}Mn_{1-x}PS_3$. The binding energies are charge corrected to the C 1*s* line (285.0 eV).

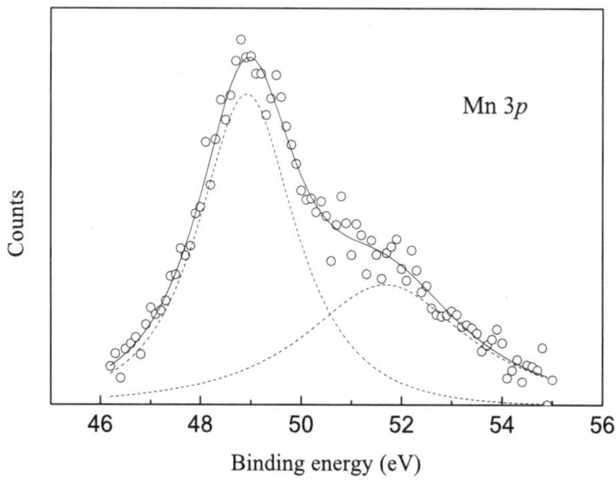

Figure 2. Mn 3*p* core level XPS spectrum in $K_{2x}Mn_{1-x}PS_3$. The binding energies are charge corrected to the C 1*s* line (285.0 eV).

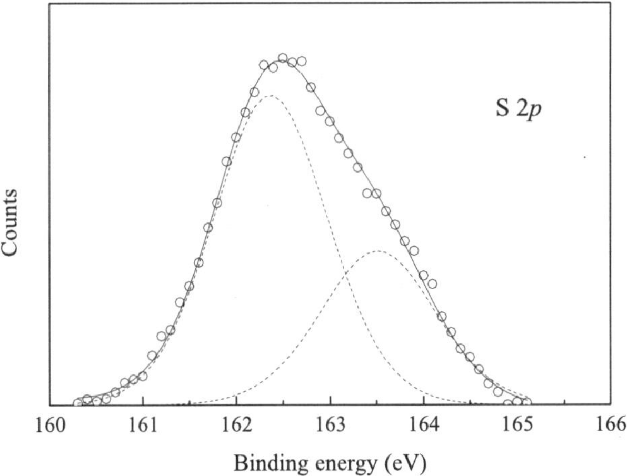

Figure 3. S $2p$ core level XPS spectrum in $K_{2x}Mn_{1-x}PS_3$. The binding energies are charge corrected to the C $1s$ line (285.0 eV).

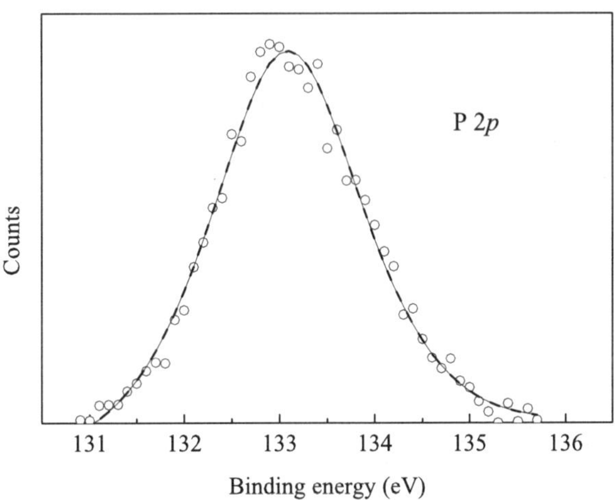

Figure 4. P $2p$ core level XPS spectrum in $K_{2x}Mn_{1-x}PS_3$. The binding energies are charge corrected to the C $1s$ line (285.0 eV).

As shown in Figure 1, owing to the spin-orbit coupling, the Mn $2p$ core levels are split into the $2p_{3/2}$ and $2p_{1/2}$ components with an energy separation of 11.8 eV. The same happens for the K $2p$ core levels (see Figure 5) whose spin-orbit components are separated of 2.8 eV. These observed spin-orbit splittings are in good agreement with the values reported in other manganese and potassium compounds respectively [8,9]. With respect to the sulphur $2p_{3/2} - 2p_{1/2}$ spin-orbit components (see Figure 3), they are weakly resolved with an energy separation of 1.1 eV.

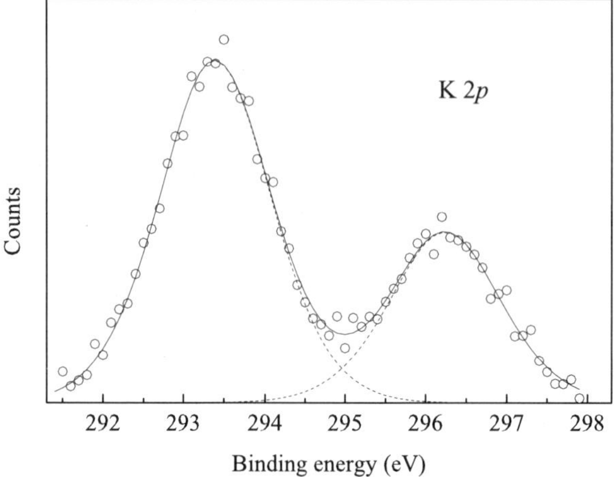

Figure 5. K $2p$ core level XPS spectrum in $K_{2x}Mn_{1-x}PS_3$. The binding energies are charge corrected to the C $1s$ line (285.0 eV).

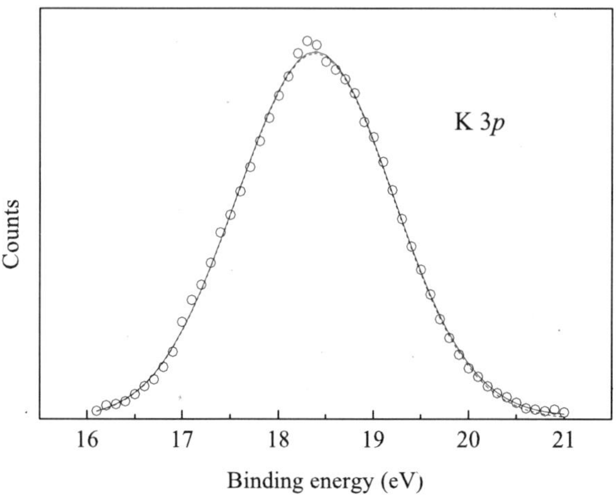

Figure 6. K $3p$ core level XPS spectrum in $K_{2x}Mn_{1-x}PS_3$. The binding energies are charge corrected to the C $1s$ line (285.0 eV).

This spin-orbit splitting value is in good agreement with one existing in other sulphur compounds [8, 9]. As results from Table 1, the magnitude of the observed spin-orbit splittings agrees well to that observed for the same levels in $MnPS_3$ too [10]. On the other hand, the Mn $3p$, P $2p$ and K $3p$ doublets, shown in Figures 2, 4 and 6 respectively, are not resolved. Moreover, the K $3p$ doublet shows negligible dispersion and strong intensity when compared to the valence band states.

Table 1. Binding energies (in units of eV) of the Mn, P, S and K core levels in $K_{2x}Mn_{1-x}PS_3$. All energies refer to the C $1s$ line (285.0eV). In the round brackets the binding energies of the same levels in $MnPS_3$ are reported for comparison.

Core-level	Mn	P	S	K
$2p$	641.8 653.6	133.1	162.4 163.5	293.4 296.2
	(641.9 653.8)[a]	(132.9)[a]	(162.4 163.4)[a]	
$3p$	48.9			18.4
	(48.8)[a]			

[a] Reference [10]

4. Discussion

As one can see in Figures 1-6 (dashed lines), all investigated core levels reveal single components indicating that in $K_{2x}Mn_{1-x}PS_3$ the atoms of Mn, P, S and K are in equivalent sites.

Table 2. Mn $2p_{3/2}$ core level binding energy position in $K_{2x}Mn_{1-x}PS_3$ and in other manganese compounds. All positions are in eV.

Compound	$2p_{3/2}$
$K_{2x}Mn_{1-x}PS_3$	641.8
Mn	639.0[a]
α-MnS	641.9[a]
MnI_2	641.9[a]
$MnCl_2$	642.0[a]

[a] Reference [8]

As shown in Table 2 the Mn $2p_{3/2}$ core level in $K_{2x}Mn_{1-x}PS_3$ is located at a binding energy which is greater than the one in the Mn metal but it is close to the binding energy observed in other ionic manganese compounds such as α-MnS, MnI_2 and $MnCl_2$ where Mn is present as a bivalent cation [8]. This similarity allows us to assert that in $K_{2x}Mn_{1-x}PS_3$ Mn is in a 2^+ formal oxidation state just as happens in $MnPS_3$. This result is also confirmed by the comparison with the Mn $2p$ XPS spectrum in $MnPS_3$ which emphasizes no remarkable change on both the binding energy positions (see Table 1) and shape [10]. As regards the binding energy of the P $2p$ core level, reported in Table 1, it is higher than that for pure phosphorus [11] but it is very close to that observed in $MnPS_3$ (see Table 1). The same happens for the S $2p$ core level binding energy which even if it is lower than the one reported for pure sulphur [12] is similar to that observed in $MnPS_3$ (see Table 1). These findings agree well with the intercalation by substitution process, which is based on an ion transfer and not on an electron transfer [13].

Regarding the binding energy position of the K $2p$ core level, it is lower than the one reported for the K element but it is close to the binding energy observed in other potassium compounds [8] where K is in a 1^+ formal oxidation state. This behavior is typical of potassium and the heavier alkali metals which exhibit the core level binding energies higher for the metallic state than for the oxidized state [14]. This fact is a further evidence that the intercalation by cationic substitution does not involve any electron transfer from the guest specie (K^+) to the host lattice ($MnPS_3$).

A careful look at Figure 1, where the Mn $2p$ core level XPS spectrum is reported, emphasizes the presence of a satellite feature on the high binding energy side of each of the spin orbit components. This can be also noted in the Mn $3p$ XPS spectrum (see Figure 2) but not at the sulphur, phosphorous or potassium core levels. These extra structures, which are observed at 645.2 and 658.1 eV in the Mn $2p$ XPS spectrum (Figure 1) and at 51.7 eV in the Mn $3p$ one (Figure 2) in $K_{2x}Mn_{1-x}PS_3$, are due, according to the literature [15, 16], to ligand-to-metal charge transfer *shake-up* transitions. This presence [16] supports the idea that $K_{2x}Mn_{1-x}PS_3$ is a large-gap insulating Mn compound like $MnPS_3$.

Confirmation that the K^+ - $(P_2S_6)^{4-}$ bond is weak is also obtained from the analysis of the unresolved K $3p$ doublet. In fact this doublet (Figure 6) is well localized and core like. Moreover, the binding energy position of this level (see Table 1) agrees well with the observed one in other K^+ compounds [17]. All this supports the presence of K^+ in $K_{2x}Mn_{1-x}PS_3$ and the idea that the intercalation process involves only a cationic substitution.

5. Conclusions

We have analyzed room temperature $K_{2x}Mn_{1-x}PS_3$ core levels with the aid of x-ray photoemission spectroscopy. All core levels investigated exhibit single components: equivalent atoms of Mn, P, S and K are present in this intercalate. The observed binding energies of the Mn $2p$, P $2p$ and S $2p$ core levels are very similar to the ones noted in $MnPS_3$ and in some ionic manganese compounds containing Mn as a bivalent cation.

In particular, the similarity with $MnPS_3$, apart from confirming the presence of Mn^{2+} in $K_{2x}Mn_{1-x}PS_3$, agrees well with the intercalation by cationic substitution process which involves an ion and not a charge transfer. This result is further supported by the analysis of the K $2p$ and $3p$ core levels whose observed binding energy positions are typical of the K^+ ion. The *shake-up* satellites at the Mn $2p$ and $3p$ core levels suggest that $K_{2x}Mn_{1-x}PS_3$ behaves as a large-gap insulating Mn compound like $MnPS_3$. The unresolved K $3p$ doublet shows a core-like character in good agreement with both the hypothesis of a weak K^+ - $(P_2S_6)^{4-}$ bond and the easy cationic exchange ability of the K^+ ion.

ACKNOWLEDGMENTS. This work is dedicated to Prof. Mario Tosi, a renowned scientist and a dear friend in the occasion of his 72nd birthday.

References

[1] W. R. CAHN, Nature **348** (1990), 389.
[2] V. GRASSO and L. SILIPIGNI, Rivista del Nuovo Cimento **25** (2002),1.
[3] R. CLEMENT, I. LAGADIC, A. LEAUSTIC, J. P. AUDIERE and L. LOMAS, Chemical Physics of Intercalation II, NATO ASI Ser. B **315** (1993).
[4] P. A. JOY and S. VASUDEVAN, J. Phys. Chem. Solids **54** (1993), 343.
[5] P. A. JOY and S. VASUDEVAN, J. Am. Chem. Soc. **114** (1992), 7792.
[6] D. ZHANG, J. QIN, YAKUSHI, Y. NAKAZAWA and K. ICHIMURA, Mater. Sci. Eng. A **286** (2000), 183.
[7] I. LAGADIC, P. G. LACROIX and R. CLEMENT, Chem. Mater. **9** (1997), 2004.
[8] J. F. MOULDER, W. F. STICKLE, P. E. SOBOL and K. D. BOOMBEN, "Handbook of X-ray photoelectron Spectroscopy", J. Chastain (ed.), Perkin-Elmer Corporation, 1992.
[9] R. GUTMANN, J. HULLIGER, R. HAUERT and E. M. MOSER, J. Appl. Phys. **70** (1991), 2648.
[10] V. GRASSO and L. SILIPIGNI, private communication.
[11] In [8], p. 59.
[12] In [8], p. 60.
[13] R. CLEMENT, In: "Hybrid organic-inorganic composites", J. E. Mark, C. Y-C Lee and P. A. Bianconi (eds.), Symposium Series, American Chemical Society, Washington, DC, 1995, Chapt. 4, p. 29.

[14] K. SZOT, F. U. HILLEBRECHT, D. D.SARMA, M. CAMPOAGNA and H. AREND, Appl. Phys. Lett. **48** (1986), 490.
[15] M. OKUSAWA, Phys. Stat. Sol. (b) **124** (1984), 673.
[16] M. PIACENTINI, F. S. KHUMALO, C. G. OLSON, J. W. ANDEREGG and D. W. LYNCH, Chemical Physics **65** (1982), 289.
[17] S. B. DICENZO, P. A. ROSENTHAL, H. J. KIM and J. E. FISCHER, Phys. Rev. B **34** (1986), 3620.

3

QUANTUM DEGENERATE GASES

Chair
Sandro Stringari

Atomic quantum gases in optical lattices

C. Fort, G. Modugno, and M. Inguscio

Despite several decades of investigation, the study of particles in periodic potentials is still one of the frontiers of modern physics. The recent achievement of quantum degeneracy in dilute atomic gases has provided a new tool to address unresolved issues of condensed matter physics, such as those related to superconductivity. The availability of ultracold bosonic and fermionic systems with tunable interactions in combination with optical lattices - perfect sinusoidal potentials - realizes a new laboratory where to study a variety of interdisciplinary phenomena, ranging from interferometry to quantum phase transitions and superfluidity.

Mario Tosi has contributed to the theory of this field in a creative and original way [1], and we hope he will appreciate our report on some of the experimental achievements in our laboratory. In recent years, at LENS we have focused our research on the basic properties of Bose-Einstein condensates (BECs) and degenerate Fermi gases in optical lattices. We have worked with 1D lattices and studied mainly the transport properties of both kinds of quantum gases, in the absence and in the presence of interactions. In this contribution we will not follow the temporal evolution of our research, but we will rather present first our results for an ideal (non-interacting) gas, to consider only at the later stage the role of the interactions, and eventually that of superfluidity in a BEC.

1. Non-interacting fermions

Ultracold fermionic atoms prepared in just one internal state are virtually non-interacting: s-wave collisions are indeed forbidden by the Pauli exclusion principle, while all other partial waves are thermally suppressed below temperatures of the order of 100 μK. This property makes an ultracold gas of fermions appropriate to study the properties of an ideal gas. On the other hand, the sinusoidal dipole potential created by two counter-propagating laser beams realizes a perfect lattice, with no defects or excitations. In a series of experiments we have actually demonstrated how this combination allows to study the fundamental transport properties of particles in lattices, such as Bloch oscillations or macroscopic conduction and insulation.

In our experiments, ^{40}K fermions are sympathetically cooled using ^{87}Rb bosons [2] in a harmonic magnetic trap. The 1D lattice is created by a retro-reflected laser beam, which is far off-resonance from any atomic transition. The typical wavelength of the laser is λ=800 nm, which determines the lattice periodicity $d=\lambda/2$ and the recoil energy $E_R=h^2/2m\lambda^2$ ($E_R \approx 300$ nK for K and 150 nK for Rb), which sets the energy scale of the problem. In particular, the energy gap between the first and second band of the lattice opens just at E_R, and for deep lattices the width of the first energy band scales as $2\delta=E_R 8s^{3/4}\exp(-2s^{1/2})/\sqrt{\pi}$, where $s=U/E_R$ is the normalized lattice depth. In order to have a single-band system one must cool the gas to a temperature comparable or below E_R. For the parameters of our fermions, this corresponds to work with a degenerate Fermi gas, since we have $E_F \approx E_R$.

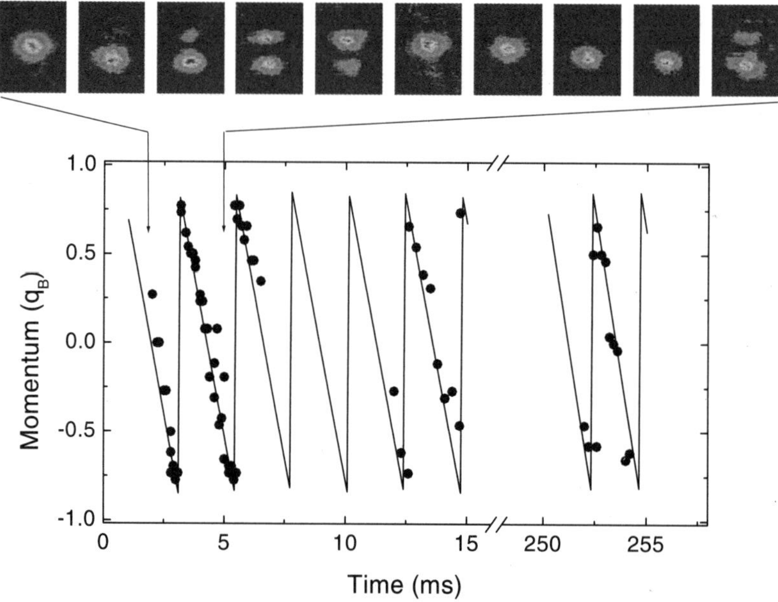

Figure 1. Time-resolved Bloch oscillations of a Fermi gas. The graph shows the measured time evolution of the peak quasi-momentum (dots), which is fitted with a sawtooth function. The absorption images show the evolution of the quasi-momentum distribution during the first Bloch period. The ability to follow many Bloch oscillations allows to measure with high sensitivity the driving force, i.e. gravity.

1.1. Bloch oscillations of an atomic Fermi gas

The simplest lattice potential to confine ultracold atoms in 3D can be created with a retro-reflected, vertically-oriented laser beam: the gaussian profile of the red-detuned beam confines the atoms in the horizontal plane whereas the lattice itself provides confinement against gravity. The motion of particles in a tilted periodic potential was studied by Bloch and Zener [3], who identified the peculiar periodic oscillations known as Bloch oscillations. This phenomenon had escaped observation in crystals since a few years ago [4], but now it has been observed in various systems, including cold atoms in optical lattices and photons in photonic crystals [5]. In all these systems however only a small number of oscillations can be seen, because interactions and defects rapidly destroy the coherence of the particles. Non-interacting fermions in optical lattices represent instead the ideal system to study this phenomenon: as a matter of fact we have been able to see hundreds of Bloch oscillations of our Fermi gas, driven by gravity [6]. Figure 1 shows the experimental data: here we study the temporal evolution of the momentum distribution of the Fermi gas, after release from the lattice. We can clearly see the displacement with constant velocity of the distribution, according to $\hbar\dot{q}=mg$, followed by a Bragg reflection at the edge of the first Brillouin zone of the lattice, $q(t)=q_B=2\pi/\lambda$. The time evolution of the central momentum can be followed for a rather long period, corresponding to hundreds of oscillations. The measured period T_B=2.32789(22) ms is not only in good agreement with the expected Bloch period, which is determined only by the force and the lattice wavelength as $T_B=2h/mg\lambda$, but can also be used for a sensitive measurement of the local acceleration of gravity. The achieved sensitivity of $10^{-4}g$, combined with a spatial resolution of the order of 10 μm, makes this system appealing for future studies of forces at small lengthscales, including forces close to surfaces. In addition, our experiment demonstrates the superiority of non-interacting fermions respect to bosons not only for this specific example of quantum interference, but more in general for high-precision atom interferometry.

1.2. Conduction and insulation in a harmonic potential

In a second series of experiments we have studied the effect of a harmonic potential on a Fermi gas in the lattice. This is the most common configuration encountered in experiments, where additional confining potential are usually present. In our case, the atoms are kept in the magnetic harmonic potential, and a 1D lattice is superimposed along the weak trap axis. We then study the response of the system to a sudden, small displacement of the magnetic trap, which excites a dipolar center-of-mass motion. While in a pure harmonic potential it proceeds undamped around the new trap minimum, with the addition of a relatively shallow lattice potential the motion is strongly perturbed. As shown in Figure 2a, the oscillation is strongly reduced in amplitude, damped and frequency shifted and, most of all, proceeds around a position far from the equilibrium position of the harmonic trap.

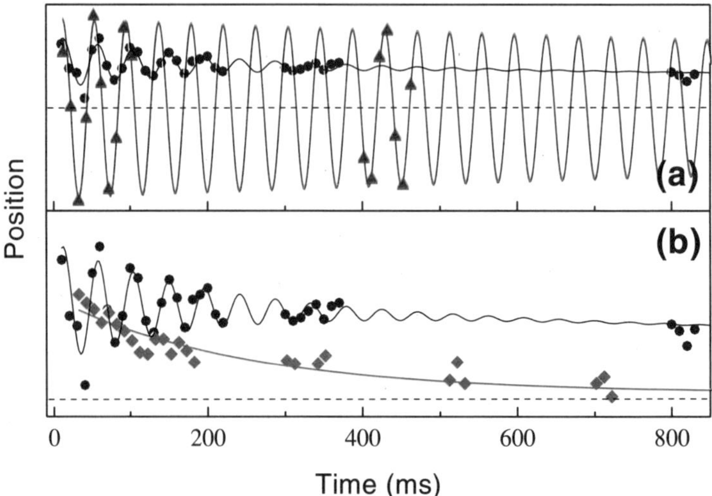

Figure 2. Dipole oscillations of a Fermi gas in a harmonic trap combined with a 1D optical lattice. a) In the presence of a lattice with U=3E_R (circles) the oscillations of a non-interacting sample are frequency-shifted, damped, and proceed with an offset from the oscillation in absence of the lattice (triangles). b) In the presence of interactions with a coexisting uncondensed bosonic sample (diamonds), the oscillations are overdamped and the center of mass decays to the equilibrium position of the trap.

This behavior can be understood by evaluating the solutions of the single-particle 1D Schrödinger equation for the combined potential [7]. While the eigenstates at the bottom of the trap resemble harmonic oscillator states, as soon as their energy grows larger than the bandwidth 2δ the states become localized on either side of the harmonic potential. These states are the analogous of the Wannier-Stark states in the tilted lattice, and they spread over just a few lattices sites: $\Delta x = 2\delta/m\omega^2 x$. This structure can be well understood also in a semiclassical approach, where the bands of the homogeneous lattice are bent along the harmonic potential. Note how in the presence of the inhomogeneous potential, a spatial forbidden region appears in addition to the usual energy gap.

Let us now consider a gas with Fermi energy lying in the first bandgap: both classes of states, delocalized across the center of the trap and localized on the sides, will be occupied. However, after the axial displacement only the atoms in delocalized states will be able to perform coherent dipolar oscillations, while the ones in localized states will just perform small incoherent oscillations on the sides. This explains both the reduction of amplitude and the offset in the oscillation center of Figure 2a. For a quantitative analysis of the dynamics we relied on a semiclassical model [8], which is able to reproduce rather closely the experimental observations, including the damping and frequency shift. The latter is simply due to the larger effective mass of the atoms oscillating in the lattice, whereas the damping is due to a dephasing of the single-particle orbits whose energy approaches the band edge.

Let us now analyze the behavior of the different portions of the Fermi gas using the language of solid-state physics. The localized states are the analogous of Wannier-Stark states, with a main difference: in the harmonic trap the extension of such states varies with the local gradient of magnetic field, and therefore is not possible to see collective Bloch oscillations. Moreover, for $E < 2\delta$ the extension of the lattice states would get longer than the harmonic oscillator states, and the states become delocalized. Now, the sudden shift of the trap is just equivalent to adding an additional linear potential, in analogy to an electric potential applied to a crystal. In our non-interacting system the single-particle eigenstates discussed above are also the eigenstates of the many-body system, and therefore atoms in localized states do not move macroscopically, while those in delocalized states can oscillate. Looking at the macroscopic motion one can therefore consider the first portion of the sample like an insulator, which does not show a current in the presence of an electric voltage, and the latter like a conductor. Note that such insulating behavior is just what one would expect for a perfect crystal, where Bloch oscillations can proceed without dissipation.

The fraction of atoms in each of the two components, insulating and conducting, clearly depends on the position of the Fermi energy with respect to the first band of the lattice. In our system it is possible to tune the ratio $E_F/2\delta$, for example by varying the depth of the lattice at fixed atom number. In Figure 3a we show the experimental results, compared with theory, for the crossover from a purely conductive Fermi gas to an almost completely insulating one, where most of the atoms are in localized states [8].

2. Interacting fermions and bosons

Let us now consider the effect of interatomic collisions in the transport phenomena described above. These collisions are normally present in

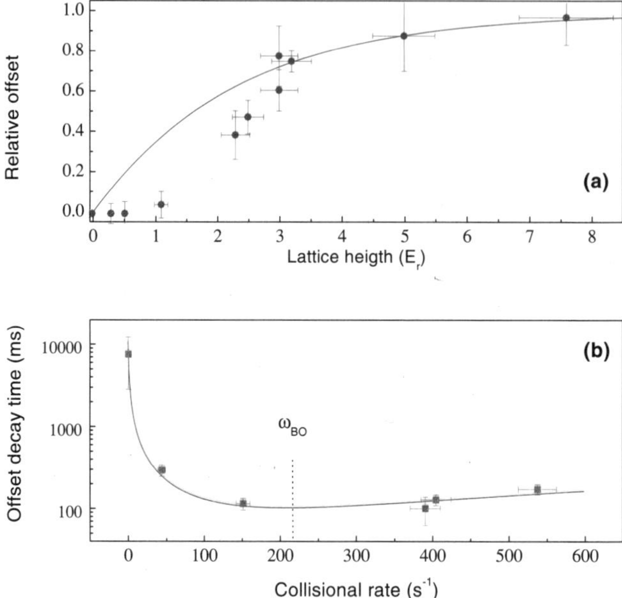

Figure 3. a) Crossover from a purely conducting to an almost completely insulating behavior of a non-interacting Fermi gas obtained by tuning the ratio $E_F/2\delta$. The relative offset of the oscillations represents the ratio between the insulating and conducting fractions. (b) Crossover from an insulator to an ohmic conductor, by tuning the collisional rate in a boson-fermion mixture. The decay time for the offset is the analogous of the resistivity of the sample.

samples of bosons, of non-identical fermions (e.g. fermions in different internal states) and in boson-fermion mixtures. At low temperatures the collisional rate $\Gamma \approx 4\pi a^2 n v$ depends just on the mean atomic density n, the thermal velocity v and the relevant scattering length a, and can be as large as $100\,\mathrm{s}^{-1}$. We find that in general the presence of collisions affects dramatically the coherent transport of particles through the lattice, causing a very rapid damping of both Bloch and dipolar oscillations. A notable exception that we will discuss in detail in the last section of this contribution is a Bose-Einstein condensate, which regains the coherence properties even in the presence of atomic interactions.

In the experiment we have studied the effect of collisions by working with an uncondensed sample of bosons, either alone or in coexistence with a sample of fermions. The latter combination allows to explore a wider range of collisional rates, since one can keep the atom number of one of the two components (the fermionic one) fixed, while the other can be varied by more than two orders of magnitude without affecting the detection of the first one.

When repeating the Bloch oscillations experiment with an interacting sample of this kind, we observed a decay of the oscillations with a time constant $\tau \approx \Gamma^{-1}$, as expected: collisions between particles lead to a dephasing between individual particles and the macroscopic oscillations are degraded. This phenomenon is analogous to what happens in real crystals, where Bloch oscillations are hardly detectable. In our vertical lattice however the time between collisions $\Gamma^{-1} \approx 50\,\text{ms}$ is long compared to T_B, due to a relatively low atomic density, and therefore we can still see several Bloch oscillations in interacting samples.

To make a full analogy with solid-state systems, it is more instructive to study the dipolar oscillations in the magnetic trap in the presence of the lattice [9]: here the strong radial confinement allows to have $\Gamma^{-1} \approx 1\div 10\,\text{ms}$, which is even larger than the local T_B due to the weak longitudinal harmonic confinement. As shown in Figure 2b collisions change drastically the behavior of the oscillations of fermions, and in particular we observe an exponential decay of the center of the oscillations, initially displaced, to the absolute minimum of the potential. Collisions now allow scattering of particles between localized states and the system can relax towards equilibrium. In the solid-state approach one interprets the decay of the position as the current induced by collisions. The perfect crystal, which has an insulating behavior, is indeed turned into a "real" crystal, where collisions provide the mechanism for a macroscopic transport. Making now an analogy between the time needed to the Fermi gas to relax to the trap center and the resistivity of a crystal, Figure 3b shows how it is possible to scan the whole crossover from an insulator to an ohmic conductor, where the conduction is hindered by collisions. In the crossover region, where the collisional rate is comparable to the Bloch oscillation frequency, one can clearly see a decrease of the resistivity for increasing collisional rates. This regime of anomalous conduction was accessed so far only in semiconductor superlattices, which however do not allow to explore the whole crossover.

3. Superfluid Bose Einstein condensate

In this section we describe a series of experiments performed with a Bose Einstein condensate (BEC) moving in a 1D optical lattice. The main characteristic to be considered now is the momentum spread of the atomic sample compared to the width of the Brillouin zone. The momentum spread Δp of a Bose-Einstein condensate is inversely proportional to its size R ($\Delta p \sim \hbar/R$). Considering an optical lattice with a spatial periodicity of $d \sim 400\text{nm}$, the width of the Brillouin zone is given by $2\hbar k = h/d$. The typical size of a BEC is $R \sim 100\mu\text{m}$ resulting in

$\Delta p \sim 10^{-3}\hbar/d$. In contrast to Fermi gases, the BEC occupy a very small portion of the Brillouin zone and can be considered as an ideal probe to study the energy bands of the periodic potential. Actually, many experimental works on condensates in optical lattices have used the Bloch approach to interpret the experimental results [10–12].

However, the Bloch picture can be complicated and enriched by the presence of interactions between the atoms, that produce a non-linearity. In the regime of high lattice heights this can even induce a quantum phase transition to a Mott insulator as demonstrated in [13]. In the other limit of low lattice heights, one of the most interesting problem following from the presence of a non-linear term is the stability of the condensate moving in the optical lattice. The first source of instability arises from the analysis of the energy spectrum of the system and takes place when the velocity of the condensate is larger than the sound velocity (energetic or Landau instability). Another mechanism that can play a role in a repulsive condensate trapped in a periodic potential is the dynamical instability. This phenomenon takes place when the spectrum of the excitations of the system exhibits complex frequencies and small perturbations can grow exponentially eventually destroying the initial state. Dynamical instability is a very general phenomenon of non-linear wave equations and has been already studied in the frame of nonlinear optics, fluid dynamics and plasma physics [14]. More recently, dynamical instability has been theoretically studied in the context of Bose-Einstein condensates in optical lattices [15] and to explain the formation of solitons in attractive condensates [16].

In the following we first describe an experiment where the condensate is loaded in an optical lattice after release from the magnetic trap [17]. In the expanded condensate interactions decrease and the dynamic can be interpreted mostly in terms of stable Bloch waves. Then we report briefly a recent experiment we have performed using a trapped condensate in order to have a high enough density to enter regimes of instability when the BEC is loaded in a moving optical lattice [18].

3.1. Band Spectroscopy with an expanding condensate: Linear Regime

In this experiment we first produce a ^{87}Rb condensate in a magnetic trap, we then switch off the magnetic potential and let the condensate expand for 2 ms. After this time the interaction energy reduces to 40% of its initial value and the density is low enough to consider the condensate as a linear probe of the periodic potential. At this point, we switch on an optical lattice moving at constant velocity v_L. Switching on adiabatically the moving lattice we load the condensate in a state with well defined quasi-momentum $q = mv_L/\hbar$ and band index n ($n = 1$ for $0 < q < q_B$, $n = 2$ for $q_B < q < 2q_B$ and so on). After 12 ms of expansion in the moving optical lattice, we take an absorption picture of the condensate extracting the center of mass position and the radii of the condensate. From the center of mass position we can extract the velocity of the condensate in the frame of the periodic potential. The results are shown in Figure 4A where the velocity v of the condensate in the moving frame of the lattice is given as a function of the quasi-momentum for optical lattice heights of $s = 1.3$ and $s = 3.8$. The obtained data perfectly reproduce the corresponding Bloch velocities obtained from band structure calculations. Furthermore, from these data we can also extract the effective mass of the condensate as $m^* = \hbar(\Delta v/\Delta q)^{-1}$ where Δv is the measured variation of the center of mass velocity corresponding to a variation of Δq in the quasimomentum. In Figure 4B we report both the value of the effective mass extracted from the measured velocity, and the one calculated from the Bloch theory (continuous line) as a function of the quasimomentum in the first two bands. Again we have a very good agreement between the experimental measurements and the theory.

From the measurement of the radii of the condensate after the expansion in the moving optical lattice we can also extract information on how the effective mass affects the expansion of the BEC. In Figure 5 we display the measured axial and radial size of the condensate as a function of the lattice velocity (or, equivalently, as a function of the condensate quasimomentum). The experimental observations are compared with the numerical simulation of the effective 1D Gross-Pitaevskii equation discussed in [19]. In particular, in the region of negative effective mass we observe a contraction of the axial size and correspondingly an increase in the BEC radial size.

3.2. Non-linear dynamics of a trapped Bose Einstein condensate: Dynamic and Energetic Instability

If the condensate is loaded in the 1D optical lattice leaving on the magnetic confinement, the non-linear density related term is big enough to study energetic and dynamical instability. Both the energetic and the dynamical instability are expected to deplete the initial atomic state, and we choose to characterize the timescales of these two phenomena by measuring the number of atoms remaining in the initial condensate as a function of thc time spent in the moving optical lattice.

The experimental results for dynamical instability are analyzed according to the model described in [20] based on a linear stability analysis

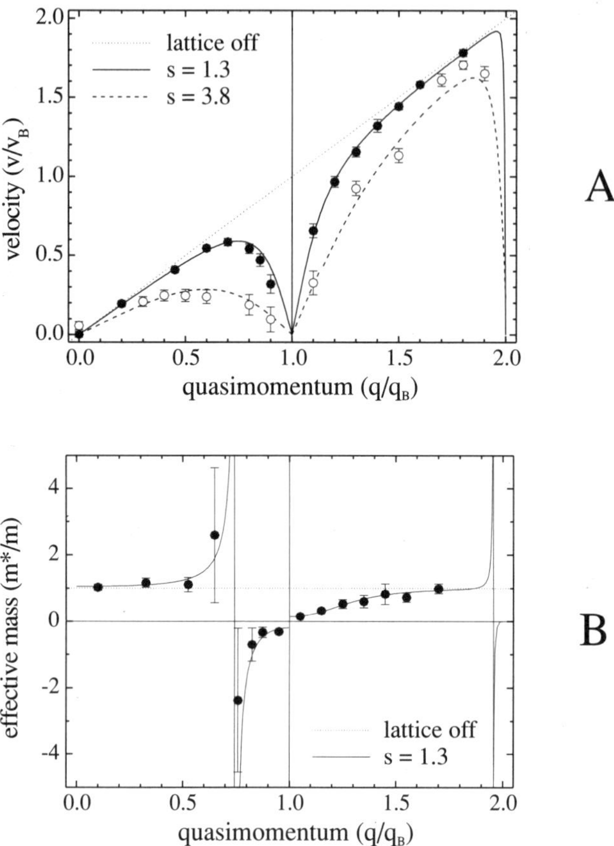

Figure 4. A) Velocity of the condensate loaded in the optical lattice as a function of its quasimomentum in the lattice frame for the first and second band. The experimental data points correspond to lattice heights of 1.3 E_R (filled circles) and 3.8 E_R (open circles). The continuous and dashed lines are the corresponding Bloch velocity calculated from the band theory while the dotted line represents the velocity of the condensate with the lattice off. B) Effective mass of the condensate loaded in the optical lattice (1.3 E_R) for different quasimomenta. The data points are extracted from the measured condensate velocities and show a good agreement with the Bloch theory calculation (continuous line).

of the Non Polynomial Schrödinger equation (NPSE) [21], a simplified 1D model that includes an effective radial-to-axial coupling. The theory yields an estimate of the dynamical instability thresholds and of the growth rates of the most unstable modes. In the experiment, we introduce a radiofrequency (RF)-shield that removes the thermal component during the measurements. The RF-shield simply consists in leaving on the RF field we use to produce the condensate.

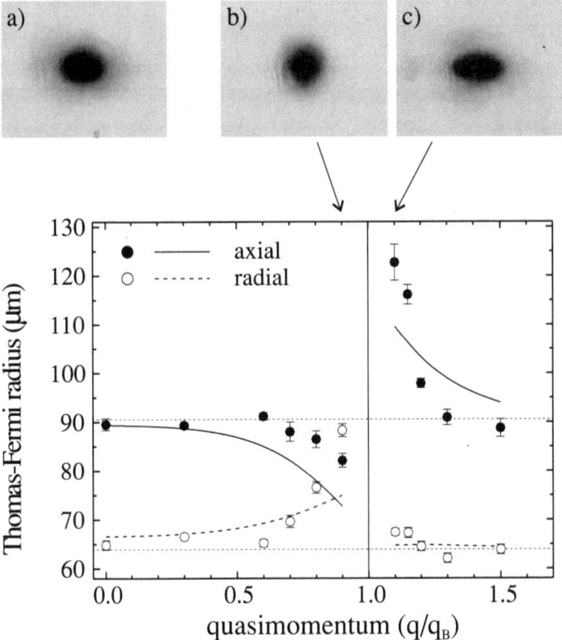

Figure 5. Absorption images of the expanded condensate. From left to right: (a) normal expansion of the condensate without the lattice; (b) axial compression in a lattice of 2.9 E_R moving with $v_L = 0.9\ v_B$; (c) enhanced axial expansion in a lattice of 2.9 E_R moving with $v_L = 1.1\ v_B$ (where $v_B = \hbar q_B/m$). In the lower part: axial and radial dimensions of the condensate after expansion in an optical lattice with a height of 2.9 E_R as a function of the quasimomentum. The experimental points (filled and open circles) show the Thomas-Fermi radii of the cloud extracted from a 2D fit of the density distribution. The dotted lines show the dimensions of the expanded condensate in the absence of the optical lattice. The continuous and dashed lines are theoretical calculations obtained from the 1D effective model presented in [19].

When the optical lattice moves at different velocities, we observe a strong dependence of the BEC lifetime on its quasimomentum. We have performed a detailed investigation of the system in the lowest three energy bands for two different lattices ($s = 0.2$ and $s = 1.15$). The results are shown in Figure 6 where we plot the loss rates (defined as the inverse of the lifetimes) together with the calculated growth rates of the most dynamically unstable mode for each value of the quasimomentum obtained using the theoretical approach described in [20]. The experimental loss rates show a dependence on the quasimomentum closely related to the behaviour of the calculated growth rates, clearly demonstrating that the variation of the condensate loss rates is caused by the dynamical instabil-

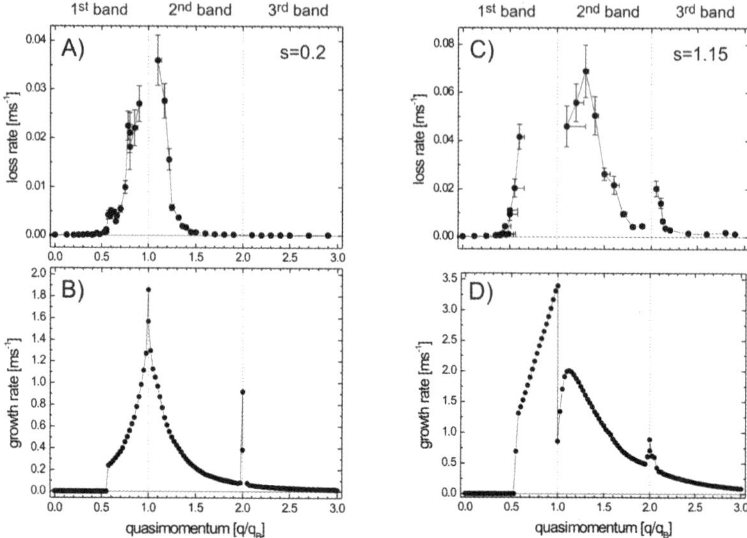

Figure 6. A) Experimental loss rates for a BEC loaded into the first three energy bands of a moving optical lattice (0.2 E_R). B) Theoretical growth rates of the most dynamically unstable modes obtained from a linear stability analysis of the NPSE [20] for 0.2 E_R. C) Experimental loss rates for a BEC loaded into the first three energy bands of a moving optical lattice (1.15 E_R). D) Theoretical growth rates of the most dynamically unstable modes obtained from a linear stability analysis of the NPSE for 1.15 E_R.

ity even if these two quantities can not be quantitatively compared since they refer to two different physical parameters. In the region where dynamical instability affects the system, we also observed that the expanded density profile is dramatically modified and density modulation appears suggesting fragmentation of the Bloch wave and creation of phase domains. This structure becomes more evident for higher lattice heights and near the zone boundaries, where the growth of the instability is faster.

We repeated the measurements in the presence of a small thermal component. The results are illustrated in Figure 7 where we show a series of pictures of the expanded density distribution taken after 15 seconds in the moving optical lattice with $s = 0.2$ for a range of BEC quasimomenta $0 < q < 0.2\, q_B$. The upper row of pictures refers to the situation where the RF-shield frequency is kept low enough to maintain a pure condensate. The number of atoms in the condensate does not change in the range of quasimomenta investigated, as this is the region where the condensate is dynamically stable (see Figure 6). The situation illustrated by the lower row of Figure 7 is completely different. In this case the RF-shield value has been chosen in order to have a mixed cloud with 30% of the atoms in

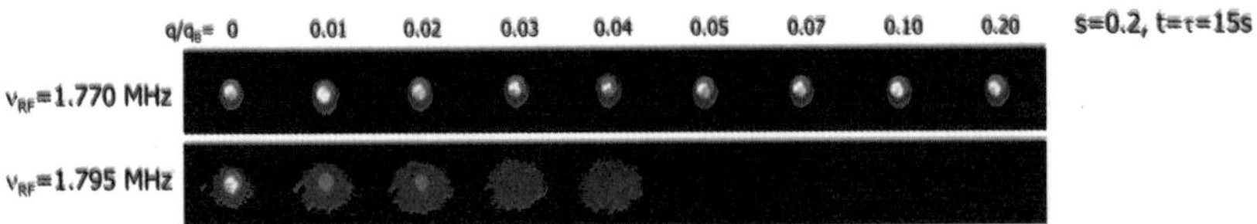

Figure 7. Density profile of the expanded condensate after 15 s spent in the combined trap realized adding to the magnetic trap an optical lattice with a height of 0.2 E_R. The different pictures refer to different BEC quasimomenta from 0 q_B to 0.2 q_B. In the upper row the experiment has been performed with a quasi pure condensate, while the lower row shows pictures obtained when the experiment has been performed with a mixed cloud (30% of the atoms in the thermal fraction).

the thermal component. As soon as $q > 0$ the lifetime reduces strongly and the condensate fraction is completely depleted after 15 seconds for $q = 0.04 q_B$. We attribute this behavior to the onset of energetic instability. The thermal fraction provides the dissipative mechanism necessary to trigger the energetic instability and the system becomes unstable if the velocity exceeds the speed of sound. To explain why this happens even at very small velocities one should consider the fact that the speed of sound is proportional to the density of the sample and that the trapped condensate has a density distribution with very low density regions in the outer shell. Knowing the BEC density distribution one can calculate the fraction of atoms having a sound velocity smaller than the center of mass velocity. This fraction is expected to become unstable. Following this reasoning we have investigated in a more quantitative way the onset of energetic instability as reported in [22].

In particular we observed that the timescale needed for the energetic instability to destroy the condensate ground state is hundreds of milliseconds, much longer than the dynamical instability timescale. This means that, in the regions of parameters where both the mechanism can be activated the faster and thus predominant one, is the dynamical instability.

4. Conclusions

In conclusion, we have presented a series of experiments on the fundamental transport properties of non-interacting, interacting and superfluid atomic samples in 1D optical lattices.

These findings open a number of research directions, which we will possibly follow in the future. On one side, fermionic Bloch oscillators can be employed for a sensitive measurement of forces with high spatial resolution; they might have important applications to the study of forces in proximity of surfaces, such as the Casimir-Polder force, or even to study of gravity at short distances. On the other side, it would be interesting to extend our studies of non-interacting Fermi gases in lattices to interacting Bose-Fermi mixtures. A number of novel quantum phases, including super-solids, is indeed expected in such systems, when prepared at low temperature in a 3D optical lattice. Furthermore the study of energetic and dynamical instability demonstrated that these effects are crucial when a condensate moves in a periodic potential. This phenomena have to be taken into account in all the possible implementations of BEC in optical lattice including quantum computing schemes. Very recently we have also started to study the properties of a condensate in a random (instead of periodic) potential created by light. This system will open the way to a new range of experiments including the possibility to add randomness to a BEC in a periodic potential. This will allow to study new quantum phases such as the Bose glass. This demonstrate how the combination of optical potential and quantum degenerate gases is still a very rich and open field of investigations.

ACKNOWLEDGMENTS. We thank the researchers at LENS who give their fundamental contribution to the experiments reported here: F. S. Cataliotti, J. Catani, E. De Mirandes, L. De Sarlo, L. Fallani, F. Ferlaino, J. E. Lye, M. Modugno, H. Ott, G. Roati, R. Saers.

References

[1] A. MINGUZZI, S. SUCCI, F. TOSCHI, M. P. TOSI and P. VIGNOLO, Phsyics Reports **395** (2004).

[2] G. ROATI, F. RIBOLI, G. MODUGNO and M. INGUSCIO, Phys. Rev. Lett. **89** (2002), 150403.

[3] C. ZENER, Proc. R. Soc. London Ser. A **145** (1934), 523; G. H. WANNIER, Phys. Rev. **117** (1960), 432.

[4] C. WASCHKE *et al.*, Phys. Rev. Lett. **70** (1993), 3319.

[5] For a recent review see: T. HARTMANN, F. KECH, H. J. KORSCH and S. MOSSMANN, New Journal of Physics **6** (2004), 2.

[6] G. ROATI, E. DE MIRANDES, F. FERLAINO, H. OTT, G. MODUGNO and M. INGUSCIO, Phys. Rev. Lett. **92** (2004), 230402.

[7] H. OTT, E. DE MIRANDES, F. FERLAINO, G. ROATI, G. MODUGNO and M. INGUSCIO, Phys. Rev. Lett. **93** (2004), 120407.

[8] L. PEZZE, L. PITAEVSKII, A. SMERZI, S. STRINGARI, G. MODUGNO, E. DE MIRANDES, F. FERLAINO, H. OTT, G. ROATI and M. INGUSCIO, Phys. Rev. Lett. **93** (2004), 120401.

[9] H. OTT, E. DE MIRANDES, F. FERLAINO, G. ROATI, G. MODUGNO and M. INGUSCIO, Phys. Rev. Lett. **92** (2004), 230402.

[10] J. H. DENSCHLAG, J. E SIMSARIAN., H. HÄFFNER, C. MCKENZIE, A. BROWAEYS, D. CHO, K. HELMERSON, S. L. ROLSTON and W. D. PHILLIPS, J. Phys. B **35** (2002), 3095 .
[11] O. MORSCH, J. H. MÜLLER, M. CRISTIANI, D. CIAMPINI and E. ARIMONDO, Phys. Rev. Lett. **87** (2001), 140402.
[12] M. GREINER, I. BLOCH, O. MANDEL, T. W. HÄNSCH and T. ESSLINGER, Phys. Rev. Lett. **87** (2001), 160405.
[13] M. GREINER, O. MANDEL, T. ESSLINGER, T. W. HÄNSCH and I. BLOCH, Nature **415** (2002), 39; T. STÖFERLE, H. MORITZ, C. SCHORI, M. KÖHL and T. ESSLINGER, Phys. Rev. Lett. **92** (2004), 130403.
[14] A. HASEGAWA and Y. KODAMA, "Solitons in optical Communications", Clarendon Press, Oxford, 1995.
[15] B. WU and Q. NIU, Phys. Rev. A **64** (2001), 061603R; B. WU and Q. NIU, New Journ. Phys. **5** (2003), 104; A. SMERZI, A. TROMBETTONI, P. G. KEVREKIDIS and A. R. BISHOP, Phys. Rev. Lett. **89** (2002), 170402; C. MENOTTI, A. SMERZI and A. TROMBETTONI, New Journ. Phys. **5** (2003), 112; M. MACHHOLM, C. J. PETHICK and H. SMITH, Phys. Rev. A **67** (2003), 053613; F. KH. ABDULLAEV, B. B. BAIZAKOV, S. A. DARMANYAN, V. V. KONOTOP and M. SALERNO, Phys. Rev. A **64** (2001), 043606; V. V. KONOTOP and M. SALERNO, Phys. Rev. A **65** (2002), 021602R.
[16] L. SALASNICH, A. PAROLA and L. REATTO, Phys. Rev. Lett. **91** (2003), 080405.
[17] L. FALLANI, F. S. CATALIOTTI, J. CATANI, C. FORT, M. MODUGNO, M. ZAWADA and M. INGUSCIO, Phys. Rev. Lett. **91** (2003), 240405.
[18] L. FALLANI, L. DE SARLO, J. E. LYE, M. MODUGNO, R. SAERS, C. FORT and M. INGUSCIO, Phys. Rev. Lett., **93** (2004),140406.
[19] P. MASSIGNAN and M. MODUGNO, Phys. Rev. A **67** (2003), 023614.
[20] M. MODUGNO, C. TOZZO and F. DALFOVO, 2004, Phys. Rev. A **70** (2005), 043625.
[21] L. SALASNICH, Laser Physics **12** (2002), 198; L. SALASNICH, A. PAROLA and L. REATTO, Phys. Rev. A **65** (2002), 043614.
[22] L. DE SARLO, L. FALLANI, J. E. LYE, M. MODUGNO, R. SAERS, C. FORT and M. INGUSCIO, Phys. Rev. A **72** (2005), 013603.

Towards a quantitative investigation of sonic black-hole analogies in condensed matter systems

F. Federici, C. Cherubini, and S. Succi

The scattering process of a sound-wave perturbation impinging on a rotating acoustic black-hole is numerically investigated using the mathematical tools recently developed in numerical relativity for rotating Kerr black holes. Preliminary results indicate that modern methods from numerical relativity can be successfully adapted to the numerical investigation of analogies between gravitational black holes and condensed matter phenomena. However, subtle effects, such as stimulated emission from the black-hole ergo-sphere (super-radiance), require very careful and well resolved simulations.

In the recent past, there has been a growing interest in exploring analogue models of gravitational physics in condensed matter systems. The rationale behind such models traces back to a seminal observation by Unruh [1], who noted a close analogy between sound wave propagation on a background inhomogeneous flow and scalar fields propagation in a curved space-time. The analogy goes on by observing that, much like superfluid hydrodynamics is only a large-scale effective theory of microscopic superfluids, field theory on a curved space-time can be regarded as a large-scale limit of a putative microscopic theory of quantum gravity. The interesting point is that, while microscopic theories of quantum gravity are mostly a matter of speculation, the microscopic theory of superfluids is well developed. As a result, it can be hoped that the wide body of knowledge available for the latter can be brought to the benefit of the former. For instance, assessing the mechanisms of sound radiation from 'terrestrial' black holes beyond the hydrodynamic picture may in principle offer new information about the microscopic origin of cosmic black hole radiance, the Hawking effect, and other cosmological phenomena. This is a long-term research program which requires several quantitative steps. One of these is the study of scattering and radiance phenomena from black holes whose background space-time can be parallelled to the flow field configurations associated with typical fluid excitations, such as fluid vortices. Among others, a distinguished model of fluid flow which seems particularly well suited to pursue the 'analogue gravity' program is provided by the so-called "draining bathtub"geometry [2]. Such a model can be regarded as a three-dimensional flow with a sink (vortex) at the origin. The flow field induced by the vortex can be shown to associate with an acoustic metric with a non-trivial event horizon and ergo-sphere, which is why the model bears a special relevance to the physics of black holes. The main scope of this paper is to analyse and develop the formal apparatus to pursue the draining bathtub analogy on quantitative grounds via numerical simulations.

1. The model

It can be shown [1, 2] that sound propagation in a draining bathtub fluid is described by a wave equation that, under the hypothesis of long-wavelength perturbations, can be mapped onto a Klein-Gordon equation associated with an effective relativistic curved space-time background called acoustic metric. We focus our attention on the acoustic metric associated to the draining bathtub model introduced by Visser [2], for a rotating acoustic black hole. Such model consists of a $(2 + 1)$-dimensional flow with a sink at the origin. The flow is assumed to be vorticity-free (apart from a possible δ function contribution at the vortex core) and the angular momentum is conserved. Such constraints imply that the density of the fluid, the background pressure and the speed of sound c are constant throughout the flow. Therefore the background velocity potential must have the form

$$\psi(r, \phi) = -ca \log(r/a) + \Omega a^2 \phi \tag{1.1}$$

where a is a length scale associated with the "radius" of the vortex horizon and Ω is the constant rotation frequency.

Such a velocity potential, being discontinuous on going through 2π radians, is a multi-valued function. Therefore it must be interpreted as being defined patch-wise on overlapping regions around the vortex core at $r = 0$. The velocity of the fluid is then given by

$$\vec{v} = \nabla\psi(r, \phi) = \frac{-ca\hat{r} + \Omega a^2 \hat{\phi}}{r}. \tag{1.2}$$

The acoustic metric associated to this configuration is given by

$$\begin{aligned} ds^2 = &-\left(c^2 - \frac{a^2c^2 + a^4\Omega^2}{r^2}\right) dt^2 + \frac{2ca}{r} dt\, dr \\ &-2\Omega a^2 dt\, d\phi + dr^2 + r^2 d\phi^2 + dz^2. \end{aligned} \tag{1.3}$$

The metric (1.3) has an event horizon located at $r = a$ and an ergo-sphere. i.e. a region from which black-hole rotational energy can be transferred to the reflected wavepacket, at $r = r_{erg} = a\sqrt{1 + a^2\Omega^2/c^2}$. It can be seen from Eq. (1.2) that, since the acoustic event horizon forms once the radial component of the fluid velocity exceeds the speed of sound, the radius of such horizon is $r_{\text{horizon}} = a$. Linear perturbations of the velocity potential $\psi^{(1)} \equiv \Psi$ satisfy the massless Klein-Gordon scalar wave equation $\nabla^\mu \nabla_\mu \Psi = 0$ on this background, i.e.

$$\Bigg[-\frac{1}{c^2}\frac{\partial^2}{\partial t^2} + \frac{2a}{cr}\frac{\partial^2}{\partial t \partial r} - \frac{2a^2\Omega}{c^2r^2}\frac{\partial^2}{\partial t \partial \phi} + \left(1 - \frac{a^2}{r^2}\right)\frac{\partial^2}{\partial r^2}$$

$$+\frac{2a^3\Omega}{cr^3}\frac{\partial^2}{\partial r \partial \phi} + \frac{c^2r^2 - a^4\Omega^2}{c^2r^4}\frac{\partial^2}{\partial \phi^2} + \frac{\partial^2}{\partial z^2} + \frac{r^2 + a^2}{r^3}\frac{\partial}{\partial r} \tag{1.4}$$

$$-\frac{2a^3\Omega}{cr^4}\frac{\partial}{\partial \phi}\Bigg]\Psi = 0\,.$$

This partial differential equation is strongly hyperbolic, meaning by this that its characteristics exhibit strong spatial variations. As a result, in order to perform a successful integration, modern tools of differential geometry and general relativity must be resorted to.

The line element (1.3) is most conveniently analysed by recasting it in the following $(3 + 1)$-form:

$$ds^2 = -\alpha^2 dt^2 + \gamma_{ij}(dx^i + \beta^i dt)(dx^j + \beta^j dt)\,. \tag{1.5}$$

The three-metric is simply the flat Euclidean space in cylindrical coordinates $\gamma_{ij} = \text{Diag}(1, r^2, 1)$. Latin letters are used for spatial indices that are raised and lowered with γ_{ij} and its inverse $\gamma^{ij} = \text{Diag}(1, r^{-2}, 1)$, whereas Greek letters indicate four-dimensional quantities. The shift is given by $\beta^i = (acr^{-1}, -a^2\Omega r^{-2}, 0)$, consequently $\beta_i = (acr^{-1}, -a^2\Omega, 0)$. The lapse is simply $\alpha = c$. Equation (1.4) is completely separable and can be analysed in the frequency domain by imposing $\Psi = e^{-i\omega t}e^{im\phi}e^{ikz}P(r)$, where the integers m and k are the azimuthal and axial wavenumbers, respectively.

Within this representation it can be shown that the case $m = 0$ can be mapped into a Schroedinger-like equation with the following potential:

$$V(r) = k^2 - \frac{5}{4}\frac{a^4}{r^6} - \frac{1}{4}\left(\frac{1 + 4k^2a^2}{r^2}\right) + \frac{3}{2}\frac{a^2}{r^4}. \tag{1.6}$$

It can also be shown that wave-frequencies in the range $0 < \omega < m\Omega$ for $\omega \in \mathcal{R}$ exhibit the so-called phenomenon of super-radiance, whereby the back-scattered wave gains energy from the vortex background. For details, see [3–6].

2. Theoretical formulation of the scattering process

In order to solve Eq. (1.4) numerically in the time domain, we resort to some of the mathematical prescriptions developed in [7] for the numerical integration of the massless scalar field perturbations on a rotating Kerr black hole background. The starting point is to introduce the following conjugate quantities of the scalar field Ψ:

$$\Phi_i = \frac{\partial \Psi}{\partial x^i}\,, \qquad \Pi = -\frac{1}{\alpha}\left(\frac{\partial \Psi}{\partial t} - \beta^i \Phi_i\right), \tag{2.7}$$

by substituting them into Eq. (1.4), one obtains a first-order, symmetric, hyperbolic system.

In order to limit the computational demands, we factorise the z and ϕ dependence as follows:

$$\Psi\ = \psi_1(t,r)e^{im\phi}e^{ikz}\,, \quad \Pi\ = \pi_1(t,r)e^{im\phi}e^{ikz}\,, \tag{2.8}$$

$$\Phi_1 = \phi_1(t,r)e^{im\phi}e^{ikz}\,, \quad \Phi_2 = im\Psi\,, \qquad \Phi_3 = ik\Psi\,. \tag{2.9}$$

The initial wave equation (1.4) is then replaced by the following first-order set of coupled partial differential equations:

$$\begin{aligned}
&\frac{\partial \psi_1}{\partial t} - c\frac{\partial}{\partial r}\left(\frac{a}{r}\psi_1\right) = \left(\frac{ac - ima^2\Omega}{r^2}\right)\psi_1 - c\pi_1 \\
&\frac{\partial \phi_1}{\partial t} + c\frac{\partial}{\partial r}\left(\pi_1 - \frac{a}{r}\phi_1\right) = \frac{2ia^2m\Omega}{r^3}\psi_1 - \frac{im\Omega a^2}{r^2}\phi_1 \\
&\frac{\partial \pi_1}{\partial t} + c\frac{\partial}{\partial r}\left(\phi_1 - \frac{a}{r}\pi_1\right) = \left(\frac{ac - ima^2\Omega}{r^2}\right)\pi_1 \\
&\qquad\qquad + c\left(k^2 + \frac{m^2}{r^2}\right)\psi_1 - \frac{c}{r}\phi_1
\end{aligned} \tag{2.10}$$

associated with the constraint

$$|C| = |\partial_r \psi_1 - \phi_1| = 0 \tag{2.11}$$

as $C = 0$ corresponds to the definition of ϕ_1. Deviations of $|C|$ from zero provide a crucial indicator of the quality of the numerical results.

The horizon is such that nothing can escape from the surface $r = a$. As a result, in order to account for one-way inward propagation from the horizon, a proper in-going radiation condition on this surface is required, as discussed in the next section.

3. Numerical results

The numerical integration of wave equations in a cylindrical geometry is often affected by problems due to reflected waves close to the internal boundary [8]. Something similar happens for black holes, where theoretically correct boundary conditions are frequently in need of 'ad-hoc' modifications. The best way to solve the problem is to use horizon-penetrating coordinates or, better yet, the Kerr in-going ones, which permit to extend the field integration inside the horizon, via the so called "excision"technique [9]. Since inside the horizon the light cones point towards the singularity, the region inside the vortex is causally disconnected from the outside region. Thanks to this property, one can impose a generic well-behaved boundary condition inside the horizon and sidestep the problem of constraint violations (see below) inside the acoustic black hole. This procedure is computationally very demanding as it relies upon a very fine discretization around the sonic horizon. Indeed, due to requirements of high resolution, long integration times and distant outer boundaries, even the simple Klein-Gordon equation in Kerr black-hole geometry, has been only recently successfully implemented using a powerful multiprocessor code [7]. Following [7], we have implemented the excision technique (details shall be presented in a future publication [10]).

According to the standard prescription for scattering processes in Kerr black holes [6], we choose as initial data a Gaussian pulse modulated by a monochromatic wave

$$\psi_1(0,r) = A\exp[-(r-r_0+ct)^2/b^2 - i\sigma(r-r_0+ct)/c]|_{t=0} \quad (3.12)$$

together with the associated conditions for π_1 and ϕ_1 given above. The corresponding power spectrum is $P(\omega) = P_{\max}\exp\left[-(\omega-\sigma)^2 b^2/4\right]$. The modulated pulse leads to a Gaussian frequency distribution centered in $\omega = \sigma$ and the range $0 < \sigma < m\Omega$ corresponds to thesuperradiant regime.

We choose units in which $a = c = 1$, together with an angular frequency equal to $\Omega = 1.4$, corresponding to 10^6 Hz in physical units. The numerical integration is performed in the space-time domain $r \in [0.5, 150.5]$ and $t \in [0, 115]$.

The amplitude of the initial Gaussian pulse is $A = 0.3$, and the pulse is centered at $r_0 = 10$, corresponding to an envelope of harmonics centered at $\sigma = 0$ with variance $b = 0.5$. The axial wavenumber is $k = 0.02$, and for the azimuthal wavenumber we analyse the cases $m = 0$, $m = 1$. In order to control the reliability of the numerical simulation. the constraint C is monitored at both the outer and inner boundaries all along the simulation span.

Figure 1 shows a density plot of the real part of ψ_1 in the interval $r \in [0.5, 20]$ and $t \in [0, 10]$. The initial gaussian pulse moves towards the vortex horizon placed at $r = 1$, its trajectory is bent by the potential outside the horizon. The bending of the pulse trajectory, with light cones heading towards the horizon, is consistent with similar findings in numerical relativity [11], as shown in the diagram of Figure 1. The same behaviour is observed in Figure 2 for the case $m = 1$.

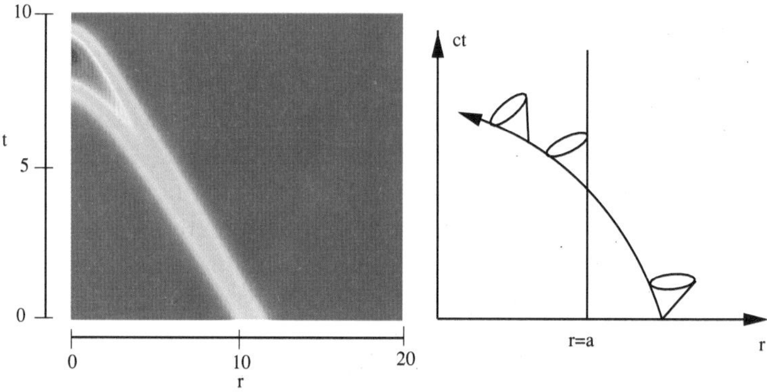

Figure 1. Density map of the real part of ψ_1 in the $r - t$ plane for the case $m = 0$. Due to the curved background, the wave trajectory leans towards the sonic horizon at $r = a$. Note that since no signal can escape from the horizon, the light-cone of the outward propagating signals is parallel to the $r = a$ axis.

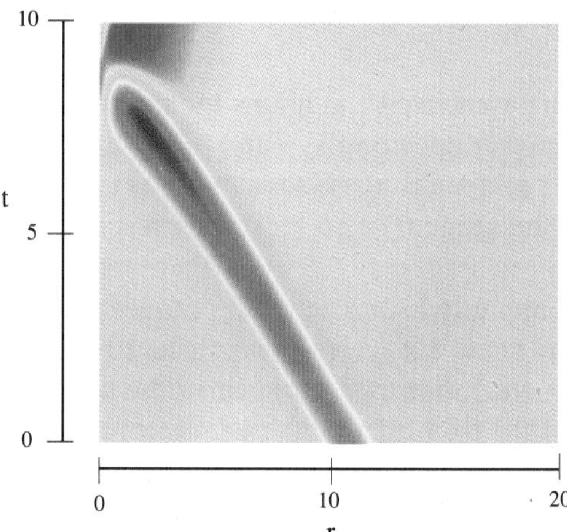

Figure 2. A density plot of the real part of ψ_1 in the $r - t$ plane is shown for the case $m = 1$. The dark spot on top of the figure corresponds to the scattered wavepacket.

Figure 3 and Figure 4 show snapshots of the real part of ψ_1 as a function of r for $t = 0, 5, 7, 10$, for the case $m = 0$ and $m = 1$ respectively. The peak in the signal at $t = 0$ is the in-going pulse. At $t = 5$ the initial pulse starts to be affected by the potential and around $t = 7$ for $m = 0$, and $t = 10$ for $m = 1$, respectively, the radiation is backscattered. The small reflected wave then relaxes towards steady-state ($Re\Psi_1 = 0$) after the characteristic ringing modes oscillations have settled down.

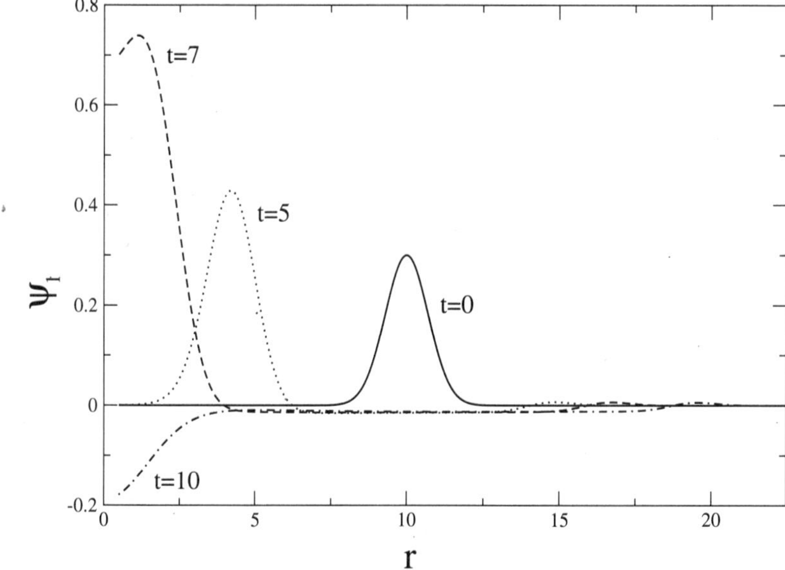

Figure 3. Snapshots of the time evolution of the perturbation ψ_1 at $t = 0$ (solid line), $t = 5$ (dotted line), $t = 7$ (dashed line), $t = 10$ (dot-dashed line), as a function of the distance r from the the vortex center for the case m=0. The graph is plotted with $r \geqslant 0.5$, being $r = 0.5$ the position of the internal boundary. The solid line shows the initial gaussian pulse travelling towards the vortex core.

In Figure 5 and Figure 6 we plot the evolution of $|C|$ as a function of t, for $m = 0$ at the inner and outer boundaries, respectively. These figures show that the boundary conditions preserve the constraints to a very good degree of accuracy for the case $m = 0$. For the case $m = 1$ (see Figure 7 and Figure 8) the evolution of $|C|$ confirms that, as found in [7], constraint violations at the outer boundary exhibit a time-asymptotic secular growth of the violation. This is still an open problem for the numerical simulation of black-hole dynamics. However, for the simulation span considered here, such violations are still well below the physical signal, so that they do not spoil the interpretation of the physical results.

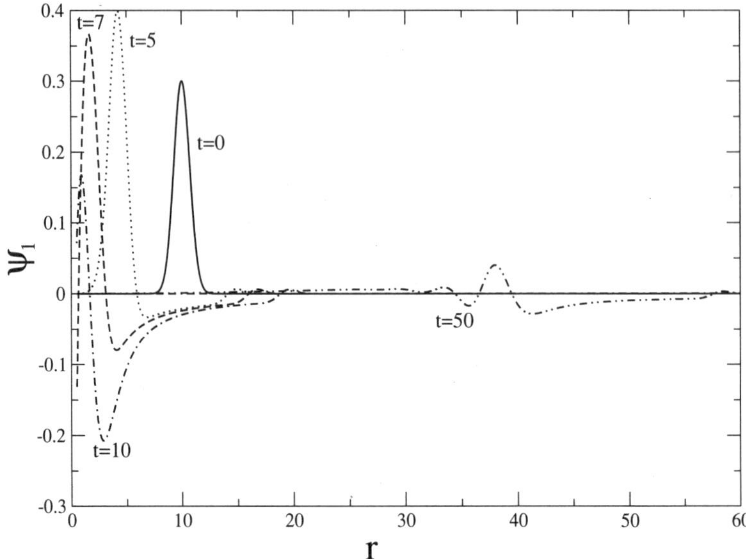

Figure 4. Snapshots of the time evolution of the perturbation ψ_1 at $t = 0$ (solid line), $t = 5$ (dotted line), $t = 7$ (dashed line) and $t = 10$ (dot-dashed line), as a function of the distance r from the the vortex center for the case $m = 1$. The graph is plotted with $r \geqslant 0.5$, $r = 0.5$ being the position of the internal boundary.

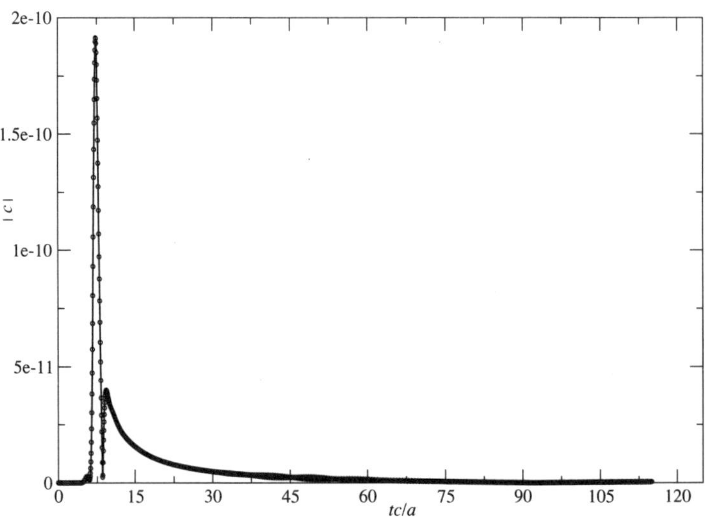

Figure 5. Time evolution of the constraint $|C|$ at the horizon for the case $m = 0$.

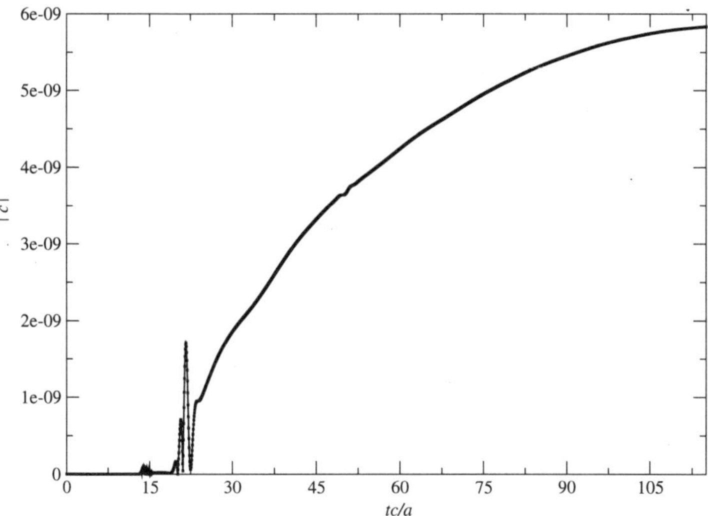

Figure 6. Time evolution of the constraint $|C|$ at $r = 150$ for the case $m = 0$.

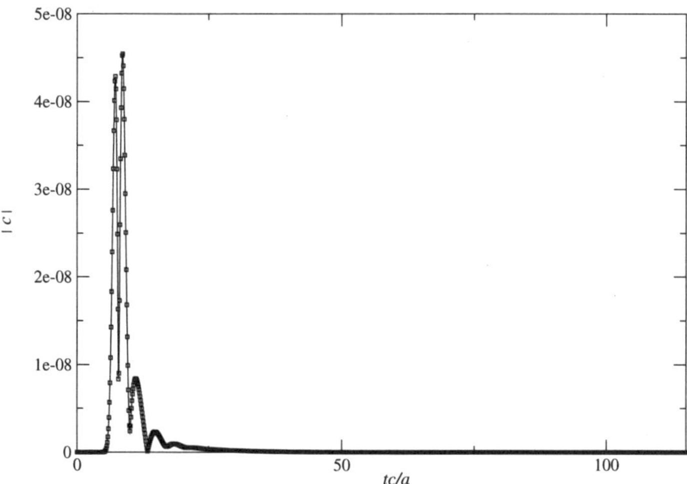

Figure 7. Time evolution of the constraint $|C|$ at the horizon for the case $m = 1$.

3.1. Summary

Summarizing, we have presented a numerical investigation of the scattering process of a sound wave perturbation impinging on a rotating acoustic black-hole. To this purpose, mathematical tools of modern numerical relativity of rotating Kerr black-holes have been adapted to the case of

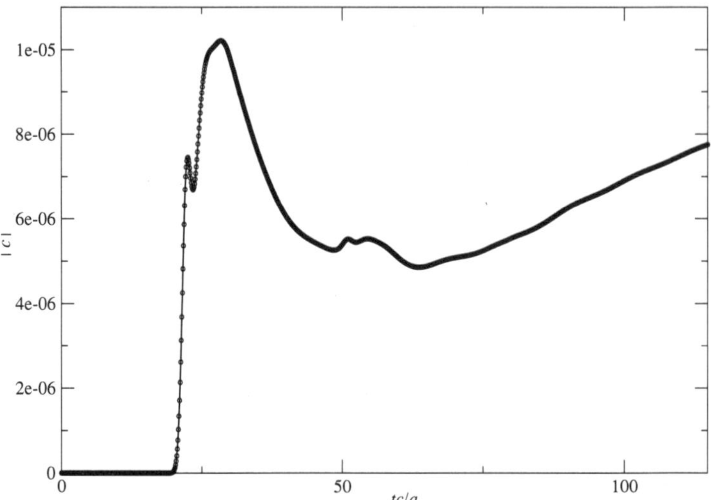

Figure 8. Time evolution of the constraint $|C|$ at $r = 150$ for the case $m = 1$.

the sound wave perturbation in a draining bathtub background geometry. The numerical results appear physically consistent, in that the indicators of numerical inaccuracies stay well below the signal amplitudes. However, subtle physical effects, such as super-radiance, are definitely in need of more systematic and accurate investigations.

An important direction for future research is the extension of the present formalism to the case of realistic BEC experiments, for which a new conceptual and numerical treatment of superfluidity and vortex quantization needs to be developed.

ACKNOWLEDGMENTS. One of the authors (SS) is warmly thankful to Prof. Mario Tosi for several years of enlightening and joyful cooperation.

References

[1] W. G. UNRUH, Physical Review Letters **46** (1981), 1351.

[2] M. VISSER, Class. Quant. Grav. **15** (1998), 1767.

[3] S. BASAK and P. MAJUMDAR, Class. Quant. Grav. **20** (2003), 2929.

[4] S. BASAK and P. MAJUMDAR, Class. Quant. Grav. **20** (2003), 3907.

[5] E. BERTI, V. CARDOSO and J. P. S. LEMOS, Physical Review D **70** (2004), 124006.

[6] N. ANDERSSON, P. LAGUNA and P. PAPADOPOULOS, Physical Review D **58** (1998), 087503.

[7] M. A. SCHEEL, A. L. ERICKCEK, L. M. BURKO, L. E. KIDDER, H. P. PFEIFFER and S. A. TEUKOLSKY, Physical Review D **69** (2004), 104006.
[8] D. KOYAMA, In: Proceedings of the 12th International Conference on Domain Decomposition Methods, 2001.
[9] J. THORNBURG, Physical Review D **59** (1999), 104007.
[10] C. CHERUBINI, F. FEDERICI, S. SUCCI and M. P. TOSI, in preparation.
[11] G. CALABRESE, L. LEHNER and M. TIGLIO, Physical Review D **65** (2002), 104031.

Solid state methods applied to a condensate travelling through one-dimensional arrays

Z. Akdeniz and P. Vignolo

We calculate within a Bose-Hubbard tight-binding model combined with a scattering matrix approach the matter-wave flow of condensed ^{87}Rb atoms driven by a constant force through various types of quasi-onedimensional arrays of potential wells. Different interference effects are observed when double-periodicity is induced by creating an energy minigap or when quasi-periodicity governed by the Fibonacci series induced the fragmentation of the spectrum.

1. Introduction

A Bose-Einstein condensate (BEC) is a gas in which a macroscopic number of massive particles reside in the same quantum state (see [1] for a review of work done on quasi-pure BEC's produced since 1995). Experiments aimed at revealing the coherence of a BEC have demonstrated its matter-wave properties. In particular, condensate interferometry can be realized by splitting a BEC into two parts with a definite phase relationship, these parts being then brought into overlap and interference as for an optical laser beam that has gone through a beam splitter. Coherent splitting of a BEC has been achieved by optically induced Bragg diffraction [2] and a number of ingenious methods have been devised to extract a collimated beam of atoms from a BEC (see *e.g.* [3]).

Solid-state like systems as a condensate or a quantum degenerate Fermi gas moving in a lattice can be realized by superposing to the atomic confinement periodic or quasi-periodic optical potentials. For instace, a quasi-onedimensional (1D) array of potential wells is created by the interference of two optical laser beams which counterpropagate. Such an optical lattice provides an almost ideal periodic potential and has allowed the study of Bloch and Josephson-like oscillations [4–7].

When the lattice is modified by introducing further periodicities, in momentum space it exists several pathes for Bloch oscillations and the number of condensed atoms leaving the lattice under the action of a constant force reflects the interference between the different pathes [8, 9]. In presence of quasi-periodicity the spectrum becomes fragmented, the simple picture for Bloch oscillations breaks down and localization effects appear [9, 10].

In this paper we discuss the use of solid-state approaches applied to the study of the condensate transport. In particular, we compare the transport properties of a condensate travelling through a single-period, a double-period and a quasi-periodic Fibonacci 1D array under the action of a constant external force. In Section 2 we discuss how to realize a BEC in a single-period, a double-period and a quasi-periodic 1D array and in Section 3 we give the Hamiltonian and the density of states corresponding to each system. The calculation of the bosonic current throughout the arrays and the numerical results are given in Section 4. Section 5 contains some concluding remarks.

2. The system

A single-period 1D lattice for an atomic BEC can be realized by superposing two optical laser beams which counterpropagate along the z axis, say, and are superposed on a highly elongated magnetic trap. This is shown in the left panel of Figure 1. The light potential for the atoms has the form

$$U(z) = U_0 \sin^2(kz)\,, \tag{2.1}$$

where U_0 is the potential well depth and k is the laser wavenumber determining the distance d of adjacent wells as $d = \pi/k$. The 1D optical lattice can be modified by means of auxiliary laser beams [8] (see right panel of Figure 1). In particular, its periodicity can be doubled by adding two beams that are rotated by angles of 60° and 120° with respect to the z axis. For a suitable choice of the phases the potential seen by the BEC atoms takes the shape

$$U(z) = U_0[\sin^2(kz) + \delta^2 \sin^2(kz/2)]\,, \tag{2.2}$$

where δ^2 is the relative energy difference between adjacent wells. In solid-state terminology, the doubling of the period when $\delta^2 \neq 0$ causes the opening of a minigap in the energy spectrum as a function of δ^2.

A schematic drawing of the set-up of optical lasers that would create an atomic Fibonacci wave guide is shown in Figure 2. Here, two pairs of counterpropagating laser beams create a square optical lattice. The projection of this lattice on a line at an angle $\alpha = \arctan(2/(\sqrt{5}+1))$

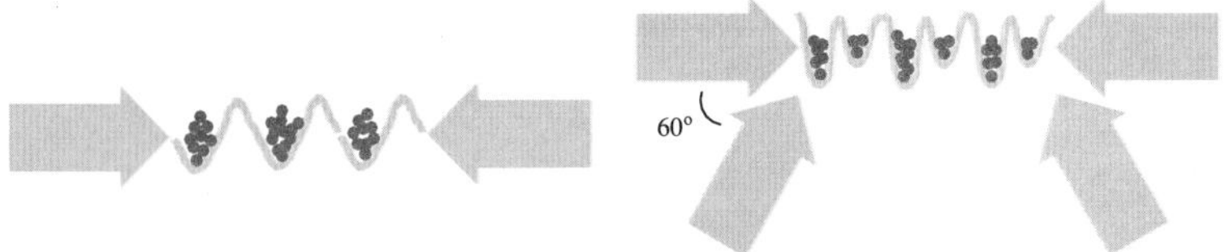

Figure 1. Optical realization of a single (left panel) and double (right panel) period 1D lattice.

relative to the lattice creates a quasi-periodic sequence of bond lengths, and hence of hopping energies, which obey the Fibonacci rule [10]. The atoms can be made to travel along the sequence by pointing along this direction a hollow beam (for a description of the latter see for example the work of Xu *et al.* [11]).

A simple way to create and control a constant external drive in a Fibonacci array, as well as in a single-period and double-period lattice, it would be to tilt the whole set-up by an angle β relative to the vertical axis (see Figure 2). In this case the external force acting on the condensate atoms of mass m is $F = mg\cos\beta$.

3. The Hamiltonian

To study the effect of the various types of arrays on the bosonic transport, we build a 1D tight-binding Hamiltonian for N condensed bosons and use a Green's function approach to evaluate their transport coefficient through the lattice. The Bose-Hubbard Hamiltonian for the bosons is

$$H = \sum_{i=1}^{n_s} [E_i |\, i\rangle\langle i\,| + \gamma_i (|\, i\rangle\langle i+1\,| + |\, i+1\rangle\langle i\,|)] \,. \tag{3.3}$$

Here, the parameters E_i and γ_i depend on the number of bosons in the lattice well labelled by the index i and represent site energies and hopping energies, respectively.

We proceed to a 1D reduction of the Hamiltonian by introducing the transverse width $\sigma_\perp$ of the bosonic wave functions in a cigar-shaped harmonic trap, together with a 1D condensate wave function $\phi_i(z)$ in the $i-th$ cell. In a tight-binding scheme $\phi_i(z)$ is a Wannier function for the bosons in the potential $U(z)$ and, according to the early work of Slater [12], can be written as a Gaussian function,

$$\phi_i(z) = \phi_i(0)\exp[-(z-z_i)^2/(2\sigma_z^2)], \tag{3.4}$$

where $|\phi_i(0)|^2$ is the number of bosons in the lattice well i. The determination of the widths $\sigma_\perp$ and σ_z is carried out variationally.

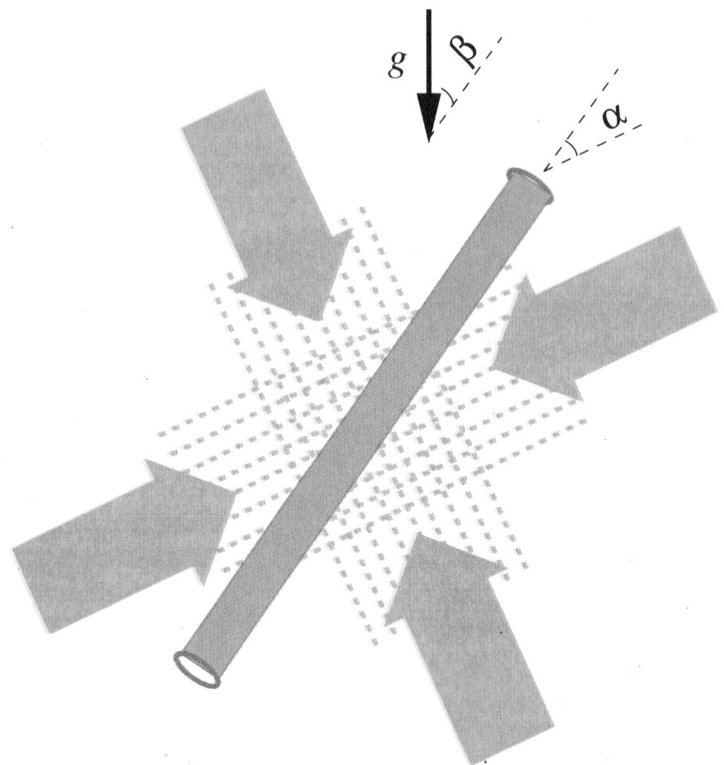

Figure 2. Schematic representation of a five laser-beam configuration to create a quasi-one-dimensional Fibonacci array. Four beams generate a square optical lattice and a hollow beam confines the condensate to a strip with slope $\alpha = \arctan(2/(\sqrt{5}+1))$ relative to an axis of the lattice. The angle β between the hollow beam and the vertical direction determines the driving force as $F = mg\cos\beta$.

We can now evaluate the parameters entering the effective Hamiltonian. The site energies are given by

$$E_i = \int dz\, \phi_i(z) \left[-\frac{\hbar^2\nabla^2}{2m} + U(z) + \frac{1}{2} g_{bb} |\phi_i(z)|^2 + maz + C \right] \phi_i(z) \quad (3.5)$$

where m is the bosonic mass, $a = F/m$ is the acceleration due to a constant external force F acting on the bosons, C is a factor which takes in account the reduction of dimensionality. The parameter g_{bb} is the strength of the 1D boson-boson interaction and is given by

$$g_{bb} = \frac{4\pi\hbar^2}{m} \frac{a_{bb}}{2\pi\sigma_\perp^2} \quad (3.6)$$

with a_{bb} the boson-boson scattering length. Consistently with the model, the hopping energies γ_i are given by

$$\gamma_i = \int dz\, \phi_i(z) \left[-\frac{\hbar^2 \nabla^2}{2m} + U(z) + \frac{1}{2} g_{bb} |\phi_i(z)|^2 + C \right] \phi_{i+1}(z). \quad (3.7)$$

A schematic representation of the Bose-Hubbard Hamiltonian in absence of an external force for a single-periodic, a double-periodic and a Fibonacci array is shown in Figure 3.

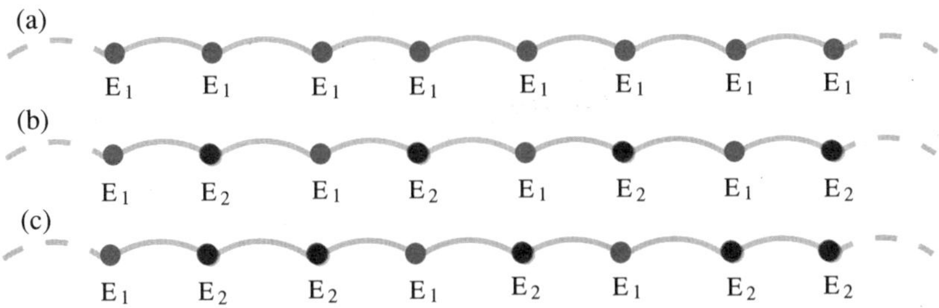

Figure 3. Schematic representation of the tight-binding Hamiltonian for a condensate in (a) a single-period lattice, (b) a double-period lattice, (c) a Fibonacci array.

By applying an external force, the energy spectrum depends on the spatial position. A constant acceleration provokes a tilt of the spectrum and the condensed bosons, which at rest are in the lowest energy level, start travelling through the lattice and at the same time through the spectrum. This is shown in Figure 4 where we have plotted the local density of states (DOS) $\rho_i(E)$ given by

$$\rho_i(E) = -\frac{1}{\pi} \lim_{\varepsilon \to 0} \operatorname{Im} \langle i | (E + i\varepsilon - H)^{-1} | i \rangle \,, \quad (3.8)$$

as a function of position and energy for a single-periodic, double-periodic and a Fibonacci array. The arrows indicate the bosonic trajectory in space and through the spectrum.

In the next section we shall see how the various spectra modify the transport properties of the condensed bosons travelling through the arrays.

4. The steady-state current

The steady-state particle current passing through the lattice can be calculated by exploiting Landauer's approach adapted to a bosonic gas with

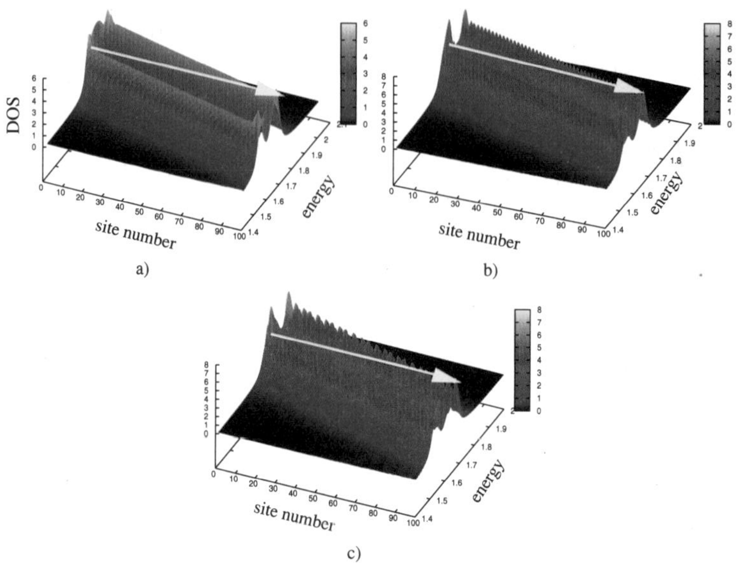

Figure 4. Local DOS as a function of position and energy for a mono-periodic, a double-periodic and a Fibonacci array.

out-of-equilibrium leads [13]. The current j can be written

$$j = \frac{N}{\pi \hbar n_s} \int \mathrm{d}E[f(E-\mu_1) - f(E-\mu_2)]\mathcal{T}(E) \tag{4.9}$$

where $f(E-\mu)$ is the Bose-Einstein distribution and $\mu_{1,2}$ are the chemical potentials of the reservoirs which act as the source and the sink of the bosons, and $\mathcal{T}(E)$ the transmittivity coefficient that we shall discuss below.

Note that the normalization N/n_s of the bosonic wavefunction plays the role of the electric charge in the Landauer formula [14]. In the zero-temperature limit we have

$$f(E-\mu_1) - f(E-\mu_2) \approx \frac{\partial f(E-\mu)}{\partial E}(\mu_2 - \mu_1) \tag{4.10}$$

and the density probability $\partial f(E-\mu)/\partial E$ is a delta function $\delta(E-\mu)$. Then Eq. (4.9) gives

$$j = \frac{N}{\pi \hbar n_s} \mathcal{T}(\overline{\mu}) \Delta\mu \tag{4.11}$$

where $\Delta\mu = \mu_2 - \mu_1$ is the difference in chemical potentials and $\overline{\mu} = (\mu_1 + \mu_2)/2$ is their average value.

The transmittivity can be evaluated by connecting the array to an incoming and to an outgoing lead, the first mimiking the continuous replenishing of the condensate and the latter playing the role of the coupling with the continuum. Within this frame the transmittivity coefficient is defined as

$$\mathcal{T}(E) = \frac{\lim_{n\to+\infty}\langle n|\phi^{\tilde{\kappa}}_{\text{out}}\rangle\langle\phi^{\tilde{\kappa}}_{\text{out}}|n\rangle v_{\text{out}}}{\lim_{m\to-\infty}\langle m|\phi^{\kappa}_{\text{in}}\rangle\langle\phi^{\kappa}_{\text{in}}|m\rangle v_{\text{in}}} \tag{4.12}$$

with v_{in} and v_{out} the velocities of the incoming and outgoing wavefunctions $\phi_{\text{in}}(z)$ and $\phi_{\text{out}}(z)$ of the incoming and outgoing leads.

The outgoing wavefunction $\phi_{\text{out}}(z)$ can be expressed as a function of the incoming one $\phi_{\text{in}}(z)$ throught the scattering matrix formalism. This leads to the formula

$$\mathcal{T}(E) = 4\frac{|T(E)_{1,N_{\text{out}}}|^2}{\gamma_L^2}\sin\left(\kappa\frac{\lambda}{2}\right)\sin\left(\tilde{\kappa}\frac{\lambda}{2}\right). \tag{4.13}$$

where $T(E)_{1,N_{\text{out}}}$ is the scattering matrix element of coherence between the first site and the site N_{out} where the bosons reach the highest energy point of the spectrum and can be transmitted towards the continuum. The full details for the evaluation of the scattering matrix $T(E)$ have been given elsewhere [8].

As a numerical application, we have chosen to focus on the typical experimental parameters used at LENS [15] for a condensed gas of ^{87}Rb, setting $N = 10^5$, $n_s = 200$, $U_0 = 3.5Er$, $E_r = \hbar^2k^2/2m$ being the recoil energy, and $\delta^2 \simeq 10^{-2}$. Our results are shown in Figure 5 where we have plotted the bosonic transmittivity throughout a single-periodic, a double-periodic and a Fibonacci array as a function of the acceleration a.

For the case of a single-period lattice, the intensity of the steady-state current increases monotonically with the acceleration. A BEC driven through a double-period lattice by a constant force is coherently split by a combination of Bragg diffraction and of tunnelling through the minigap. The interference peaks, appearing on the outgoing particle flow, have been shown to be periodical as a function of the ratio between the Bloch period and the tunnelling time through the minigap [8, 9]. In a Fibonacci array the simple picture for Bloch oscillations breaks down. Some peaks are still presents analogously with the double-period case, but the peaks seem to be randomly distributed and the number of points with vanishing transmittivity increases [9, 10]. The null transmittivity reflects the localization of the condensate in the array.

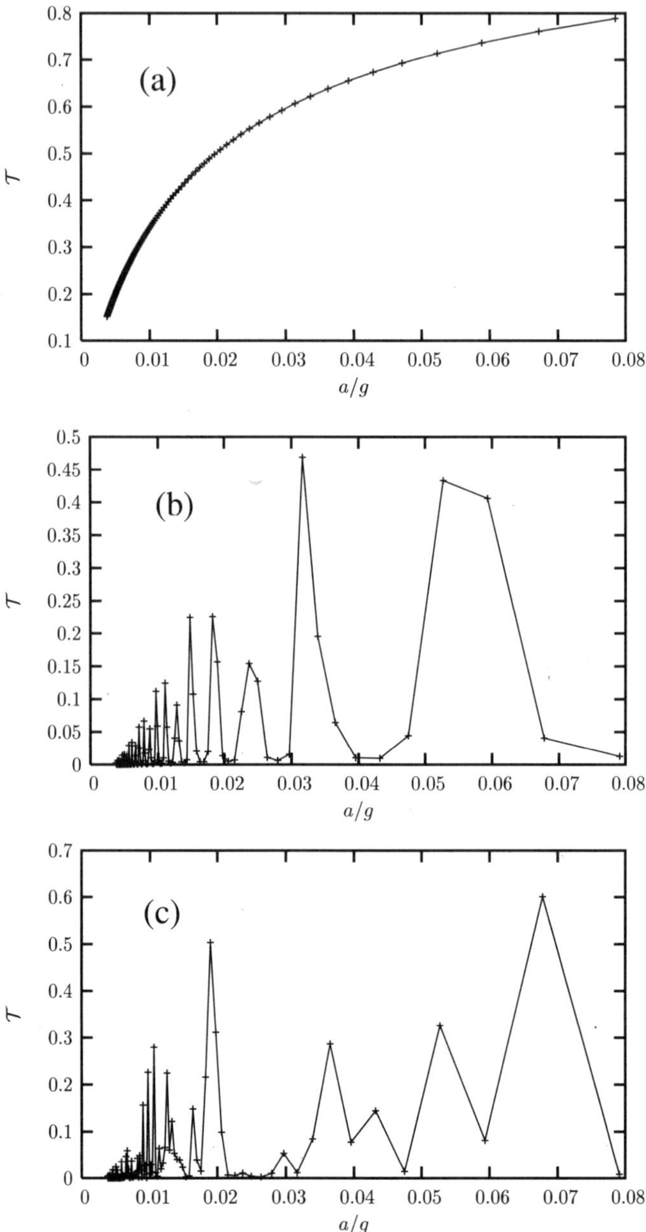

Figure 5. The transmittivity $\mathcal{T}$ as a function of the acceleration a (in unit of the gravity acceleration g) throughout a single-periodic (a), double-periodic (b) and a Fibonacci array (c).

5. Concluding remarks

In this paper we have used an out-of-equlibrium Green's function approach to evaluate the bosonic steady-state current through various typologies of onedimensional arrays. We have found that when single-periodicity is broken by a higher-order periodicity or by a quasi-periodicity the behaviour of the current is not monotonical with the applied force and interference phenomena appear. In the particular, in a quasi-periodic array, coherence and disorder effects coexist analogously to the case of quasi-periodic crystals.

ACKNOWLEDGMENTS. This work was partially supported by INFM through the PRA-Photonmatter.

Z.A. acknowledges support from TUBITAK and from the Research Fund of Istanbul University under Project Number YÖP-316/13082004.

References

[1] A. MINGUZZI, S. SUCCI, F. TOSCHI, M. P. TOSI and P. VIGNOLO, Phys. Rep. **395** (2004), 223.
[2] M. KOZUMA, L. DENG, E. W. HAGLEY, J. WEN, R. LUTWAK, K. HELMERSON, S. L. ROLSTON and W. D. PHILLIPS, Phys. Rev. Lett. **82** (1999), 871.
[3] S. INOUYE, T. PFAU, S. GUPTA, A. P. CHIKKATUR, A. GÖRLITZ, D. E. PRITCHARD and W. KETTERLE, Nature **402** (1999), 641.
[4] M. B. DAHAN, E. PEIK, J. REICHEL, Y. CASTIN and C. SALOMON, Phys. Rev. Lett. **76** (1996), 4508.
[5] B. P. ANDERSON and M. A. KASEVICH, Science **282** (1998), 1686.
[6] F. S. CATALIOTTI, S. BURGER, C. FORT, P. MADDALONI, F. MINARDI, A. TROMBETTONI, A. SMERZI and M. INGUSCIO, Science **293** (2001), 843.
[7] G. ROATI, E. DE MIRANDES, F. FERLAINO, H. OTT, G. MODUGNO and M. INGUSCIO, Phys. Rev. Lett. **92** (2004), 230402.
[8] P. VIGNOLO, Z. AKDENIZ and M. P. TOSI, J. Phys. B **36** (2003), 4535.
[9] Y. EKSIOGLU, P. VIGNOLO and M. P. TOSI, Opt. Comm. **243** (2004), 175.
[10] Y. EKSIOGLU, P. VIGNOLO and M. P. TOSI, Laser Phys. **15** (2005), 356
[11] X. XU, K. KIM, W. JHE and N. KWON, Phys. Rev. A **63** (2001), 063401.
[12] J. C. SLATER, Phys. Rev. **87** (1952), 807.

[13] M. Paulsson, cond-mat/0210519
[14] R. Landauer, Philos. Mag. **21** (1970), 863.
[15] G. Roati, F. Riboli, G. Modugno and M. Inguscio, Phys. Rev. Lett. **89** (2002), 150403.

Dynamics of trapped ultracold boson-fermion mixtures

P. Capuzzi

1. Introduction

The experimental achievement of Bose-Einstein condensation of dilute atomic trapped gases about ten years ago triggered a renewed interest in the field of atomic gases [1–3]. Since then, an enormous amount of research has been devoted to the properties of Bose-condensed gases and latter on to statistical mixtures of bosons and fermions (see, e.g., refs. [4] and [5] for a review). The atoms are confined by magnetic and optical potentials and cooled down to temperatures where quantum degeneracy effects are well noticeable. The versatility of experiments and the diluteness of the trapped gases provided a new, easyly accesible, ground in which quantum theories for finite systems could be tested.

At the beginning the studies on trapped Bose-Fermi mixtures where just a natural continuation of those performed in pure Bose-Einstein condensates of single- and multi-component gases. Many authors investigated the spatial density profiles and excitations as functions of tunable experimental parameters such as number of particles, strength and sign of the interparticle interactions, and geometry of the confinement. More recently, the experiments [6–8] have been focused on strongly interacting fermions aiming at identifying the presence of a BCS superfluid or of a condensate of molecules.

In this article we will discuss the dynamics of boson-fermion mixtures with special emphasis on the effects of collisions. The paper shall not attempt to be an exhaustive but rather an illustrative description of some phenomena that may be observed and analysed in actual experiments in trapped degenerate boson-fermion mixtures.

The paper is organised as follows. In the first section we introduce the trapped boson-fermion mixtures and review several approximation for the calculation of the ground-state densities. The second section deals with the dynamics of the system and how the frequency of collisions influences it. In particular, we describe three regimes, namely, hydrodynamic, collisionless and crossover regimes and illustrate some of their features with experiments and theoretical calculations. Finally, Section 4 contains the summary of the paper.

2. The trapped mixture

We study a mixture of two species of neutral atoms, one species are bosons and the other spin-polarised fermions. The particles are confined by magnetic and/or optical external potentials which can be well represented by

$$V_{\text{ext}}^{\sigma}(\mathbf{r}) = \frac{1}{2} m_\sigma \left(\omega_{x,\sigma}^2 x^2 + \omega_{y,\sigma}^2 y^2 + \omega_{z,\sigma}^2 z^2\right), \tag{2.1}$$

with $\sigma = F$ for fermions and $\sigma = B$ for bosons. Here m_σ is the mass of atoms of species σ and $\omega_{j,\sigma}$ the angular trap frequencies along the axis x_j. Our system is at ultralow temperatures and hence for a dilute system collisions occur predominantly in the s-wave. The interaction can then be modelled by means of contact potentials $V_{\text{int}}(\mathbf{r} - \mathbf{r}') = g\,\delta(\mathbf{r} - \mathbf{r}')$, where $g > 0$ corresponds to repulsions whereas $g < 0$ indicates attractive potentials. These give rise to the mean-field potentials

$$V_{\text{mf}}^{B}(\mathbf{r}) = g_{BB}\,\rho_B(\mathbf{r}) + g_{BF}\,\rho_F(\mathbf{r}), \tag{2.2}$$

for bosons and

$$V_{\text{mf}}^{F}(\mathbf{r}) = g_{BF}\,\rho_B(\mathbf{r}), \tag{2.3}$$

for fermions. Notice that in Eq. (2.3) we have included only the interaction between bosons and fermions since for spin-polarised fermions fermion-fermion interactions are suppressed due to the Pauli principle. The coupling strengths g_{ij} can be written in terms of the s-wave scattering lengths as $g_{ij} = 2\pi\, a_{ij}\hbar^2/m_{ij}$, with m_{ij} the reduced mass of the pair (i, j). The external potentials and interactions define the mean-field Hamiltonians

$$H_B = \frac{p^2}{2\,m_B} + V_{\text{ext}}^{B} + g_{BB}\,\rho_B + g_{BF}\,\rho_F, \tag{2.4a}$$

$$H_F = \frac{p^2}{2\,m_F} + V_{\text{ext}}^{F} + g_{BF}\,\rho_B. \tag{2.4b}$$

The equilibrium densities of the mixture can be obtained by minimising the total energy functional

$$E[\rho_B, \rho_F] = \int d^3r \left(\kappa_B(\mathbf{r}) + V^B_{\text{ext}}(\mathbf{r})\rho_B + \frac{g_{BB}}{2}\rho_B^2\right) + \int d^3r \left(\kappa_F(\mathbf{r}) + V^F_{\text{ext}}(\mathbf{r})\rho_F\right) + \int d^3r \, g_{BF}\, \rho_B\, \rho_F \quad (2.5)$$

for given number of fermions (N_F) and bosons (N_B). In Eq. (2.5), $\kappa_i(\mathbf{r})$ is the kinetic-energy density of component i. For bosons in the Gross-Pitaevskii approximation [9] the kinetic-energy density for vortex-free configurations reads

$$\kappa_B(\mathbf{r}) = -\frac{\hbar^2}{2m_B}|\nabla\sqrt{\rho_B}|^2. \quad (2.6)$$

On the other hand, fermions are often treated in two different approximations according to the collisionality of the gas. Assuming that the fermions can be described by independent particles in a Hartree-Fock (HF) fashion, we can write

$$\rho_F(\mathbf{r}) = \sum_i |\psi_i|^2\, \Theta(\mu_F - \varepsilon_i) \quad (2.7)$$

where μ_F is the chemical potential of fermions, and ε_i and $\psi_i(\mathbf{r})$ are the single-particle energy levels and eigenfunctions of the Hamiltonian (2.4b). In its turn, the kinetic-energy density is

$$\kappa_F(\mathbf{r}) = \frac{\hbar^2}{2m_F}\sum_i |\nabla\psi_i|^2\, \Theta(\mu_F - \varepsilon_i). \quad (2.8)$$

On the contrary, if the atoms are in local equilibrium [10], the kinetic-energy density of the fermions takes the form

$$\kappa_F(\mathbf{r}) = \frac{5}{3}A\rho_F^{5/3} + \xi_F(\mathbf{r}), \quad (2.9)$$

where $A = \hbar^2(6\pi^2)^{2/3}/2m_F$. In Eq. (2.9) the first term in the rhs is the volume ideal-gas contribution to the kinetic energy and $\xi_F(\mathbf{r})$ represents surface and higher order terms.

If the number of particles of each species is large enough, the most common approximation consists in neglecting completely the kinetic energy of bosons and keeping only the volume contribution for fermions. These are called Thomas-Fermi (TF) approximations. Under isotropic harmonic confinement and for number of fermions $N_F \sim 10^4$ as in current experiments we have verified that the fermion equilibrium profiles are well described by the solution of a generalised Thomas-Fermi approximation which includes the surface kinetic energy effects in the form of Von Weizsäcker [11]. We showed that the results of this are practically indistinguishable from the full solutions of the HF equations.

Depending on the strength of the interspecies interaction g_{BF} the mixture may be in a well mixed phase or in a fully demixed phase. Moreover, for strongly attractive interactions the system may become unstable and collapse. These two possibilities are illustrated in Figure 1 where we plot the boson and fermion densities of a mixture of ^{7}Li and ^{6}Li atoms at several values of the coupling g_{BF} and with the other parameters as in the experiment of Schreck *et al.* [12]. The demixing and instability of the mixture may lead to a decrease in the efficiency of the cooling process which might in turn prevent experimentalists from exploring other regimes at even lower temperatures. The collapse of a boson-fermion mixture has been experimentally observed in an system of ^{87}Rb-^{40}K at LENS [13]. In that experiment the collapse of the mixture manifested as the disappearance of trapped fermions that resulted in a more stable system.

3. Collisions and dynamics

The dynamics of these mixtures is strongly influenced by the frequency of collisions. If collisions occur very rarely we expect the dynamics to be governed by the mean-field Hamiltonians (2.4). On the contrary, if collisions are very frequent the system is constantly in local equilibrium and its motion resembles that of a standard fluid. Finally, in a general situation one has to resort to kinetic equations to follow the dynamics more accurately. The Boltzmann treatment of the kinetic problem [14] constitutes a common choice that works very well for a gas in a semiclassical regime.

The analysis of the dynamical properties of mixture has been the most frequently used tool to investigate and quantum-mechanically characterise trapped degenerate systems. These studies have been generally focused on the behaviour of the density fluctuations, induced e.g. by deformations of the confining potential, and on the free expansion of the trapped gas after the confinement has been switched off.

The transition from a collisionless to a hydrodynamic behaviour as the collisionality increases has been experimentally [15] and theoretically [16] explored in mixtures. A clear signature of this transition is the locking of the motion of both species to a common oscillation with-

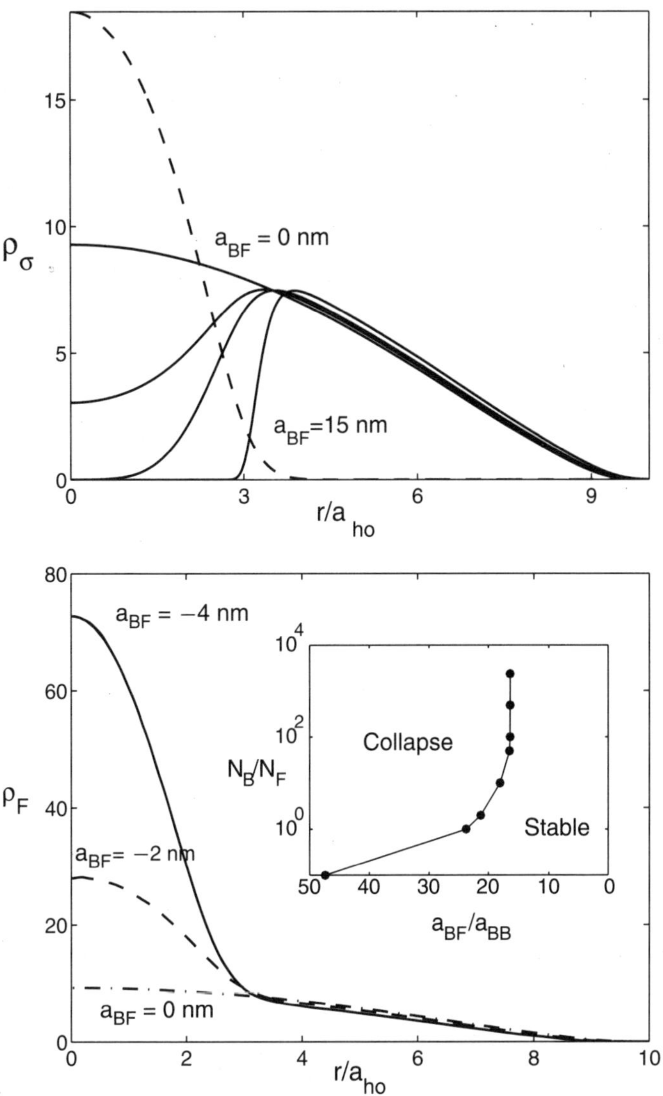

Figure 1. Density profiles (in arbitrary units) of a Li mixture in an isotropic trap of frequency $\omega_0 = 2\pi \times 1000\ \text{s}^{-1}$ as functions of r (in units of the oscillator length $a_{\text{ho}} = \sqrt{\hbar/m_F\omega_0}$). Top panel: Demixing of the mixture for $a_{BF} = 0$, 1, 2, and 15 nm. Solid lines correspond to the fermionic density profiles and the dashed line to the bosonic one. Bottom panel: Approaching of the collapse instability showing the increase of the fermionic density at the trap centre. The inset shows the phase diagram of the Li mixture in terms of the relative number of particles N_B/N_F (in log scale) *versus* the ratio of scattering lengths a_{BF}/a_{BB}.

out damping. While in the collisionless regime each species oscillates at a, *a priori*, renormalised trap frequency; in the fully hydrodynamic case there is only one oscillatory motion well defined. In Figure 2 we illustrate the collisionless to collisional transition as demonstrated in experiments with fermion-fermion and boson-fermion mixtures.

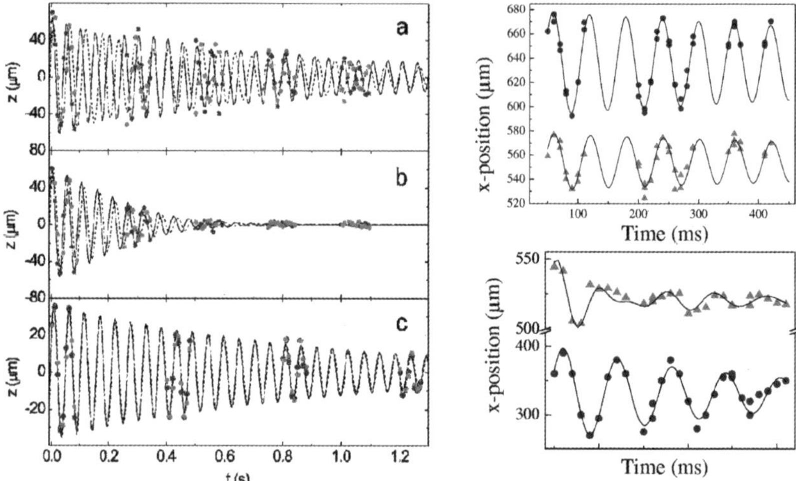

Figure 2. Transition from a collisionless to a collisional regime in the sloshing mode of a binary mixture. Left panels: Transition in a ^{40}K mixture (fermions) taken from Ref. [15]. The top, middle and bottom panels show the oscillatory motion of two species of fermions in the collisionless, crossover and hydrodynamic regimes, respectively. The different linestyles indicate the two species. Right panels: Transition in a ^{87}Rb-^{40}K mixture from Ref. [17]. The top (bottom) panel illustrates the oscillation of bosons and fermions in the hydrodynamic (crossover) regime.

3.1. Hydrodynamic approach

When a mixture evolves while maintaining its local equilibrium ensured by collisions, its evolution can be described in terms of the equation of continuity of mass for the particles density $\rho_\sigma(\mathbf{r}, t)$ and current $\mathbf{j}_\sigma(\mathbf{r}, t)$

$$\frac{\partial \rho_\sigma}{\partial t} + \nabla \cdot \mathbf{j}_\sigma = 0\,, \tag{3.10}$$

plus the equation of conservation of momentum or Euler equation, which in the linear regime reads

$$m_\sigma \frac{\partial \mathbf{j}_\sigma}{\partial t} + \nabla \cdot \Pi^\sigma + \rho_\sigma \nabla \widetilde{V}_\sigma = 0, \tag{3.11}$$

where Π^σ is the kinetic-stress tensor and $\widetilde{V}_\sigma = V^\sigma_{\rm ext} + V^\sigma_{\rm mf}$ is the total effective potential experienced by species σ. The system of equations (3.10) and (3.11) is closed when a form of Π^σ in terms of the densities and currents is specified. This is equivalent to specifying the equation of state of the atoms. In the Thomas-Fermi approximation we have $\Pi^B_{ij}=0$ for Bose-condensed atoms and $\Pi^F_{ij}=2A\,\rho_F^{5/3}\delta_{ij}/5$ for spin-polarised fermions.

If we put the system slightly out-of-equilibrium by creating perturbations on the density, the induced oscillations contain information on the collective excitation modes of the system. Performing a small oscillations analysis of Eqs. (3.10) and (3.11) we obtain the coupled equation for the collective modes of the mixture [11]

$$m_\sigma\, \omega_i^2\, \delta\rho_\sigma(\mathbf{r}) = \nabla \cdot (\rho_\sigma \delta\mathbf{F}_\sigma) - g_{BF}\, \nabla \cdot (\rho_\sigma \nabla \delta\rho_{\bar{\sigma}})\,, \qquad (3.12)$$

where $\mathbf{F}_\sigma$ is the total effective force acting on species σ (see Ref. [11] for the detailed expressions) and $\bar{\sigma}$ is the species different from σ. In addition, in *spherically* confined mixtures the collective modes can be labelled according to their total angular momentum ℓ and z-projection ℓ_z. We have examined the behaviour of the low-lying collective modes with $\ell = 0$ (monopolar) with varying the strength of the interaction. In Figure 3 we show the monopolar modes of a mixture of ^{6}Li and ^{7}Li atoms as functions of the interspecies repulsion a_{BF}. The approach of the demixing, depicted in Figure 1, is reflected in the monopolar modes as a clear lowering of the *fermionic* modes followed by a sharp upturn. These have been interpreted [11] as the appearance of a dynamical instability and the formation of a hard shell of fermions where sound propagates.

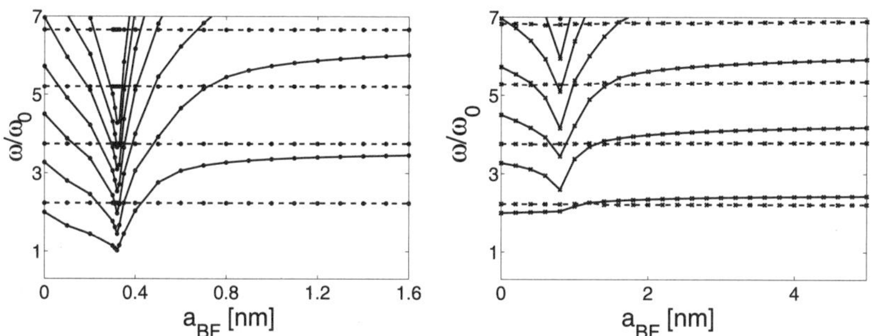

Figure 3. Frequencies of $\ell = 0$ collective modes (in units of the trap frequency ω_0) as functions of the boson-fermion scattering length a_{BF} (in nanometres). The left and right panels correspond to mixtures with $N_F = 10^4$ fermions and $N_B = 2.4 \times 10^7$ and 10^6 bosons respectively. Each panel show the low-lying fermionic (solid lines) and bosonic (dashed lines) modes from the numerical solution of Eq. (3.12).

Another experiment [13] that has recently attracted some attention is the collapse of a mixture of ^{87}Rb and ^{40}K atoms with attractive interactions. The behaviour of the collective modes of such mixtures shows that on approaching critical values of a_{BF} and N_B a set of collective modes rapidly soften to zero when the mixture collapses. This is illustrated in Figure 4 for the same parameters that the experiment performed at LENS and for the ^{7}Li-^{6}Li mixture. Even though this softening is more dramatic than that observed during the demixing, for the current experimental parameters, it might be quite difficult to observe due to the narrow window in a_{BF} and N_B where the effect is noticeable.

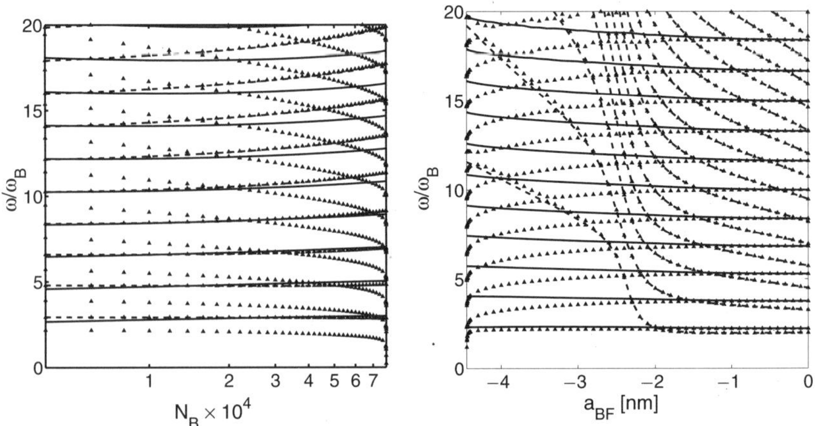

Figure 4. Monopolar collective modes of a boson-fermion mixture with attractive interactions. Left panel: Frequencies of the collective modes (in units of the trapping frequency of the bosons) as functions of the number of bosons N_B (in log scale) for the ^{87}Rb-^{40}K mixture of the LENS experiment [13]. Right panel: Frequencies of the collective modes (in units of the trapping frequency of the bosons) as functions of a_{BF} (in nanometres) for the ^{6}Li-^{7}Li mixture.

3.2. Collisionless approach

In the collisionless regime the dynamics of the mixture can be treated in the time-dependent Hartree-Fock approximation of the mean-field Hamiltonians (2.4). This is equivalent to the Random Phase approximation (RPA) [18] and considers the bosonic and fermionic fluctuations on the same basis, satisfying e.g. the f-sum rule and generalised Kohn theorems [19]. Since it neglects correlations between the fluctuations, the RPA is reliable in the dilute limit specified by the inequalities $\rho_B\, a_{BB}^3 \ll 1$ and $k_F\, |a_{BF}| \ll 1$, where $k_F = (6\pi^2 \rho_F)^{1/3}$ is the Fermi wave number.

Within the RPA it is customary to study the dynamics of systems in the linear regime by means of the density-density response functions. For multi-component systems these are defined by

$$\bar{\chi}_{\sigma\,\sigma'}(\mathbf{r},\mathbf{r}',t-t') = \frac{\delta\rho_\sigma(\mathbf{r},t)}{\delta U_{\sigma'}(\mathbf{r}',t')}$$

$$\stackrel{\text{def}}{=} -\,i\,\theta(t-t')\,\langle[\delta\rho_\sigma(\mathbf{r},t),\,\delta\rho_{\sigma'}(\mathbf{r}',t')]\rangle,$$

where $\delta U_\sigma(\mathbf{r},t)$ represents the perturbing potential, and σ and σ' can take the values F or B. The RPA provides a coupled set of equations to be solved for the four density-density response functions. The Fourier transform of the RPA equations with respect to time can be recast in terms of the density fluctuations $\delta\rho_\sigma(\mathbf{r},\omega)$ as [20]

$$\delta\rho_F(\mathbf{r},\omega) = \int d^3r'\,\bar{\chi}^{0F}(\mathbf{r},\mathbf{r}',\omega)\left[\delta U_F(\mathbf{r}',\omega) + g_{BF}\,\delta\rho_B(\mathbf{r}',\omega)\right],$$

$$\delta\rho_B(\mathbf{r},\omega) = \int d^3r'\,\bar{\chi}^{0B}(\mathbf{r},\mathbf{r}',\omega)\left[\delta U_B(\mathbf{r}',\omega) + g_{BB}\,\delta\rho_B(\mathbf{r}',\omega)\right.$$
$$\left. + g_{BF}\,\delta\rho_F(\mathbf{r}',\omega)\right]. \tag{3.13}$$

In Eqs. (3.13) $\bar{\chi}^{0\,F}$ is the Lindhard density-density response function constructed with the HF orbitals [21] and $\bar{\chi}^{0\,B}$ is the condensate density-density response function built out of the excited states of the Gross-Pitaevskii equations [20, 22]. The solution of Eqs. (3.13) yields the density fluctuations at frequency ω from which it is possible to construct susceptibilities $\chi_{\sigma\sigma'}(\omega)$ and dynamic structure factors of the mixture [23]. These provide not only the location of the collective resonances but also the relative amplitude between them. Furthermore, we may also evaluate the static limit of the responses $\chi_{\sigma\sigma'}(\omega \to 0)$ which are related to the compressibility of the mixture. This turns out to be particularly useful to analyse the collapse and demixing instabilities of the mixture. In Figure 5 we show the fermion compressibility $\kappa \stackrel{\text{def}}{=} -\,\chi_{FF}(\omega = 0)$ for the ^{6}Li-^{7}Li mixture and the same parameters that in Sec. 3.1

The main information contained in Figure 5 concerns the transition to demixed or collapsed phases. For attractive couplings ($a_{BF} < 0$), we observe that the compressibility decreases fast to zero reflecting an increased difficulty to compress the fermions since the density at the centre of the trap rises on approaching the collapse point. On the other hand, for repulsions ($a_{BF} > 0$) we find first a slight increase of the compressibility at the point where the frequencies of the collective modes have a clear peak and then it follows a slow decrease in agreement with the sharp upturn of the collective modes.

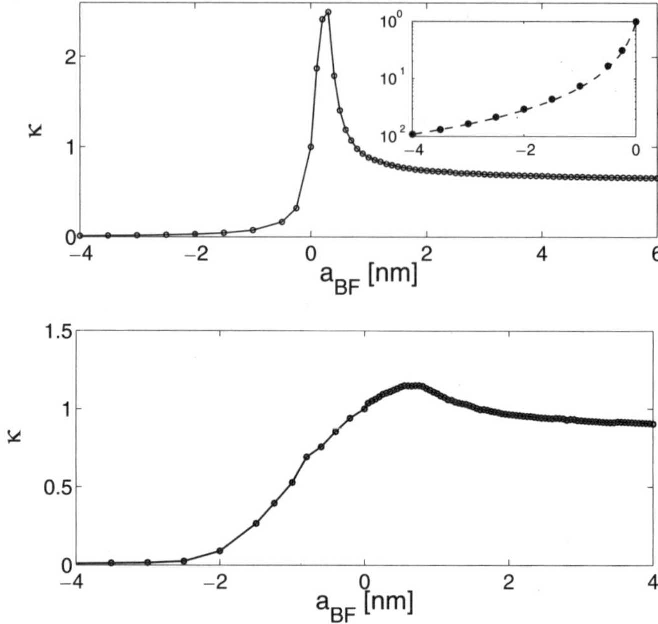

Figure 5. Generalised compressibility κ (in arbitrary units) of the ^{6}Li-^{7}Li mixture as a function of a_{BF} (in nanometres). Top and bottom panels correspond to a system with $N_F = 10^4$, and $N_B = 2.4 \times 10^7$ and 10^6 respectively. The inset shows a zoom of κ in log scale for negative a_{BF}.

3.3. Crossover regime

When collisions occur often but are not enough to establish a local equilibrium, fully collisionless or hydrodynamic approaches are inadequate to describe the dynamics of boson-fermion mixtures. A possible scenario of this is during the free expansion of the gases. Even if the mixture is in a hydrodynamic regime at the beginning of the expansion, on switching off the confinement the densities start melting and eventually the system acquires a collisionless behaviour. It is therefore evident that a situation like this needs a description which accounts for all the intermediate collisional cases, an approach that describes the crossover from one regime to the other. A simple way of dealing with this is to study the dynamics by means of a Boltzmann equation for the one-body distribution functions $f_\sigma(\mathbf{r}, \mathbf{p})$ [24]. The inclusion of the mean-field interactions in the Boltzmann equation gives rise to the so-called Vlasov-Landau equation

$$\frac{\partial f_\sigma}{\partial t} + \nabla \widetilde{V}_\sigma(\mathbf{r}) \cdot \nabla_{\mathbf{p}} f_\sigma - \frac{\mathbf{p}}{m} \cdot \nabla_{\mathbf{r}} f_\sigma = C[f_F, f_B] \qquad (3.14)$$

where C is the collision integral which may include, *a priori*, collisions between bosons and between bosons and fermions. However, since fermions are spin-polarised and at very low temperature, collisions between them are not permitted by the Pauli principle.

We have developed a numerical procedure based in particle-in-cell plus Monte Carlo sampling techniques [16, 25] which allow us to solve Eq. (3.14) down to a temperature $T = 0.1\,T_F$, where T_F is the Fermi temperature. The deviation from pure collisionless or hydrodynamic regimes manifests in the appearance of non-negligible damping rates in some collective modes [26,27]. In particular, the dipolar modes of a mixture show a clear increase of the damping rates at intermediate collisionalities. Also the aspect ratio of the density profiles after the free expansion of the mixture from an anisotropic trap gives useful information about the collisionality of the system and has been extensively used to determine its state before the expansion. In Figure 6 we show the final aspect ratio $R_\perp/R_\parallel$ of fermions in a ^{87}Rb-^{40}K mixture confined in anisotropic traps with anisotropy $\lambda = \omega_\perp/\omega_\parallel$ obtained from the numerical solution of Eq. (3.14) and compare with the expectations from fully collisionless or hydrodynamic approaches. The results indicate that if the traps are highly anisotropic, e.g. $\lambda \ll 1$, the aspect ratio of the fermionic density profiles tends to one irrespective of the initial collisional state. We have also shown [28] that the magnitude of the maximum deviation of the aspect ratio from one depends on the number and temperature of the bosons.

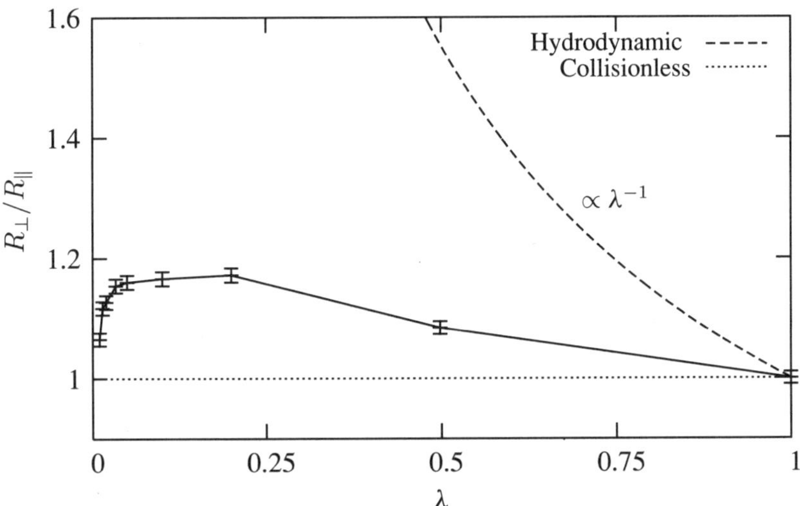

Figure 6. Aspect ratio $R_\perp/R_\parallel$ of the fermions in a ^{87}Rb-^{40}K mixture as a function of $\lambda = \omega_\perp/\omega_\parallel$.

4. Summary and concluding remarks

We have studied some aspects of the dynamics of boson-fermion mixtures in confined systems. We showed that depending on the strength and sign of the interparticle interaction and on the frequency of collisions different regimes may be reached.

In the hydrodynamic regime we explored the behaviour of the monopolar collective modes on varying the strength of the boson-fermion coupling. We found that the demixing transition is accompanied by a clear decrease of the low-lying monopolar modes followed by a sharp upturn. On the other hand, the transition to the collapse is signalled by a softening of the same modes to zero in a narrow range of the boson-fermion coupling and number of bosons.

The collisionless approach also allowed us to characterise both transitions quite clearly, we have illustrated these in the plots of suitable fermion compressibilities which attained negligible values approaching the collapse and have a peak around the demixing point.

The crossover regime, where neither hydrodynamic nor collisionless approaches are adequate, is characterised by the appearance of damping in some collective modes. Moreover, its importance is more evident during the free expansion of the mixture where the system may go from hydrodynamic to collisionless regimes. In this case, we showed that for highly anisotropic traps the final aspect ratio of the fermions tends to one irrespective of the initial collisional state.

Most of this work has been done in collaboration with Anna Minguzzi, Patrizia Vignolo and Mario P. Tosi.

References

[1] M. H. ANDERSON, J. R. ENSHER, M. R. MATTHEWS, C. E. WIEMAN and E. A. CORNELL, Science **269** (1995), 198.

[2] K. B. DAVIS, M. O. MEWES, M. R. ANDREWS, N. J. VAN DRUTEN, D. S. DURFEE, D. M. KURN and W. KETTERLE, Phys. Rev. Lett. **75**, (1995), 3969.

[3] C. C. BRADLEY, C. A. SACKETT, J. J. TOLLETT and R. G. HULET, Phys. Rev. Lett. **75** (1995), 1687.

[4] M. INGUSCIO, S. STRINGARI and C. WIEMAN (EDS.), "Bose-Einstein condensation in atomic gases", Proc. Int. School 'Enrico Fermi', Amsterdam, 1999.

[5] A. MINGUZZI, S. SUCCI, F. TOSCHI, M. P. TOSI and P. VIGNOLO, Phys. Rep. **395** (2004), 223.

[6] K. M. O'HARA, S. L. HEMMER, M. E. GEHM, S. R. GRANADE and J. E. THOMAS, Science **298** (2002), 2179.
[7] M. GREINER, C. A. REGAL and D. S. JIN, Nature **426** (2003), 537.
[8] C. CHIN, M. BARTENSTEIN, A. ALTMEYER, S. RIEDL, S. JOCHIM, J. HECKER DENSCHLAG and R. GRIMM, Science **305** (2004), 1128.
[9] F. DALFOVO, S. GIORGINI, L. PITAEVSKII and S. STRINGARI, Rev. Mod. Phys. **71** (1999), 463.
[10] The local equilibrium could be achieved not only by frequent collisions with themselves or other species but also by including a *small* [25,29] fraction of heavy impurities against whom the fermions may collide.
[11] P. CAPUZZI, A. MINGUZZI and M. P. TOSI, Phys. Rev. A **67** (2003), 053605.
[12] F. SCHRECK, L. KHAYKOVICH, K. L. CORWIN, G. FERRARI, T. BOURDEL, J. CUBIZOLLES and C. SALOMON, Phys. Rev. Lett. **87** (2001), 080403.
[13] G. MODUGNO, G. ROATI, F. RIBOLI, F. FERLAINO, R. J. BRECHA and M. INGUSCIO, Science **297** (2002), 2240.
[14] K. HUANG, "Statistical Mechanics", Wiley, New York, 1987.
[15] S. D. GENSEMER and D. S. JIN, Phys. Rev. Lett. **87** (2001), 173201.
[16] F. TOSCHI, P. VIGNOLO, S. SUCCI and M. P. TOSI, Phys. Rev. A **67** (2003), 041605.
[17] F. FERLAINO, R. J. BRECHA, P. HANNAFORD, F. RIBOLI, G. ROATI, G. MODUGNO and M. INGUSCIO, J. Opt. B **5** (2003), S3.
[18] D. PINES and P. NOZIÈRES, The Theory of Quantum Liquids, vol. I, Benjamin, New York, 1966.
[19] J. F. DOBSON, Phys. Rev. Lett. **73** (1994), 2244.
[20] P. CAPUZZI and E. S. HERNÁNDEZ, Phys. Rev. A **64** (2001), 043607.
[21] P. CAPUZZI and E. S. HERNÁNDEZ, Phys. Rev. A **63** (2001), 063606.
[22] A. MINGUZZI and M. P. TOSI, J. Phys: Condens. Matter (UK) **9** (1997), 10211.
[23] A. L. FETTER and J. D. WALECKA, "Quantum Theory of Many-Particle Systems", McGraw-Hill, New York, 1971.
[24] D. GUÉRY-ODELIN, F. ZAMBELLI, J. DALIBARD and S. STRINGARI, Phys. Rev. A **60** (1999), 4851.
[25] P. CAPUZZI, P. VIGNOLO, F. TOSCHI, S. SUCCI and M. P. TOSI, Phys. Rev. A **70** (2004), 043623.

[26] L. VICHI, J. Low. Temp. Phys. **121** (2000), 177.
[27] X.-J. LIU and H. HU, Phys. Rev. A **67** (2003), 023613.
[28] P. CAPUZZI, P. VIGNOLO and M. P. TOSI, Phys. Rev. A **72** (2005), 013618.
[29] M. AMORUSO, I. MECCOLI, A. MINGUZZI and M. P. TOSI, Eur. Phys. J. D **7** (1999), 441.

4

PHASE TRANSITIONS

Chair
Attilio Rigamonti

Phase transitions - An introductory overview dealing with the major achievements by M. Tosi in the field

A. Rigamonti

I would like to thank the organizers for inviting me here and asking to chair the section on phase transitions. Actually I do thank them, most because they gave me the opportunity to attend such a nice and stimulating meeting and to be present when we all greet Mario Tosi.

At the same time, asking me to give a quick overview of the major achievements in the field the organizers put me somewhat in trouble. Obviously I was aware of scientific milestones that Tosi had set. Since the late 1950s in Pavia his juvenile activity on the cohesion mechanism in ionic solids was already well known. Later, I encountered him at meetings or at schools, as at Collegio Ghislieri in Pavia or at ISI in Villa Gualino in Torino, or even before when the young researchers belonging to the Gruppo Nazionale di Struttura della Materia of CNR were requested to illustrate their projects for grants and Mario Tosi was one of the most "careful" referees.

Still, I felt that what I knew was not enough to yield a proper introductory overview and then I decided to take a look to the papers by Tosi all along fifty years. That originated the trouble: an impressive list of works directly involving or at least related to phase transitions! How to frame all that in a few minutes? Special concern was about inter-related arguments, since a good scientific paper usually has impact on more than one area and positively affects several issues.

An example of that kind is found in the studies of fundamental statistical mechanics carried out by Mario in the 1960s, most in Argonne and in collaboration with Singwi and Sjolander. For the *generalized dielectric function* $\varepsilon(k, \omega)$ for the model system of electron fluid imbedded in a neutralizing uniform background the inadequacy of the RPA and of the Hubbard's formulation was emphasized and the inclusion of exchange and correlation through the *local field factor* $G(k, \omega)$, as in

$$\varepsilon(k,\omega)=1-\frac{(4\pi e^2/k^2)\chi^0(k,\omega)}{1+(4\pi e^2/k^2)\chi^0(k,\omega)G(k,\omega)} \quad (\chi^0(k,\omega) \text{ Lindhard function})$$

was shown to be crucial for the accurate description of the collective properties and of the charge and spin susceptibilities.

A group of works [Singwi, Tosi and Sjolander "Il Nuovo Cimento" 54, B 160 (1968); Singwi, Tosi, Land and Sjolander, Phys. Rev. 176, 589 (1968); Lobo, Singwi and Tosi, Phys. Rev. 186, 470 (1969); Singwi and Tosi, in "Solid State Physics" 36, 177 (1981)] that later on turned out to have deep consequences in subsequent developments, including Quantum Monte Carlo simulations, computational "experiments" and the density-functional theory of Wigner crystallization [Senatore and Pastore, Phys. Rev. Lett. 64,303 (1990)].

Let us begin the overview starting from the most classical phase transition, namely the *gas to liquid transition* along the critical isochore [March and Tosi "Atomic Dynamics in Liquids" (Dover 1991)]. After the progresses related to the scaling approach and to renormalization group techniques the first steps towards complete structural theories were made [Senatore and March, J. Chem. Phys. 80, 5242 (1980); March, Perrot and Tosi, Molecular Physics 93, 355 (1998)].

From the equation for pressure

$$p = \rho k_B T - (\rho^2/6) \int \underbrace{g(r)}_{\text{pair correlation function}} r [\partial \overbrace{\phi(r)}^{\text{pair potential}} / \partial r] dr$$

by taking into account the relationship of $g(r)$ to the three-atom correlation function $g^{(3)}$ and then to $S_{k \to 0}(k, 0)$ (or to the density dependence of $g(r)$):

i) the structural properties at the critical point were obtained;
ii) three structural integrals were expressed in terms of the compressibility ratio and determined to good numerical accuracy (particularly for classical rare gas);
iii) in turn, the relationships involving the force $-[\partial\phi(r)/\partial r]$, the density derivatives of $g(r)$ and the corrected Orstein-Zernike factor $c(k) \simeq 1/[1 + \alpha k^{2-\eta}]$ were derived.

Future directions of further studies were significantly anticipated.

Special mention is worthy to give to the problem of the gas-liquid transition and of the coexistence line for charged fluids or electrolitic solutions [Rovere, Parrinello, Tosi *et al.*, Phys. Chem. Liq. 9, 11 (1979); Ballone, Senatore and Tosi, Nuovo Cimento Lett. 31, 619 (1981); Badirkhan and Tosi, Phys. Chem. Liq. 21,177 (1990)]. The theoretical findings in the early works remained as landmarks for more than twenty years and qualitative and quantitative progress could be achieved only when simulation techniques became possible. M. Fisher was impressed and has been working with enthusiasm on the subject in the last years.

[See also March and Tosi, Phys. Chem. Liq. 11, 89 (1981);______ 11, 79(1981); Rovere and Tosi, Solid State Comm. 55, 1109 (1985); Tosi, Physica Scripta T29, 277 (1989)]

Another line of work evidencing the achievements by Mario Tosi involves the fascinating field of the *two-dimensional (2D) electron fluid* (nowadays experimentally realized in semiconductor heterostructures and possibly in electron-doped calcium hexaborides).

A very rich phenomenology characterizes this system, particularly for low areal density when

$$< E_{e-e} > \geq < E_{\text{kinetic}} > \approx E_{\text{Fermi}} \,,$$

e.g. for average electron-electron distance $\geq 10\,a_0$, with localization, spin polarization, dimensionality-related magnetic phase transitions and enormous response to external field.

That strong-coupling many body system is a forefront area in theoretical condensed matter physics and involves the fundamental aspects of itinerant magnetism. Various descriptions have been proposed (ranging from melting of a Wigner solid from the insulating (I) side to the formation of one from the metallic (M) side; from superconductivity to quantum percolation; from semiclassical one-electron with no M-I transition to non-Fermi liquid scenario). While each of these descriptions is possibly suitable to explain one or another set of experimental observations, a comprehensive picture is still missing [See Castellani, Di Castro and Lee, Phys. Rev. B 57, R9381 (1998); Abrahams, Krachenko and Sarachik, Rev. Mod. Phys. 73, 251 (2001)].

The electron gas with the Coulomb Hamiltonian and neutralizing background is a useful reference model and some of the relevant contributions driven by Mario Tosi are:

i) the density-wave theory of the classical Wigner crystallization [Rovere and Tosi, J.Phys. C 18, 3445 (1985)] and the liquid structure and the related freezing process of the 2D classical electron fluid [Ballone, Pastore, Rovere and Tosi, J. Phys. C 18, 4011 (1985)]. In particular it was shown that the theoretical description by Ramakrishnan and Youssuf of the solid described in terms of the static correlation functions of the liquid, had to involve the three-body correlation functions, in view of the long-range Coulomb interactions.

ii) the vibrational properties of the Wigner electron lattices near the melting transition: from density functional theory the dispersion relations and the elastic constants have been derived, showing that exchange and correlation lead to major anharmonic softening of the transverse phonon branches [Tosi and Tozzini, Europhysic Lett. 23, 433 (1995); Tozzini and Tosi, J. Phys.: Condensed Matter 8, 8121 (1996); Ferconi and Tosi, Europhysics Lett. 14, 797 (1991);_____ J. Phys: Condensed Matter 3, 9943 (1991)];

iii) derivation of the charge and spin susceptibilities with the analytical parameterisation of the local field factors $G_{+,-}$ embedding exchange and correlation as in

$$\chi_s(k) = -\mu_B^2 \chi^0(k)/[1 + V_k \chi^0(k) G_{-+}(k)]$$

with thermo dynamical sum rules (both in 3D and 2D) and predictions about the magnetic phase transition [Polini and Tosi, Phys. Rev. B 63, 045118 (2001); Davoudi, Polini, Giuliani and Tosi, Phys. Rev. B 64, 233110 (2001); Davoudi, Polini, Giuliani and Tosi, Phys. Rev. B 64, 153101 (2001); Davoudi, Polini, Sica and Tosi, Solid State Communications 121, 295 (2002); Davoudi and Tosi, Physica B- Condensed Matter 322, 124 (2002)].

Other issues related to *phase transitions* are:

i) the study of the collapse instability in trapped boson-fermions mixtures [Capuzzi, Minguzzi and Tosi, Phys. Rev. A 69, 053615 (2004)] and the phase separation for ultra-cold mixtures under confinement [Akdeniz and Tosi, Z. Phys. Chem. 217, 927 (2003)];

ii) the derivation of the magnetization and entropy across the melting curve in a classical 2D plasma [March, Capuzzi and Tosi, Phys. Lett. A 327, 226 (2004)] evidencing the central role of the magnetization of the Coulomb liquid, yet being entirely due to particle repulsions;

iii) liquids just above freezing and "hot" crystals approaching melting, in terms of phonons renormalization and broadening and evidencing interesting analogies and differences [Tosi and Tozzini, Phil. Mag. B 69, 833 (1994) and Phil. Mag. B 72, 577 (1995); Ferconi and Tosi, J. Phys.: Condensed Matter 3, 9943 (1991)];

iv) liquid-solid transition for elemental semiconductors in the bond particle model and freezing of liquid alkali metals as screened ionic plasmas [Badirkhan, Rovere and Tosi, Phil. Mag. B 65, 921 (1992) and J. Phys.: Condensed Matter 3, 1627 (1991)]

and others that here is possible only to list, such as liquid metals near freezing, orientational disorder and melting transition in solid halogens and H_2.

Finally I would like to mention that the *metal -nonmetal transition* has been successfully studied by Tosi in a variety of systems and conditions (such as in solutions of metals in molten salts, in naturally layered crystals, in quantum wires and in graphite intercalation compounds). Thermo dynamical, structural and dynamical properties were derived, the electrons accompanying the metallic component being treated as inert neutralizing background and the electron-electron correlations playing more relevant role in case of confinement. [Miesenbock and Tosi, Z. Phys. B 78, 225 (1990); Chabrier, Senatore and Tosi, Nuovo Cimento 3 D,730 (1984); Fantoni and Tosi, Physica B 217, 35 (1996); Tosi, J. Mat. Science and Technology 14, 1 (1998); March and Tosi, Phys. Chem. Liq. 30, 103 (1995); Yurdabak, Akdeniz and Tosi, Nuovo Cimento 16 D, 307 (1994); Akdeniz and Tosi, Nuovo Cimento 18 D, 613 (1996); for more see March and Tosi, Adv. Phys. 44, 299 (1995)].

Again, an impressive amount of work was performed by Mario Tosi, with outstanding results that will remain relevant in the field of theoretical condensed matter physics.

The participants to this meeting and in particular the speakers of the session is about to initiate, are happy to display their appreciation.

ACKNOWLEDGMENTS. A discussion and comments by Gaetano Senatore are gratefully acknowledged.

Quantum many-body approach to nonequilibrium collective phenomena

R. Stinchcombe

Relationships between nonequilibrium collective phenomena and quantum many-body problems are introduced and exploited. In particular the mapping between classical stochastic nonequilibrium particle systems and quantum spin systems is explained and illustrated with standard examples. These include particle-conserving flow processes, one of which exhibits (already in one-dimension) the simplest example of a nonequilibrium steady state transition. This is a quantum phase transition in the associated spin model. A generalisation of the flow models, including pair creation and annihilation effects is also described, together with its exact solution for a special one-dimensional case by mapping, via a quantum spin system, to a free fermion model. We further discuss the relationship of this special case to the approach, by domain coarsening, to the low-temperature equilibrium state in the Ising chain with Glauber kinetics. The generalisation of this latter phenomenon to domain nucleation and growth, eg after reversal of an external field at low temperatures and in higher dimensions, is more difficult, and recent analytic and numerical advances exploiting the quantum spin Hamiltonian description are outlined. Finally a discussion is presented of new features occurring when allowance is made for modifications of Glauber kinetics still consistent with the detailed balance condition necessary to ensure approach to equilibrium.

1. Introduction

This paper is dedicated to Mario Tosi on the occasion of his 72nd birthday. Professor Tosi's leadership in research, and his critical encouragement to so many others, at ICTP Trieste and elsewhere, has been an inspiration to us all.

In this contribution we have tried to echo the progression of his interests by carrying forward many body and quantum condensed matter concepts and approaches to a developing area, which is here nonequilibrium collective behaviour. Consequently there will be contact with the types of many-body and quantum condensed matter phenomena and approaches surveyed in the books edited by March and Tosi [1], and by Lunqvist, March, and Tosi [2], and developed in others of his books. Aspects of classical and quantum spin models, and of phase transitions [3, 5] (including quantum transitions [4]) will play a special role. This is because of mappings, to be described below, of nonequilibrium and particle systems to quantum and spin ones. Such relationships allow techniques developed for quantum many-body systems to be exploited for the typically much less explored area of nonequilibrium collective phenomena.

In § 2 such connections are presented in a general way, followed by the particular mapping of classical stochastic nonequilibrium particle systems to quantum spin systems. Subsequent sections will introduce and develop specific examples. These include in § 3 brief discussions of the simplest (driven) system showing a nonequilibrium steady state phase transition (corresponding to a quantum transition in the mapped model), and of its generalisation including particle pair creation and annihilation. The quantum spin system equivalent to this generalised problem includes as special cases the Ising and Heisenberg models and it can be solved by mapping to a free fermion problem for the case where the four rate parameters occurring are related by one condition. It will be shown in § 4 that this simplified problem is related to the approach to the low temperature equilibrium state of the zero field Ising model with Glauber kinetics. As will be outlined in § 4, such problems involving the approach to the equilibrium state rather than to a more general nonequilibrium steady state are subject to kinetics with detailed balance. Even though they are much simpler than general nonequilibrium problems (without detailed balance) the dynamics can be rich and difficult if they start far from equilibrium.

An example of this which will be developed in §§ 4, 5 is droplet or domain nucleation and coarsening in gas-liquid or magnetic systems. This problem has been much investigated for many years, largely for Glauber kinetics on simple lattices. That however is only one of the possible processes which satisfy the detailed balance condition set by the final thermodynamic equilibrium state. § 5 outlines how recent progress has been made on alternative kinetic rules, and on the nonzero field generalisation, and on more general lattices, by exploiting the Hamiltonian formulation resulting from the quantum spin mapping. The concluding section (§ 6) gathers together the main points and puts them into context.

2. Nonequilibrium processes as quantum problems

Collective nonequilibrium phenomena range from nucleation, shock phenomena, glassy dynamics, ageing, self organisation, and pattern formation etc. in classical physical systems to traffic flow and financial markets in a wider context. They are typically modelled [6–8] by some appropriate many body assembly (particles, spins, vehicles, orders, agents) with specified interactions and dynamic rules. For the physical systems, but also in restricted models of traffic [9] and financial markets [10–12] leaving aside the very interesting problems arising from agents' choice, these systems can in general be described by the following master equation:

$$dP_C/dt = \sum_{C'} (W_{C' \to C} P_{C'} - W_{C \to C'} P_C). \tag{2.1}$$

This gives [13] the evolution in time t of the probability P_C of configurations C of the system in terms of the basic specified rates W, eg that $(W_{C \to C'} P_C)$ for C to go to C'.

The general relationship to a quantum problem occurs because this equation has obvious analogies to a Schroedinger equation in imaginary time, with P_C playing the role of state vector and with Hamiltonian H such that

$$H_{CC'} = W_{C' \to C} - \delta_{CC'} \sum_{C''} W_{C \to C''}. \tag{2.2}$$

The signs are chosen so that the related evolution operator is e^{-Ht}. Then the equation implies that the eigenvalues of H (which is non-hermitian in general) have non-negative real parts. In particular, the steady state of the nonequilibrium process is the right eigenstate corresponding to the "lowest" (zero) eigenvalue of H. This state is often far from obvious, and it can take different forms, eg change its symmetry, as parameters, eg rate parameters or externally imposed currents, etc, are changed. This corresponds to a nonequilibrium steady state transition in the original system, and to a quantum transition in the related one (see e.g. § 3).

The quantum description using (2.2) of any nonequilibrium problem provides the time-dependent history-averaged quantities through the operators θ representing them in an expression of the form

$$\langle \theta \rangle = \langle S | \theta e^{-Ht} | P(0) \rangle. \tag{2.3}$$

Here $|P(0)\rangle$ is the state corresponding to the initial probability of the system configurations, and $\langle S|$ is the state corresponding to the equally weighted sum over all configurations. This is by definition stationary and therefore a zero-eigenvalue left eigenstate of H, that is $\langle S|H = 0$. Consequently

$$d\langle\theta\rangle/dt = \langle[H, \theta]\rangle \tag{2.4}$$

and we can utilize Heisenberg equations of motion for any operator (see e.g. § 4).

A very important class of nonequilibrium model comprises the particle exclusion models [6–8]. In these an effective hard core repulsion is modelled by confining the particles to sites of a lattice which may only be occupied by at most one particle at a time. In addition dynamic processes and their associated rates are prescribed. Some examples are shown in Figure 1(a) and developed further below. Like the lattice gases of equilibrium statistical mechanics, these are minimal models containing the two basic ingredients for collective behaviour, namely interactions and many particles.

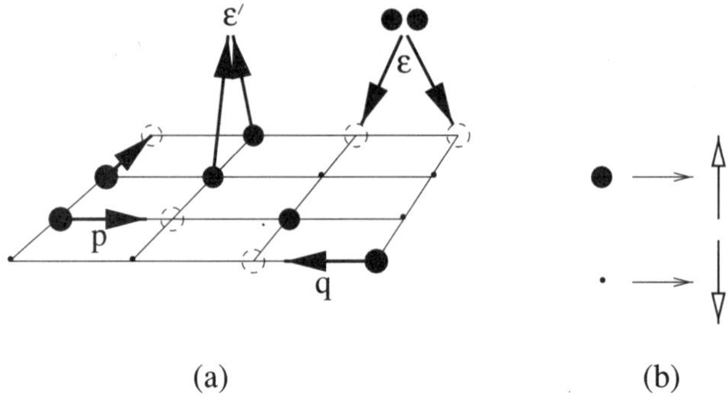

Figure 1. (a) Representative stochastic 'exclusion' processes of hard core particles on a square lattice: asymmetric nearest neighbour hopping, at rates p, q; pair creation/deposition and annihilation/evaporation, at rates $\varepsilon, \varepsilon\prime$. (b) pseudospin representation of particle/vacancy as spin up/down.

As with the lattice gases the two possible states at each site l (occupied or vacant) can be represented by a pseudospin σ_l^z up or down, as shown in Figure 1(b). In the nonequilibrium case the resulting spin model is governed by a Hamiltonian H, as in (2.2), which captures the prescribed processes. Since these processes typically place and remove particles at sites, site-specific spin flip operators $\sigma_l^{\pm}$ are involved and the Hamiltonian is therefore a quantum spin Hamiltonian. In this approach, therefore, the original nonequilibrium classical particle problem maps to a quantum spin one [7, 14–16].

3. Examples: non equilibrium and quantum phase transitions; approximate and exact solutions

A simple example of a collective nonequilibrium process is hopping of hard core particles on a lattice. This is one of the processes in Figure 1(a). Hopping of a particle from site l to site $l\prime$ at rate p (i.e. flipping corresponding pseudospins at these sites down and up respectively) is produced by a term $p\sigma_l^- \sigma_{l\prime}^+$ in H. This corresponds to a contribution to the first term on the right hand side (RHS) of (2.2). The contribution corresponding to the second term is $pP_l^+ P_{l\prime}^-$ where

$$P_l^{\pm} \equiv \frac{1}{2}(1 \pm \sigma_l^z) \tag{3.5}$$

are projection operators for the appropriate spin up or down configurations. Allowing also for hopping at rate q in the opposite direction, and summing over all bonds then produces the quantum spin Hamiltonian $H(p, q)$ corresponding to asymmetric hard core hopping with rates p, q. For symmetric hopping ($p = q$) this becomes the spin-1/2 Heisenberg ferromagnet, and the many known results for this quantum system provide analogous ones for the nonequilibrium classical process. For example the spin wave modes associated with the Goldstone symmetry of this spin model correspond to diffusion modes of the classical particle process.These, like the spin waves are not linearly superposable but they are amenable to treatment by standard methods for quantum spin systems like the Bethe ansatz [17] in one dimension. It is important to mention that, though we will not develop them further here, there also exist methods applying directly to the original nonequilibrium model, such as the operator algebra methods for one dimensional cases, in either their original form for steady states [6, 18], or in the general form applying also to dynamics [7, 8, 19]. This general form has been successfully applied to provide new exact results for the symmetric hard core hopping process just described [19].

The fully asymmetric version ($q = 0$) of hard core hopping is the ASEP (asymmetric exclusion process) which even in one dimension with nearest neighbour coupling shows a nonequilibrium steady state phase transition. This has been exactly treated by the steady state operator algebra referred to above [6, 18], but we now use a simpler approximate approach [20] to arrive at it.

The phase transition is evident in a change in the character of the steady state average density profile ρ_l as the average steady state current $J_{l,l+1}$ on any bond $l, l+1$ goes through a critical value $J_c = p/4$. Here

$$\rho_l = \langle n_l \rangle, \; J_{l,l+1} = \langle p n_l (1 - n_{l+1}) \rangle , \tag{3.6}$$

where n_l is the fundamental fluctuating occupation variable, ie the number of particles (=1 or 0) at site l, p is the hopping rate, in just the forward direction, and $\langle ... \rangle$ denotes the average over histories in the steady state.

The transition is easy to discuss within a mean field approximation in which the average is factorised to give $J_{l,l+1} \sim p\rho_l(1 - \rho_{l+1})$, since particle conservation makes $J_{l,l+1} = J$, independent of l in the steady state. The resulting relation for ρ_{l+1} in terms of ρ_l and J is a map with different characteristics for $J > p/4$, $J < p/4$, resulting in two different types of profile ρ_l, labelled A and B in Figure 2(a), for the two phases. The steady state transition produced by this mean field theory [20] is actually confirmed by the exact solution [6, 18] using the steady state operator algebra techniques. The steady state transition corresponds to a boundary-field driven quantum (T=0) transition in the quantum spin system with Hamiltonian $H(p, 0)$. This contains Heisenberg and imaginary staggered Dzyaloshinsky-Moriya interactions. These support a helical spin configuration whose twist differs qualitatively in the two phases, as depicted in Figure 2(b).

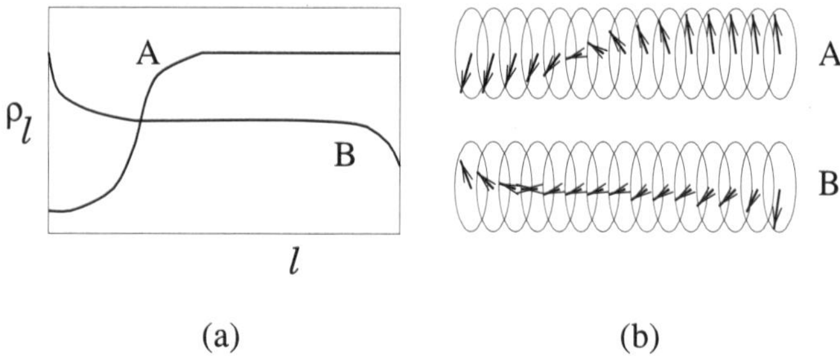

Figure 2. (a) Density profiles in the two phases A and B of the 1d ASEP (hard core asymmetric hopping), and (b) (classical representation of) corresponding spin configurations in the equivalent quantum spin model.

A generalisation of the model [21–23] has as well as partially asymmetric hopping at rates p, q, pair creation and annihilation of particles at rates ε and $\varepsilon\prime$, say (e.g. through deposition and evaporation [15]). All these processes are shown in Figure 1(a). For the special case where $p = q$, $\varepsilon = \varepsilon\prime$ (in any dimension) the corresponding quantum spin system is the anisotropic Heisenberg model and this further reduces to the Ising one if $p = \varepsilon$ (and to the isotropic Heisenberg one for the special case $\varepsilon = 0$, symmetric hopping, discussed above) [21]. Well known results e.g. for the spectrum and correlation functions for all these spin models carry over to eg the decay and correlations of the nonequilibrium particle system.

Another special case is when the following rate relationship applies:

$$p + q = \varepsilon + \varepsilon\prime. \tag{3.7}$$

Then the Hamiltonian H of the quantum spin version of this model becomes bilinear in spin raising and lowering operators and linear in σ^z's. That makes it exactly soluble [22] in 1d (one dimension) since the Jordan-Wigner transformation [24] takes the Hamiltonian into the free fermion form

$$H_{FF} = \sum_l [p c_l c_{l+1}^+ + q c_l^+ c_{l+1} + \varepsilon c_l^+ c_{l+1}^+ + \varepsilon\prime c_l c_{l+1} - ...]. \tag{3.8}$$

Here c_l^+ and c_l are fermion creation and annihilation operators at site l; terms indicated by ... , which are linear combinations of fermion number operators at l and $l + 1$, are the transformed version of the contribution corresponding to the probability-conserving terms, cf the second term on the RHS of (2.2). H_{FF} can be diagonalised by going to wave-vector (k) space and carrying out a Bogoliubov transformation [25]. The spectrum of the resulting fermion-conserving model is [22]

$$E_k = p + q + (\varepsilon - \varepsilon\prime)\cos k + i(p - q)\sin k. \tag{3.9}$$

This spectrum implies ballistic motion and decay in the original particle problem, and it has a gap proportional to the smaller of ε, $\varepsilon\prime$ so the decay is exponentially fast unless ε or $\varepsilon\prime$ vanishes, when power law decays occur.

This model will be seen to have a further relationship to nonequilibrium spin processes discussed in the next section.

4. Approach to equilibrium and detailed balance; Ising kinetics

The approach to equilibrium can be described by a master equation having rates W satisfying detailed balance conditions [13]

$$W_{C\prime\to C} P_{C\prime}^E = W_{C\to C\prime} P_C^E. \tag{4.10}$$

These ensure that the Boltzmann equilibrium distribution $P_C^E \equiv e^{-\beta E_C}$ solves the steady state master equation.

A particular example of detailed balance dynamics is Glauber single-spin-flip kinetics for the Ising model in contact with a heat bath [26].Since $W_{C\to C\prime}$ involves initial and final configurations which differ only by the flip of a single spin, e.g. that at site r, it follows that the quantum Hamiltonian (2.2) becomes [30]

$$H_G = -\sum_r [R_r^+(\sigma_r^+ - P_r^-) + R_r^-(\sigma_r^- - P_r^+)], \tag{4.11}$$

where $R_r^\pm$ are appropriate rates satisfying the detailed balance condition (4.10). The standard choice has the individual rates dependent only on the energy difference involved, which is

$$E_C - E_{C\prime} = \pm 2(h + \sum_{r\prime} J\sigma_{r\prime}^z) \equiv \pm\psi(r)/\beta \tag{4.12}$$

where J is the (nearest neighbour) exchange coupling and h is the external field, $\sigma_{r\prime}$ are the spins neighbouring site r, and the two signs refer to flip up and down. Then

$$R_r^\pm = (1 + e^{\mp 2\psi_r})^{-1}. \tag{4.13}$$

For Glauber kinetics the fundamental Heisenberg equation is, using (4.11),

$$\mathrm{d}\sigma_l^z/\mathrm{d}t = [H, \sigma_l^z] = R_l^+\sigma_l^+ - R_r^-\sigma_l^-. \tag{4.14}$$

This, together with (2.4), (4.13) and the identity $\langle S|\sigma_l^\pm = \langle S|P_l^\mp$, implies for Glauber dynamics

$$\mathrm{d}m_l/\mathrm{d}t \equiv \mathrm{d}\langle\sigma_l^z\rangle/\mathrm{d}t = 1/2\langle(\sigma_l^z + \tanh\psi_l)\rangle \tag{4.15}$$

which is a well known result normally obtained by manipulations starting from the master equation.

For the 1d case in zero field using (4.12) the last equation easily reduces to

$$\mathrm{d}m_l/\mathrm{d}t = (1/4)[2m_l + (m_{l+1} + m_{l-1})\tanh(2\beta J)]. \tag{4.16}$$

This describes the space-time evolution of the local magnetisation, and has a spectral gap proportional to $1 - \tanh(2\beta J)$, vanishing in the low temperature limit.

The processes underlying the last results are domain wall motion and nucleation. To illustrate this and bring various things already mentioned together we continue with the simple one-dimensional example of Glauber dynamics, but including nonzero field. In that case the up and down flips of spin l described by (4.11) have rates $R_l^\pm(n)$ which only depend (through ψ_r in (4.13)), (4.12)) on n, the combined σ^z's of the two neighbouring spins (generalising to z neighbouring spins in higher-dimensional lattices of coordination number z):

$$R^{\pm}(n) = (1 + e^{\mp 2\beta(h+nJ)})^{-1}. \tag{4.17}$$

In 1d the neighbours provide four possible environments for spin l and give rise to eight possible rates $R^{\pm}(n)$ for the flip of spin l, as depicted in Figure 3.

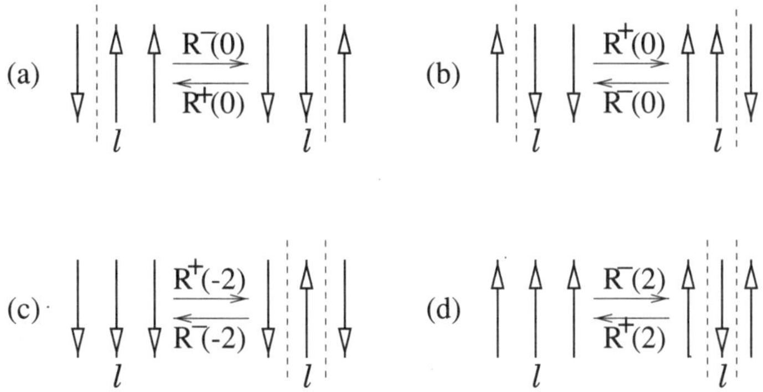

(a) $R^-(0)$ / $R^+(0)$ l l (b) $R^+(0)$ / $R^-(0)$ l l

(c) $R^+(-2)$ / $R^-(-2)$ l l (d) $R^-(2)$ / $R^+(2)$ l l

Figure 3. Glauber single-spin flip processes, and associated domain wall processes, for the spin at site l in each of its possible environments for $d = 1$. Here, $R^{\pm}(n)$ are the rates for up and down flips when the neighbouring σ^z's add up to n.

Also shown, by dashed lines, in Figure 3 are the walls between up and down domains. So (a), (b) correspond to asymmetric hopping of domain walls and (c), (d) to pair creation and annihilation of domain walls in down spin or up spin backgrounds respectively (nucleating and destroying single spin domains). In the hopping processes (a), (b) the bias occurs because the field favours the growth of domains aligned in its direction.

Now, if we think of the domain walls as particles, the processes (a), (b), (c), (d) are those of the generalised particle model discussed at the end of § 3. Moreover, since $R^-(0) + R^+(0) = 1 = R^+(-2) + R^-(-2)$ the Glauber kinetics of domain wall motion and nucleation and antinucleation in a background of down spins satisfies a condition equivalent to (3.7)) in the particle model (and likewise for an up spin background). Indeed, the free fermion Hamiltonian (3.8)) fully describes the domain coarsening process in the 1d Ising model with Glauber dynamics in nonzero field, and the free fermion solution becomes for $p = q$ equivalent to the earlier solution for $h = 0$ [26] (see also [27]). In particular, at very low temperatures, where nucleation is extremely unlikely, the rate ε vanishes in the corresponding particle or free fermion model which makes the spectrum gapless. That can also be seen in (4.16)) (for $h = 0$), which becomes a discrete diffusion equation for $\tanh(2\beta J) \to 1$.

5. Nucleation and generalised dynamics

The results concerning low temperature kinetics and nucleation given in the previous section are not new, though the application of the Hamiltonian method to it and the interconnections are. But we turn now to more difficult aspects of Ising domain nucleation and coarsening, concerning in particular general lattices and fields, and generalised dynamics.

The basic problem in low temperature nuclcation [28] concerns the characteristic time to reach equilibrium having started from a state far from the eventual equilibrium state. A representative example shown in Figure 4 might concern the evolution at low temperatures T from an initial state of mostly spins down (say) to a final equilibrium state (with large up-domains) after abrupt reversal of the magnetic field.

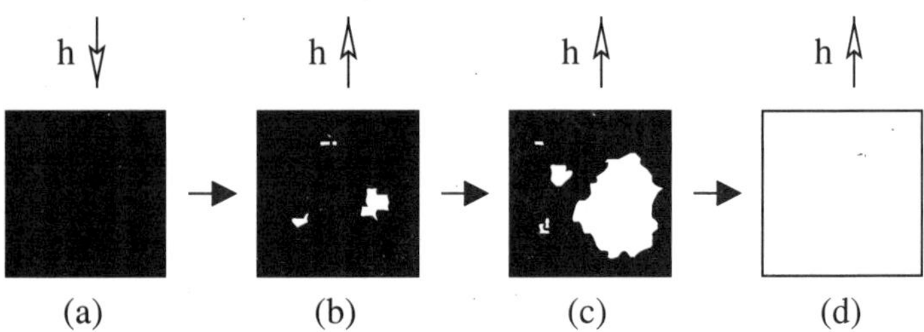

Figure 4. Domain nucleation and evolution in low temperature Ising system following reversal of small external field ($J >> T, h$): (a) after equilibrating in down field, (b) small up-spin domains soon after field reversal, (c) up-domain of critical size has formed, (d) rapid subsequent growth has given new equilibrium state.

The associated "barrier", Γ, is defined in terms of the characteristic time τ to carry out the slow stages of the nucleation by $1/\tau = Ae^{-\beta\Gamma}$. Analytic work [30] starting from (4.11) gives $\Gamma(h, T)$ at low temperatures and high fields in any lattice. For example it turns out that for Ising Glauber nucleation on a lattice of coordination number z, $\Gamma = 0$ for $h > zJ$ and $\Gamma = 2(zJ - h)$ for $zJ > h > (z - 2)J$ (which comes from $-\beta^{-1} \ln R^{+}(-z)$). This is because, in this latter range, once a single-spin domain (in the new field direction) has nucleated, which is slow at low T if $zJ > h$, the growth involves no more slow rates [29]. The 1d example considered in § 4 is atypical in that this applies also for arbitrarily small h.

In the next range $(z-2)J > h > ...$ (occurring for $z > 2$) $R^+(-(z-2))$ comes in and $\Gamma = -\beta^{-1}[\ln R^+(-z) + M \ln R^+(-(z-2))] = [2(M+1)(zJ-h) - 4MJ]$, where M is a lattice-dependent integer. For lattices of dimension $d > 1$ one can crudely estimate, by comparing field energy gain from the domain interior with its competing surface exchange energy, that for $J >> h$ very large clusters have to be formed to facilitate the nucleation. So in higher dimensions with small field and low temperatures the calculation of Γ is difficult, and simulations are very limited. The Hamiltonian provides a powerful alternative approach [30] which we now enlarge on.

The zero eigenvalue of the Hamiltonian (4.11)) corresponds to the eventual steady equilibrium state, which is the Gibbs state for the temperature parametrising the rates $R^{\pm}$. Both the analytical and the numerical work use the fact that the spectral gap to the next eigenvalue is the inverse characteristic time τ^{-1} so its logarithm gives Γ. However H_G is a non-hermitian operator, which would be expected to rule out numerical exploitation using the powerful Lanczos method [31]. But since H_G describes a dynamical process with detailed balance there exists a transformation U which takes it to a hermitian operator $H_{G'} = UHU^{-1}$. It can be shown that the transformation corresponds to rotating each spin σ_r around the z-axis with imaginary angle $-i\psi_r$. The result is

$$H_{G'} = 1/2 \sum_r [1 - \sigma_r^z \tanh \psi_r - \sigma_r^x sech \psi_r]. \tag{5.18}$$

This hermitian operator is now amenable to numerical investigation using the Lanczos method [31], and by such means new results have been obtained for the nucleation barrier Γ (and the coefficient A) for a variety of lattices and field regions [30].

Now, the Glauber single spin flip dynamics discussed above is not the only dynamics respecting detailed balance. Another is the Kawasaki process in which a neighbouring pair of oppositely oriented spins is exchanged with rates consistent with detailed balance. This process conserves total spin, which is necessary when, for example, the two spin directions denote two different species of atoms in a binary alloy. Furthermore, for all such processes (Glauber single spin flip, Kawasaki exchange, etc.) there are many alternative rate prescriptions, differing by a common factor in all rates, which satisfy detailed balance. For example, one alternative to the standard single spin flip Glauber prescription is the "modified" Glauber dynamics proposed in [32] where in place of (4.13) the rates are

$$R_r^{\pm\prime} = (1 + e^{\mp 2\beta h})^{-1}(1 + e^{\mp 2\phi_r})^{-1} \tag{5.19}$$

where $\phi_r = 2\sum_{r'} J\sigma^z_{r'}$. The Hamiltonian approach obviously provides "modified" analogues to the forms (4.14)) (4.15)) (4.16)), etc; and the Ising nucleation problem with this "modified" Glauber dynamics has been investigated for the square lattice in the master equation formalism [32] and extended to other lattices using the quantum spin Hamiltonian and the Lanczos algorithm approach [30]. In the modified dynamics, unlike the standard dynamics, the nucleation barrier is not related to energy barriers, and it remains non-zero ($= 2zJ$) at high fields.

6. Conclusion

After a general introduction the preceding sections have outlined the equivalence between nonequilibrium problems and quantum ones. This relationship applies to all nonequilibrium processes, and it allows them to be cast into quantum spin problems for cases where the nonequilibrium systems consist of a single-species of particle, or classical spins. This has been illustrated with simple examples showing collective nonequilibrium behaviour. Possibilities for analytic progress via the quantum Hamiltonian formulation have been illustrated by a free fermion solution

This more detailed account of a model related to Ising kinetics leads into recent work on domain nucleation and growth, exploiting the quantum Hamiltonian approach. The transformation to Hermitian Hamiltonians used there is always possible with detailed balance rates, making such problems amenable to attack by the Lanczos method.

There was not space to develop here other techniques (such as the Bethe ansatz) applicable after the quantum spin mapping, nor to give a balanced account of alternative approaches. These include direct attacks starting from the master equation and reformulating it, e.g. in terms of operator algebras (referred to in § 3). These have provided important advances [6–8, 33], while mean field approximations (briefly illustrated in § 3) and field theoretic approaches [34] have also proved very useful in extending our understanding of non-equilibrium phenomena.

ACKNOWLEDGMENTS. I wish to thank the NEST secretariat, and in particular Fabio Beltram and Marina Berton for organising such a stimulating and enjoyable symposium, and for the invitation to participate. This work was supported by EPSRC under the Oxford Condensed Matter Theory Grants GR/R83712/01 and GR/M04426.

References

[1] N. H. MARCH and M.P. TOSI, eds., "Polymers, Liquid Crystals and Low dimensional Solids", Plenum, New York, 1984.

[2] S. LUNQVIST, N. H. MARCH and M. P. TOSI, eds., "Order and Chaos in Nonlinear Physical Systems", Plenum, New York, 1988.

[3] R. B. STINCHCOMBE, "Polymers, Liquid Crystals and Low dimensional Solids", N. H. March and M. P. Tosi (eds), Plenum, New York, 1984, 335-400

[4] B. K. CHAKRABARTI, A. DUTTA and P. SEN, "Quantum Ising Phases and Transitions in Transverse Ising Models", Springer, Berlin, 1996.

[5] R. B. STINCHCOMBE, "Order and Chaos in Nonlinear Physical Systems", S. Lunqvist, N. H. March and M. P. Tosi (eds.), Plenum, New York 1988, 295-340.

[6] B. DERRIDA and M. R. EVANS, "Non-Equilibrium Statistical Mechanics in One Dimension", V. Privman (ed.), CUP, Cambridge, 1997, and references therein.

[7] R. B. STINCHCOMBE, Advances in Physics **50** (2001), 431 and references therein.

[8] G. M. SCHUTZ, "Phase Transitions and Critical Phenomena", C. Domb and J. Lebowitz (eds.), Vol. **19** (2001), and references therein.

[9] M. SCHRECKENBURG, A. SCHADSCHNEIDER, K. NAGEL and N. ITOH, Phys. Rev. E **51** (1995), 2339.

[10] D. CHALLET and R. STINCHCOMBE, Physica A**300** (2001), 285, preprint cond-mat/0106114.

[11] M. G. DANIEL, J. D. FARMER, J. IORI and E. SMITH, preprint cond-mat/0112422 (2001).

[12] J-P. BOUCHAUD and M. POTTERS, "Theory of Financial Risks", CUP, Cambridge, 2001.

[13] N. G. VAN KAMPEN, "Stochastic Processes in Physics and Chemistry", 2nd edition, North Holland, Amsterdam, 1992.

[14] L-H.GWA and H. SPOHN, Phys. Rev. Lett. **68** (1992), 725; Phys. Rev. A **46** (1992), 844.

[15] M. BARMA, M. D. GRYNBERG and R. B. STINCHCOMBE, Phys. Rev. Lett. **70** (1993), 1033; R. B. STINCHCOMBE, M. D. GRYNBERG and M. BARMA, Phys. Rev. E **47** (1993), 4018.

[16] F. C. ALCARAZ, M. DROZ, M. HENKEL and V. RITTENBERG, Ann. Phys. (New York) **230** (1994), 250.

[17] H. BETHE, Z. Phys.**71** (1931), 205.

[18] B. DERRIDA, M. R. EVANS, V. HAKIM and V. PASQUIER, J. Phys. A **26** (1993), 1493.

[19] R. B. STINCHCOMBE and G. M. SCHUTZ, Europhys. Lett. **29** (1995), 663; R. B. STINCHCOMBE and G. M. SCHUTZ, Phys. Rev. Lett. **75** (1995), 140.

[20] B. Derrida, E. Domany and D. Mukamel, J. Stat. Phys. **69** (1992), 667.

[21] M. D. Grynberg, T. J. Newman and R. B. Stinchcombe, Phys. Rev. E **50** (1994), 957.

[22] M. D. Grynberg and R. B. Stinchcombe, Phys. Rev. Lett. 74 (1995), 1242; Phys. Rev. Lett. **76** (1996), 851.

[23] J. Santos, G. M. Schutz and R. B. Stinchcombe, J. Chem. Phys. **105** (1996), 2399; J. Santos, J. Phys. A **30** (1997), 3249.

[24] P. Jordan and E. Wigner, Z. Phys. **47** (1928), 631.

[25] N. N. Bogoliubov, Nuov. Cim. **7** (1958), 794.

[26] R. J. Glauber, J. Math. Phys. **4** (1963), 294.

[27] F. Family and J. G. Amar, J. Stat. Phys. **65** (1991), 1235.

[28] See for example J. D. Gunton and M. Droz, "Introduction to the Theory of Metastable and Unstable States", Springer, Berlin, 1983.

[29] H. J. Hilhorst, Physica A **97** (1975), 171.

[30] M. D. Grynberg and R. B. Stinchcombe, Phys. Rev. E, to be published.

[31] See for example G. H. Golub and C. F. van Loan, "Matrix Computations", Johns Hopkins University Press, Baltimore, Berlin, 1996.

[32] K. Park, P. A. Rikvold, G. M. Buendia and M. A. Novotny, Phys. Rev. Lett. **92** (2004), 015701.

[33] M. Depken and R. B. Stinchcombe, Phys. Rev. Lett. **93** (2004), 040602.

[34] See for example J. L. Cardy, "Mathematical Beauty of Physics", J. N. Drouffe and J.-B. Zuber (eds.), Advanced Series in Mathematical Physics, Vol. 24, World Scientific, Singapore, 1997, p. 113.

The resonating valence bond wave functions in quantum antiferromagnets

A. Parola, S. Sorella, F. Becca, and L. Capriotti

Projected-BCS wave functions have been proposed as the paradigm for the understanding of disordered spin states (spin liquids). Here we investigate the properties of these wave functions showing how Luttinger liquids, dimerized states, and gapped spin liquids may be described by the same class of wave functions, which, therefore, represent an extremely flexible variational tool. A close connection between spin liquids and "frozen" superconductors emerges from this investigation.

Many years after the first proposal [1], the very existence of a spin-liquid ground state in two-dimensional (2D) spin-$1/2$ models is still a controversial issue. Short-range resonating-valence-bond (RVB) [2] phases with exponentially decaying spin-spin correlations and no broken lattice symmetry (i.e., with no valence-bond order), are conjectured to be stabilized by quantum fluctuations. However, while it is possible to show that spin-liquid ground states can be found in quantum dimer models [3], the numerical evidence in favor of disordered ground states in Heisenberg-like antiferromagnets is still preliminary [4–7]. Renewed interest on this topic is triggered by the possible realization of a frustrated Bose-Hubbard model with tunable interactions, by trapping cold atomic clouds in optical lattices [8]. A thorough analysis of the zero-temperature phase diagram of these models will probably require an outstanding numerical effort, however, it is possible to tackle the spin-liquid problem by a different, less ambitious but remarkably informative, perspective. Here, we present a detailed analysis of the properties of a class of RVB wave functions, which have been shown to represent extremely good variational states for a wide class of microscopic Hamiltonians, ranging from effective low-energy models for correlated electrons on the lattice to realistic models of atoms and molecules [4, 9, 10]. In particular, we will show that both spin-liquid states and valence-bond crystals may result from the "freezing" of Cooper pairs in the zero-doping limit of correlated electron models.

Following Anderson's suggestion [1], we define the class of projected-BCS (pBCS) wave functions in a N-site spin lattice, starting from the ground state of a suitable BCS Hamiltonian:

$$H(t,\Delta) = \sum_{i,j\sigma}(t_{ij} - \mu\,\delta_{ij})\, c^\dagger_{i,\sigma} c_{j,\sigma} - \sum_{i,j}\left[\Delta_{ij} c^\dagger_{i,\uparrow} c^\dagger_{j,\downarrow} + \Delta^*_{ij} c_{j,\downarrow} c_{i,\uparrow}\right]$$
$$= \sum_{k\sigma}(\epsilon_k - \mu)\, c^\dagger_{k,\sigma} c_{k,\sigma} - \sum_{k}\left[\Delta_k c^\dagger_{k,\uparrow} c^\dagger_{-k,\downarrow} + \Delta^*_k c_{-k,\downarrow} c_{k,\uparrow}\right], \quad (1)$$

where the bare electron band ϵ_k is real and both ϵ_k and Δ_k are even functions of k. A chemical potential μ is introduced in order to fix the number of electrons equal to the number of sites. In order to obtain a class of non-magnetic, translationally invariant, singlet wave functions for spin-1/2 models, the ground state $|\mathrm{BCS}\rangle$ of Hamiltonian (1) is then restricted to the physical Hilbert space of singly-occupied sites by the Gutzwiller projector P_G.

The first feature of such a wave function we want to discuss is the *redundancy* implied by the electronic representation of a spin state, i.e., the extra symmetries which appear when we write a spin state as Gutzwiller projection of a fermionic state. In turn, this property reflects in the presence of a local, i.e., gauge, symmetry of the fermionic problem, as already pointed out several years ago [11]. Let us consider a generic spin operator defined on a site in fermionic representation: $X_{\alpha\beta} = c^\dagger_\alpha c_\beta$. This operator acts in the Hilbert subspace of singly-occupied sites (which is left invariant under any spin Hamiltonian). It is easy to check that the three SU(2) generators $N_0 = (c^\dagger_\uparrow c_\uparrow + c^\dagger_\downarrow c_\downarrow - 1)/2$, $N_+ = c^\dagger_\uparrow c^\dagger_\downarrow$, $N_- = (N_+)^\dagger$ commute, in the singly occupied site subspace, with $X_{\alpha\beta}$. This property reflects the invariance of the operator $X_{\alpha\beta}$ under the usual $U(1)$ gauge transformation g_ϕ:

$$g_\phi : \qquad c^\dagger_\alpha \to e^{i\phi} c^\dagger_\alpha, \quad (2)$$

and also by the SU(2) rotation Σ_θ:

$$\Sigma_\theta : \qquad \begin{pmatrix} c^\dagger_\uparrow \\ c_\downarrow \end{pmatrix} \to \begin{pmatrix} \cos\theta & -i\sin\theta \\ -i\sin\theta & \cos\theta \end{pmatrix} \begin{pmatrix} c^\dagger_\uparrow \\ c_\downarrow \end{pmatrix}. \quad (3)$$

These transformations are local, i.e., can be performed on each site independently leaving every spin state invariant: they generate the SU(2) gauge symmetry group.

Let us now consider the Gutzwiller projected BCS state in a Heisenberg-like model on a lattice with an even number $N = 2n$ of sites.

$$|\mathrm{pBCS}\rangle = P_G|\mathrm{BCS}\rangle = P_G \prod_k (u_k + v_k c^\dagger_{k,\uparrow} c^\dagger_{-k,\downarrow})|0\rangle, \quad (4)$$

where the product is over all the N wave vectors in the Brillouin zone. The diagonalization of Hamiltonian (1) gives explicitly

$$u_k = \sqrt{\frac{E_k + \epsilon_k}{2E_k}} \qquad v_k = \frac{\Delta_k}{|\Delta_k|}\sqrt{\frac{E_k - \epsilon_k}{2E_k}} \qquad E_k = \sqrt{\epsilon_k^2 + |\Delta_k|^2}\,,$$

while the BCS pairing function f_k is given by:

$$f_k = \frac{v_k}{u_k} = \frac{\Delta_k}{\epsilon_k + E_k}. \tag{5}$$

Clearly, the local gauge transformations previously defined change the BCS Hamiltonian, breaking in general the translation invariance. In the following, we will restrict to the class of transformations which preserve the translational symmetry of the lattice in the BCS Hamiltonian, i.e., the subgroup of *global* symmetries corresponding to site independent angles (ϕ, θ). By applying the transformations (2) and (3), the BCS Hamiltonian keeps the same form with modified couplings:

$$\begin{aligned} t_{ij} &\to t_{ij} \\ \Delta_{ij} &\to \Delta_{ij} e^{2i\phi} \end{aligned} \tag{6}$$

for g_ϕ, while the transformation Σ_θ gives:

$$\begin{aligned} t_{ij} &\to \cos 2\theta\, t_{ij} - i \sin\theta \cos\theta\, (\Delta_{ij} - \Delta_{ij}^*) \\ &= \cos 2\theta\, t_{ij} + \sin 2\theta\, \mathrm{Im}\Delta_{ij} \\ \Delta_{ij} &\to (\cos^2\theta\, \Delta_{ij} + \sin^2\theta\, \Delta_{ij}^*) - i \sin 2\theta\, t_{ij} \\ &= \mathrm{Re}\Delta_{ij} + i\left(\cos 2\theta\, \mathrm{Im}\Delta_{ij} - \sin 2\theta\, t_{ij}\right). \end{aligned} \tag{7}$$

These relations are linear in t_{ij} and Δ_{ij} and, therefore, equally hold for the Fourier components ϵ_k and Δ_k. Note that, being Δ_r an even function, the real (imaginary) part of its Fourier transform Δ_k equals the Fourier transform of the real (imaginary) part of Δ_r. It is easy to see that these two transformations generate the full $O(3)$ rotation group on the vector whose components are $(\epsilon_k, \mathrm{Re}\Delta_k, \mathrm{Im}\Delta_k)$. As a consequence, the length E_k of this vector is conserved by the full group. In summary, this shows that there is an infinite number of different translationally invariant BCS Hamiltonians which, after projection, give the same spin state. Choosing a specific representation does not affect the physics of the state but changes the pairing function f_k before projection. Within this class of states the only scalar under rotations, which can be given some physical meaning, is the BCS energy spectrum E_k. Clearly, the projection operator will modify the excitation spectrum associated to the BCS wave function. Nevertheless, the invariance with respect to SU(2) transformations suggests that E_k may reflect the nature of the physical excitation spectrum.

Remarkably, in one dimension it is easy to prove that such a class of wave functions is able to faithfully represent both the physics of Luttinger liquids, appropriate for the nearest-neighbor Heisenberg model, and the gapped spin-Peierls state, which is stabilized for sufficiently strong frustration. In fact, it is known [12] that the simple choice of nearest-neighbor hopping t_{ij} ($\epsilon_k = -2t\ \cos k$, $\mu = 0$) and vanishing gap function Δ_{ij} reproduces the exact solution of the Haldane-Shastry model, while choosing a next-nearest neighbor hopping ($\epsilon_k = -2t\ \cos 2k$, $\mu = 0$) and a sizable nearest-neighbor pairing ($\Delta_k = 4\sqrt{2}t\ \cos k$) we recover the Majumdar-Gosh state [13], i.e., the exact ground state of the frustrated Heisenberg model when the next-nearest-neighbor exchange constant J_2 is half of the nearest-neighbor one J_1. Note that in this case, the BCS dispersion E_k is strictly positive, i.e., the BCS Hamiltonian is gapped.

A quantitative analysis of this class of wave functions can be only carried out numerically. As an example, in Figure 1 we show that the spin-spin correlations $G(r)$ have a long-range tail, i.e., $G(r) \sim 1/r$, for the known $\Delta = 0$ solution [12] but also for our pBCS state with nearest-neighbor hopping and nearest- and third-neighbor Δ, whose energy is indeed remarkably accurate [9]. Moreover, Figure 2 shows that the same class of wave functions is able to display a clear dimer ordering as soon as the BCS dispersion E_k shows a gap at the Fermi level. The agreement with the exact results (given by the Density-Matrix Renormalization Group method) for the frustrated Heisenberg model at $J_2/J_1 = 0.4$ is excellent also in this case.

Now we specialize to the 2D square lattice and we investigate whether it is possible to further exploit the redundancy of the fermion representation of a spin state in order to define a pairing function which, before projection, breaks the reflection symmetries of the lattice, while the projected state retains all the correct quantum numbers. We will show that, if suitable conditions are satisfied, a fully symmetric projected BCS state is obtained from a BCS Hamiltonian with fewer symmetries than the original spin problem.

We first introduce a set of unitary operators.

- Spatial symmetries: $R_x(x, y) = (x, -y)$; $R_{xy}(x, y) = (y, x)$. We define the transformation law of creation operators $R\, c^\dagger_{j,\sigma} R^{-1} = c^\dagger_{Rj,\sigma}$ and the action of the operator on the vacuum $R\,|0\rangle = |0\rangle$. Note that these operators map each sublattice into itself.

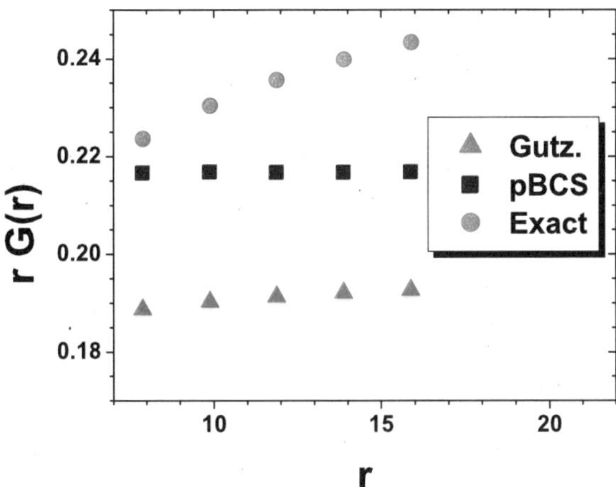

Figure 1. Spin-spin correlation function at the largest distance: $G(r) = (-1)^r \langle S_r^z S_1^z \rangle$ in a chain with L=2r sites as a function of r. Circles: exact results for the Heisenberg model, Squares: pBCS wave function with nearest-neighbor hopping and third-neighbor Δ. Triangles: Gutzwiller projected Fermi sea. A $1/r$ decay is clearly present in the variational wave functions. The exact result shows logarithmic corrections as predicted by conformal field theory.

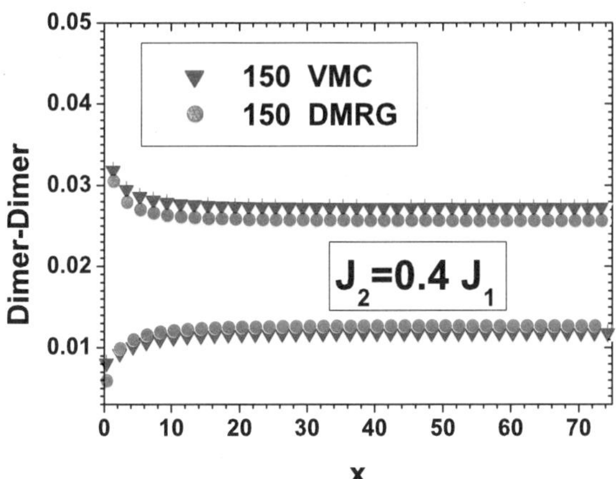

Figure 2. Dimer-dimer correlation function $\langle S_{x+1}^z S_x^z S_1^z S_0^z \rangle$ as a function of x calculated via the best variational pBCS function compared to the exact results for the J_1-J_2 modelobtained via Density-Matrix Renormalization Group (DMRG) in a 150-site chain.

- Particle-hole: $P_h c_{j,\sigma}^\dagger P_h^{-1} = i\,(-1)^j c_{j,-\sigma}$, while the action of the P_h operator on the vacuum state is $P_h|0\rangle = \prod_j c_{j,\uparrow}^\dagger c_{j,\downarrow}^\dagger |0\rangle$.
- Gauge transformation: $G\, c_{j,\sigma}^\dagger G^{-1} = i\, c_{j,\sigma}^\dagger$ and $G\,|0\rangle = |0\rangle$.

Clearly, R_x and R_{xy} are symmetries of the physical problem (e.g., the Heisenberg model). G is a symmetry because the physical Hamiltonian has a definite number of electrons and P_h leaves invariant every configuration where each site is singly occupied if the total magnetization vanishes: $N_\downarrow = N_\uparrow = n$. Therefore, the previously defined operators commute with the Heisenberg Hamiltonian and commute with each other because reflections do not interchange the two sublattices. The ground state of the Heisenberg model on a finite lattice, if it is unique, must be simultaneous eigenstate of all the symmetry operators. We want to investigate sufficient conditions which guarantee that the projected BCS state is indeed eigenstate of all these symmetries.

Let us consider a hopping term which just couples the two sublattices: $\epsilon_{k+Q} = -\epsilon_k$ and a gap function $\Delta = \Delta_s + \Delta_{x^2-y^2} + \Delta_{xy}$ with contributions from different symmetries. Moreover, we consider the case in which Δ_s and $\Delta_{x^2-y^2}$ will couple different sublattices while Δ_{xy} is restricted to the same sublattice. If this is the case, the BCS Hamiltonian $H(t, \Delta_s, \Delta_{x^2-y^2}, \Delta_{xy})$ transforms in the following way under the different unitary operators:

$$R_x H(t, \Delta_s, \Delta_{x^2-y^2}, \Delta_{xy}) R_x^{-1} = H(t, \Delta_s, \Delta_{x^2-y^2}, -\Delta_{xy})$$

$$R_{xy} H(t, \Delta_s, \Delta_{x^2-y^2}, \Delta_{xy}) R_{xy}^{-1} = H(t, \Delta_s, -\Delta_{x^2-y^2}, \Delta_{xy})$$

$$P_h H(t, \Delta_s, \Delta_{x^2-y^2}, \Delta_{xy}) P_h^{-1} = H(t, \Delta_s^*, \Delta_{x^2-y^2}^*, -\Delta_{xy}^*)$$

$$G H(t, \Delta_s, \Delta_{x^2-y^2}, \Delta_{xy}) G^{-1} = H(t, -\Delta_s, -\Delta_{x^2-y^2}, -\Delta_{xy}).$$

Starting from these transformations, it is easy to define suitable composite symmetry operators which indeed leave the BCS Hamiltonian invariant. For instance, let us consider the case in which Δ is real: in this case we select the $R_x P_h$ and R_{xy} if $\Delta_{x^2-y^2} = 0$ or $R_{xy} P_h G$ if $\Delta_s = 0$. We cannot set simultaneously $\Delta_{x^2-y^2}$ and Δ_s different form zero. The eigenstates $|\text{BCS}\rangle$ of Eq. (1) will be generally simultaneous eigenstates of these two composite symmetry operators with given quantum numbers, say α_x and α_{xy}. Let us consider the effect of projection over these states:

$$\begin{aligned}\alpha_x P_G |\text{BCS}\rangle &= P_G R_x P_h |\text{BCS}\rangle = R_x P_h P_G |\text{BCS}\rangle \\ &= R_x P_G |\text{BCS}\rangle,\end{aligned} \tag{8}$$

where we have used that both R_x and P_h commute with the projector and that P_h acts as the identity on singly occupied states. Analogously, when a $d_{x^2-y^2}$ gap is present,

$$\begin{aligned}\alpha_{xy} P_G|\mathrm{BCS}\rangle &= P_G R_{xy} P_h G|\mathrm{BCS}\rangle = R_{xy} P_h G P_G|\mathrm{BCS}\rangle \\ &= (-1)^n R_{xy} P_G|\mathrm{BCS}\rangle. \end{aligned} \tag{9}$$

These equations show that the projected BCS state with both xy and x^2-y^2 contributions to the gap has definite symmetry under reflections, besides being translationally invariant. The corresponding eigenvalues, for $n = N/2$ even, coincide with the eigenvalues of the modified symmetry operators $R_x P_h$ and $R_{xy} P_h G$ on the pure BCS state. Note that the quantum numbers we have defined, refer to the fermionic representation. It turns out that an extremely good variational wave function for the frustrated two dimensional Heisenberg antiferromagnet can be obtained by including gap functions of different symmetry [4].

An alternative, but equally interesting representation of the pBCS state can be given in terms of Slater determinants through the following argument. The pBCS wave function can be written in real space as:

$$|\mathrm{pBCS}\rangle = P_G \left[\sum_{R<X} \sum_{\sigma} f(\sigma, R; -\sigma, X) c^{\dagger}_{R,\sigma} c^{\dagger}_{X,-\sigma} \right]^n |0\rangle \tag{10}$$

where R and X run over the lattice sites and σ is the spin index. The antisymmetric pairing function $f(\sigma, R; -\sigma, X) = -f(-\sigma, X; \sigma, R)$ is simply related to the Fourier transform $f(r)$ of the previously defined f_k of Eq. (5): $f(\uparrow, R; \downarrow, X) = f(R - X)$. In order to avoid double counting, a given, arbitrary, ordering of the lattice sites has been assumed in Eq. (10). By expanding Eq. (10) and defining $f(\sigma, R; \sigma, X) = 0$ we get

$$\begin{aligned}|\mathrm{pBCS}\rangle = \sum_{\substack{R_1,\sigma_1 \cdots R_n,\sigma_n \\ X_1,\sigma'_1 \cdots X_n,\sigma'_n}} & f(\sigma_1, R_1; \sigma'_1, X_1) \cdots f(\sigma_n, R_n; \sigma'_n, X_n) \\ & \times |\sigma_1, R_1; \sigma'_1, X_1 \cdots \sigma_n, R_n; \sigma'_n, X_n\rangle \end{aligned} \tag{11}$$

where the $2n$ labels $(R_1 \cdots R_n; X_1 \cdots X_n)$ define a generic partition of the $N = 2n$ sites of the lattice. We now rearrange the creation operators in the many body state of Eq. (11)

$$|\sigma_1, R_1; \sigma'_1, X_1 \cdots \sigma_n, R_n; \sigma'_n, X_n\rangle = c^{\dagger}_{R_1,\sigma_1} c^{\dagger}_{X_1,\sigma'_1} \cdots c^{\dagger}_{R_n,\sigma_n} c^{\dagger}_{X_n,\sigma'_n} |0\rangle,$$

according to a given *site* ordering in the lattice (irrespective of the spin). This operation gives the $(S_z = 0)$ spin state $|\sigma_1 \cdots \sigma_N\rangle$ multiplied by a phase factor equal to the sign ϵ of the associated permutation. However, many distinct terms correspond to the same spin state because they differ only by the way pairs are coupled:

$$|\text{pBCS}\rangle = \sum_{\{\sigma_i\}} \sum_{P} \epsilon_P f(\sigma_1, R_1; \sigma_1', X_1) \cdots f(\sigma_n, R_n; \sigma_n', X_n)\ |\sigma_1 \cdots \sigma_N\rangle$$

Here P runs over all the possible $(N-1)!!$ partitions of the $N = 2n$ sites of the lattices into pairs, for a given spin state $|\sigma_1 \cdots \sigma_N\rangle$ uniquely identified by the variables $\{\sigma_i\}$. The weight of the spin configuration is exactly the Pfaffian of the $2n \times 2n$ antisymmetric matrix [14]:

$$\mathbf{A} = \left(\begin{bmatrix} f(\uparrow, R_\alpha; \uparrow, R_\beta) \\ f(\downarrow, X_\alpha; \uparrow, R_\beta) \end{bmatrix} \begin{bmatrix} f(\uparrow, R_\alpha; \downarrow, X_\beta) \\ f(\downarrow, X_\alpha; \downarrow, X_\beta) \end{bmatrix} \right), \tag{12}$$

where the matrix has been written in terms of $n \times n$ blocks and R_α are the positions of the up spins in the $|\sigma_1 \cdots \sigma_N\rangle$ state while X_α are the positions of the down spins. The known relation

$$[\text{Pf}\mathbf{A}]^2 = \det\mathbf{A}, \tag{13}$$

together with the fact that we considered $f(\sigma, R; \sigma, X) = 0$ imply that $\text{Pf}\mathbf{A} = \det\Big[f(R_\alpha - X_\beta) \Big]$ [15]. Therefore, the weight of the spin state $|\sigma_1 \cdots \sigma_N\rangle$ may be written as the determinant of a matrix which depends on the BCS pairing function.

When we want to represent a spin state in a 2D lattice by the pBCS wave function, special care must be given to the definition of the phases of the pairing function $f(r)$. As an example, let us consider a nearest-neighbor pairing function, i.e., we choose $|f(r)| = 1$ when r connects nearest neighbor sites in the lattice and zero otherwise. From Eq. (10) it is clear that the pBCS state is written as the sum of all possible partitions of the N-site lattice into singlets (R_i, X_i) and the amplitude of a given partition is provided by the generic term of the Pfaffian of the matrix $\mathbf{A}$:

$$\epsilon_P\, f(R_1 - X_1) \cdots f(R_n - X_n). \tag{14}$$

Here we introduced the usual convention to orient a singlet from sublattice A to sublattice B and we fix $R_i \in A$ $(X_i \in B)$, while the permutation P relates the chosen ordering of sites in the lattice and the sequence $(R_1, X_1 \cdots R_n, X_n)$. Remarkably, as proved by Kasteleyn [14], in *planar* lattices it is possible to choose the phase of the function $f(r)$ so that the products of the form Eq. (14) have all the same sign. In such a case, the pBCS wave function exactly reproduces the short-range RVB state: the equal-amplitude superposition of all possible partitions of the lattice into nearest-neighbor singlets [2]. In particular, on a rectangular lattice, the resulting pairing function has $s + id$ symmetry [16], i.e., $f(r) = 1$ on horizontal bonds and $f(r) = i$ on vertical bonds. In turn, this may be obtained from the BCS Hamiltonian (1) by considering third neighbor hopping $\epsilon_k = -2t(\cos 2k_x + \cos 2k_y)$ (with $\mu = 0$) and complex nearest-neighbor gap function $\Delta_k = 8t(\cos k_x + i \cos k_y)$.

The properties of of the pBCS wave functions have been investigated by Lanczos technique and variational Monte Carlo method in the frustrated Heisenberg model on a square lattice, defined by the Hamiltonian:

$$H = J_1 \sum_{\langle i,j \rangle} \mathbf{S}_i \cdot \mathbf{S}_j + J_2 \sum_{\langle\langle i,j \rangle\rangle} \mathbf{S}_i \cdot \mathbf{S}_j, \tag{15}$$

where the first sum runs over the nearest neighbors and the second on the next-nearest neighbors (i.e., along the diagonal). When both coupling constants are positive the model is frustrated. In Figure 3 we compare the previously introduced dimer-dimer correlations $\langle S^z_j S^z_i S^z_1 S^z_0 \rangle$ (where (i, j) and $(0, 1)$ are nearest neighbor sites) obtained via the optimized pBCS wave function and the exact Lanczos results for the Hamiltonian (15) at $J_2/J_1 = 0.55$. The figure shows that the pBCS wave function is able to capture the correct behavior of correlations also in 2D. In Figure 4 we report the size scaling of the squared dimer order parameter obtained via the optimal variational wave function in the pBCS class. The absence of dimer order is clearly suggested by the variational approach, pointing toward the existence of a spin-liquid in the 2D $J_1{-}J_2$ model.

As a final remark, we like to comment on the fate of the insulating state described by a pBCS wave function upon doping. When a limited number of holes are injected into the lattice, it is likely that the basic structure of the pBCS state is not affected by the presence of mobile charges, being largely determined by the super-exchange interaction among spins. As long as the BCS pairs present in the insulating state remain well defined even at low doping, the system is generally expected to display superconducting properties. This possibility has been explicitly verified in the so-called $t{-}J$ model [9], where the natural generalization of a pBCS wave function has been shown to give rise to off-diagonal long-range order. Real materials, like high-temperature superconductors, display a considerably richer physics and other effects may inhibit the actual realization of this scenario: charge-density waves may occur at special values of the doping or the gain in hole kinetic energy may induce a global change in the spin state causing the breaking of electron pairs. Only a detailed study of the microscopic Hamiltonian will discriminate among these and

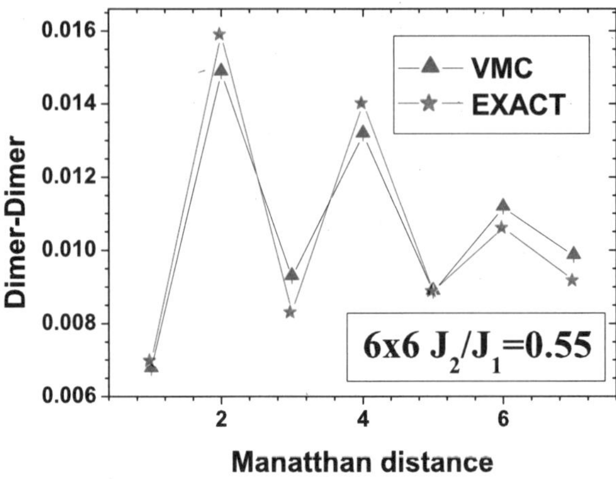

Figure 3. Dimer-dimer correlations in the 6×6 square lattice for the Hamiltonian (15) at $J_2/J_1 = 0.55$. Stars: exact Lanczos results. Triangles: optimized pBCS wave function.

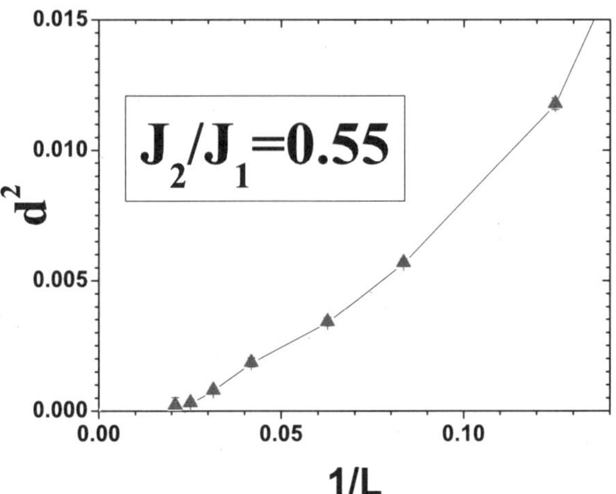

Figure 4. Size scaling of the squared dimer order parameter as predicted by the pBCS wave function.

other possibilities, and no general statement can be drawn on the basis of our analysis. However, our study suggests that a spin liquid (or even a valence bond crystal) may be thought of as an insulating state adiabatically connected to a superconducting phase which directly originates as soon as electrons are removed: this is a remarkable manifestation of the effects of electron correlations present in a gapped spin state which goes beyond standard band theory and the Fermi Liquid approach.

ACKNOWLEDGMENTS. Partial support has been provided my MIUR through a PRIN grant. F.B. is supported by INFM.

References

[1] P. W. ANDERSON, Science **235** (1987), 1196.
[2] F. FIGUEIRIDO, A. KARLHEDE, S. KIVELSON, S. SONDHI, M. ROCEK and D. S. ROKHSAR, Phys. Rev. B **41** (1989), 4619.
[3] D. S. ROKHSAR and S. A. KIVELSON, Phys. Rev. Lett. **61** (1988), 2376; R. MOESSNER and S. L. SONDHI, Phys. Rev. Lett. **86** (2001), 1881.
[4] L. CAPRIOTTI, F. BECCA, A. PAROLA and S. SORELLA, Phys. Rev. Lett. **87** (2001), 097201; Phys. Rev. B **67** (2003), 212402.
[5] J. B. FOUET, P. SINDZINGRE and C. LHUILLIER, Eur. Phys. J. B **20** (2001), 241.
[6] M. MAMBRINI and F. MILA, Eur. Phys. J. **17** (2001), 651.
[7] G. MISGUICH, C. LHUILLIER, B. BERNU and C. WALDTMANN, Phys. Rev. B **60** (2000), 1064; W. LI MING, G. MISGUICH, P. SINDZINGRE and C. LHUILLIER, Phys. Rev. B **62** (2000), 6372.
[8] M. POLINI, R. FAZIO, M. P. TOSI, J. SINOVA and A. H. MACDONALD, Laser Physics **14** (2004), 603.
[9] S. SORELLA, G. B. MARTINS, F. BECCA, C. GAZZA, L. CAPRIOTTI, A. PAROLA and E. DAGOTTO, Phys. Rev. Lett. **88** (2002), 117002.
[10] M. CASULA and S. SORELLA, J. Chem. Phys. **119** (2003), 6500.
[11] F. C. ZHANG, C. GROS, T. M. RICE and H. SHIBA, Supercond. Sci. Technol. **36** (1988), 1.
[12] F. D. M. HALDANE, Phys. Rev. Lett. **60** (1988), 635.
[13] C. K. MAJUMDAR and D. K. GOSH, J. Math. Phys. **10** (1969), 1388.
[14] See for instance, P. W. KASTELEYN, J. Math. Phys. **4** (1963), 287.
[15] The ambiguity of the sign can be resolved by a careful inspection.
[16] N. READ and B. CHAKRABORTY, Phys. Rev. B **40** (1989), 7133.

Towards determining the ground state of the 2D electron system with weak disorder

D. Neilson and D. J. W. Geldart

We reconcile the possible presence of a quantum critical point in metal-insulator transition phenomena in two-dimensional electron systems with observed electronic properties in the low resistivity region which are those of a weakly disordered Fermi liquid. The central issue of whether the ground state of an interacting 2D electron system in the presence of weak potential scattering from disorder is a metal or an insulator remains unresolved. Additional new effects which may become important at extremely low electron densities are briefly mentioned.

Information technology has radically changed our society. Behind these developments lies a very good understanding of fundamental physics. The developments have been powered by a potent combination of innovation and knowledge. However the rapid advances in technology have now reached the stage where new phenomena due to unexplained quantum effects are emerging in ultra-high quality transistors, effects that cannot be explained by existing theories.

Chip development has driven the developments in telecommunications technology. Ultra-fast transistors and semiconductor lasers based on semiconductor heterostructures play a decisive role in modern telecommunications and they form the key components in high frequency transistors. They are found in everything from mobile telephones, to CD-players, to bar-code readers, to HD television.

The first transistors consisted of a silicon crystal, oxidised to form a silicon dioxide layer on its surface. On top of this layer is placed a metal "gate" electrode. By applying a positive voltage to this gate electrode, electrons in the silicon semiconductor are attracted up to the silicon dioxide layer. The electrons cannot get through the oxide layer since it is electrically insulating, so they congregate below the oxide to form a two-dimensional (2D) sheet of electrons. Although the sheet is typically about 10 nanometres thick, it is an exact two-dimensional system because in the direction perpendicular to the insulating layer the electrons occupy one unique quantum state at low temperatures. The free electrons between the "source" and "drain" electrodes can carry a current between the source and drain, and the transistor is "on". By decreasing the attractive voltage on the "gate" the number of electrons under the oxide layer can be reduced, until the transistor is "off".

Twenty years ago it was found that transistors made from gallium arsenide (GaAs) with an aluminium gallium arsenide (AlGaAs) electrical barrier were much faster than silicon. The main reason is that the silicon dioxide layer is amorphous and contains many electrical traps. In the GaAs/AlGaAs system both sides of the interface are perfect crystals, and since their crystal structures match perfectly there are no interface states to trap charges. This property allows the electrons to move very freely ("high mobility") in thc 2D shcct along the interface formed by the two semiconductors and this gives rise to exceptionally fast transistors.

The increase in the quality of the heterostructures is exploited in low-noise high-frequency amplifiers in satellite communications and for improving the signal-to-noise ratio in mobile telephones. Frequencies up to 600 GHz have been measured in a heterotransistor, a factor of one hundred times greater than in silicon. Speeds of operation will continue to increase as will the carrier frequencies of communication, at present around 1 GHz. Wide band telecommunications will be widely connected into homes, and both mobile and fibre-optic communication will continue to expand in importance. An example of tomorrow's developments is anti-collision radar in the 70 GHz band. Radio astronomy is another example with its needs to detect ever weaker signals at higher and higher frequencies.

Thus while two-dimensional sheets of conducting electrons might appear to be abstruse, they are in fact central to the operation of many devices in use today in ultra high frequency communication networks and systems. An understanding of how they work and how we can exploit their unusual properties is extremely important for the design of tomorrow's electronic devices for communication and computing. Not only are these high quality transistors vital for advanced electronic devices, they also exhibit new and fundamental physics phenomena. The Integer and Fractional Quantum Hall Effects are well-known examples.

As transistors get smaller the number of electrons within a device decreases. Silicon Valley pioneer Gordon Moore formulated his empirical law that the number of transistors on a chip doubles every 18 months while the chip price is constant [1]. By the year 2000 there were already 1 billion transistors per chip. Moore's law also predicts that every year the number of electrons within a transistor decreases so that by 2010 there will be only one electron per device, the so-called " brick wall".

This dramatically illustrates the commercial need for a fundamental understanding of what happens in a high quality device when the number of electrons becomes very small. It is no less than crucial for continued development in the semiconductor industry.

This trend towards lower and lower electron numbers means that the interactions between the electrons are becoming extremely important. This is happening because with few electrons the Fermi energy becomes very small and the electrical forces acting between the electrons start to dominate, whereas in conventional devices there are a lot of electrons and their Fermi energy is so dominant that it kills off most effects from the electron interactions.

A corollary of Moore's law is the need to keep reducing the number of impurities in the device, since the smaller the device the greater is the effect of even a tiny number of impurities. In practice the demands on the quality of the semiconductor materials have become extraordinarily high and high purity semiconductor device material today must have less than one impurity per billion atoms, equivalent to one westerner in China.

In such systems there is a body of experimental evidence indicating new quantum effects [2–5]. One phenomenon which is still controversial suggests in devices with very low levels of impurities, that a transition can be induced in the ground state from an insulating to metallic state by increasing the electron density. Associated with this phenomenon is some evidence for an apparent new quantum critical point (QCP) that would separate the insulating and metallic ground states. There have been some demonstrations of apparent scaling behaviour of the temperature dependent resistivity with the fractional deviation of the electron density from a critical density.

The issue of the nature of the conduction-electron state in 2D goes back more than thirty years to Mott's concept of a mobility edge. Mott predicted for semiconductors with impurities that there is a mobility edge in energy. This separates those electrons of low energy that can only hop from trap to trap with the assistance of thermal excitations, from those electrons of higher energy that move freely even at zero temperature. By increasing the conduction-band electron density we move the Fermi energy through the mobility edge, and when this happens the nature of the electron ground state changes. There is a transition from an insulator with only localized states to a conductor with extended states [6]. Studies by Thouless and coworkers of the sensitivity of electronic states to changes in boundary conditions established the key role of the dimensionless conductivity in a scaling description of this transition [7]. This led to the development of a field theoretical scaling theory of the metal-insulator transition for non-interacting electrons moving in a random potential in a space of $2 + \epsilon$ dimensions where ϵ is introduced as a small expansion parameter. It is believed that a metal-insulator transition occurs for $\epsilon > 0$ (including 3D systems) and that 1D electronic systems with disorder have no metallic phase. The case of dimensionality $d = 2$ is special and has been the subject of much interest.

Twenty five years ago Abrahams *et al.* [8] proposed that the ground state for the 2D disordered electron system, at least for non-interacting electrons, should always be insulating. This result was based on a physically plausible renormalization group argument to link the perturbative treatment of the quantum-coherent enhancement of the electron back-scattering off the impurities to the presumed strong localization limit in the ground state. The coherent enhancement of electron back-scattering is due to the constructive interference of time reversed diffusion paths out to distances of the order of the quantum phase coherence length. As the temperature tends to zero the coherence length goes to infinity introducing a logarithmic divergent correction in the low temperature resistance in 2D. The continuity argument of Abrahams *et al.* [8] implies that this initially weak localization persists so that the electrons always become localized in 2D.

Neglect of the interactions between the electrons (except for linear screening effects) is a good approximation in conventional silicon devices. For these devices the nature of the electron ground state is determined purely by scattering off impurities. However this is not the case in ultra-high quality devices where the total number of impurities can be so low, and the number of electrons so few (small Fermi energy) that it is the interactions between the electrons that in fact dominates. For sufficiently dilute electron systems it is already known that new states of matter can form, such as the electron (Wigner) solid. The Wigner solid is generated by the (relatively) strong interactions between the electrons. These overcome the delocalising tendency of zero-point motion effects associated with the Fermi energy.

It is important to note that there is no consensus on an interpretation of the experimental data as a quantum phase transition. The origins of the behaviour, and the question of whether it persists in the limit of zero temperature remain subjects of debate. There are points at which the scaling theory of the QCP interpretation and experiment appear to be in conflict. Indeed, Altshuler *et al.* [9] have demonstrated that some of the more convincing experimental evidence for scaling in silicon can be explained as material dependent effects. Furthermore scaling of the resistivity data with respect to parallel magnetic fields has been reported only on the "metallic" branch and not on the "insulating" branch. And even if we accept the splitting of the resistivity as a function of temperature into two branches, with the implication of a metallic ground state, the issue of whether the effects really persist in the limit of zero temperature remains a vexed issue. Simmons *et al.* [5] have presented evidence that the down-trending resistivity curves in the "metallic" state, at least near the "critical density" n_c, show an upturn when the temperature is made lower. If this upturn were to extend to all higher densities n there would be no metallic limit and the eventual ground state would be insulating. However the observed "up-turn" could in fact be caused by a crossing of the temperature dependent melting curve. This would be consistent with only a 1% error in pinpointing the exact density for the separatrix.

In the end one must conclude that the key issue of whether the 2D interacting electron system in the presence of weak disorder can be a metal or instead is always an insulator at zero temperature remains unresolved.

Let us now recall some properties of the 2D interacting electron system in the presence of weak disorder for which there is a consensus. In the limit of very high densities any effect of the interaction between the electrons will be overwhelmed by the Fermi energy. If the electron-electron interaction is neglected entirely, the weak localization due to quantum interference is believed to be the prelude to strong localization and an insulating ground state, at least in the case of purely potential scattering without spin-orbit interactions. In the opposite limit of very low densities, the effect of the interaction between the electrons will dominate leading to crystallization of the electrons, the Wigner crystal. The crystal will be pinned by the defects, again resulting in an insulating ground state.

It is in the intermediate density region that there is little consensus. One school argues that, from Landau Fermi liquid theory for the pure system, we know that the quasiparticles are weakly interacting at sufficiently low temperatures. We recall that this result comes from a simple phase space argument. At low temperatures the thermally excited quasiparticles are confined in a narrow shell near $E_F \gg k_B T$. A quasiparticle of momentum k_1 can decay into a quasiparticle of momentum k_3, accompanied by particle-hole excitation k_2, k_4. Energy conservation requires that $|\epsilon_3| \leq \epsilon_1$, so ϵ_3 is confined to a narrow shell of width $\epsilon_1/E_F \ll 1$ near E_F. Momentum conservation $k_1 - k_3 = k_4 - k_2$ further restricts available final states by another factor also $\sim \epsilon_1/E_F$. Thus at low temperatures the inverse lifetime of the original quasiparticle varies as $(\epsilon_1/E_F)^2$, implying stability.

If the electron-electron interactions while weak are not completely negligible, then to first order in the electron-electron interactions, there is a logarithmic correction to the Drude conductivity σ_0 at low temperatures $T/T_\tau \ll 1$ given by,

$$\sigma(T) = \sigma_0 + (e^2/\pi h)(1 - F)\log(T/T_\tau) \qquad (1)$$

where τ is the elastic scattering time and F is the exchange correction [10, 11]. For $1 - F > 0$ the correction to σ_0 is negative for small T, indicating insulator-like behavior.

However the argument that interactions between electrons are negligible in the low temperature limit is based on Landau Fermi Liquid Theory for the pure system. The corresponding theory in the presence of weak disorder has not yet been resolved. The disorder introduces additional scattering channels in which quasiparticles can scatter without inducing any accompanying particle-hole excitation. Thus it is not a priori clear that treating electron-electron interactions only to first order will be adequate in the intermediate range of densities where the effects of interactions are argued to be strong.

There does exist a quantitative theory for 2D electrons in the presence of weakly disordered potential scattering, a theory which treats electron-electron scattering to all orders. This is the renormalization group method [12–14]. The renormalization group equations are based on a perturbation expansion in powers of the dimensionless resistivity $\mathcal{R} = (e^2/\pi h)R_\square \ll 1$, where $R_\square$ is the resistance per square:

$$\frac{d\mathcal{R}}{d\log\lambda^{-1}} = \alpha(\gamma_2)\mathcal{R}^2 + \ldots \tag{2}$$

$$\alpha(\gamma_2) = n_v + \left[1 + ((2n_v)^2 - 1)\left\{1 - \frac{1+\gamma_2}{\gamma_2}\log(1+\gamma_2)\right\}\right]. \tag{3}$$

n_v is the number of valleys, $\Gamma_2 = Z\gamma_2$ is the electron-hole scattering amplitude for the triplet spin state, and Z is the dynamical energy rescaling function. The variable λ describes rescaling of the energy after integrating over the energy shell specified by $\lambda k_0^2 < k^2 < k_0^2$. There are also renormalization group equations for γ_2, Z, and the particle-particle scattering amplitude $\Gamma_c = Z\gamma_c$. In Eq. (3) we have omitted γ_c since particle-particle scattering does not play an essential role in the following discussion. An interpretation of the coefficients in Eq. (3) in terms of particle-hole singlet and triplet spin states has been given using conventional many-body theory [15].

In the renormalization group approach, Eq. (2) and the equations for the other parameters are integrated upward in λ^{-1}. This leads to renormalization of the various parameters and $\mathcal{R}(T)$ develops accordingly. This generic rescaling procedure is applicable in the true quantum critical regime where the relative density shift $\delta_n = [(n - n_c)/n_c]$ is small and $T_0(\delta_n)/T \ll 1$. It is also applicable at small $\mathcal{R}$.

For weak electron-electron interactions $\alpha(\gamma_2) \simeq n_v + 1 + [(2n_v)^2 - 1](\gamma_2/2)$ is positive since $\gamma_2 \ll 1$, and so $\mathcal{R}(T)$ increases with decreasing

T, implying insulating behaviour. On the other hand when γ_2 becomes larger, we can see from inspection that the coefficient $\alpha(\gamma_2)$ will change sign leading to metallic behaviour. This strongly indicates that electron-electron interactions can compensate for the localizing effects and can stabilize metallic behaviour.

Unfortunately only the leading term in the 2D renormalization group equations is known so that Eq. (2) is strictly valid only for $\mathcal{R} \ll 1$, while the bifurcation is observed in the region $\mathcal{R} \sim 1$. The dimensionless resistivity $\mathcal{R}$ is usually small only for densities far from the critical density n_c at the bifurcation. We recall for relative density shifts δ_n that are not small compared with unity, the resistance data lie well outside the critical region, and the correlation length ξ cannot grow but remains microscopically small. Thus in this region, while the scaling equations are valid, they do not describe critical fluctuations and no critical T dependence results from rescaling.

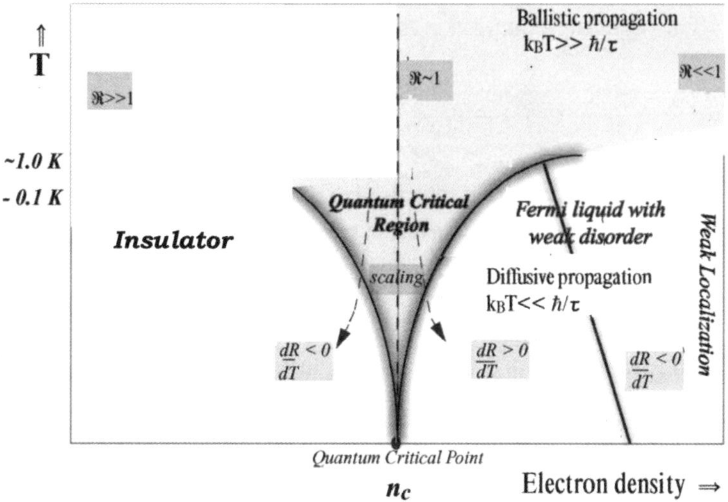

Figure 1. Schematic representation of the insulating and weakly disordered Fermi liquid domains, separated by a possible quantum critical region in the temperature-density plane. The diffusive and ballistic regions are indicated. The dimensionless resistivity $\mathcal{R} = (e^2/\pi h)R_\square$.

If a quantum critical point does exist it would be a repulsive fixed point of a renormalization group flow. For classical critical phenomena, a lowering of the temperature can cause the system variables to enter the critical regime from outside it. But for quantum critical phenomena, if the flow starts from outside the critical regime, then reducing the temperature cannot induce the flow to enter the critical regime. Unless $\mathcal{R} \sim 1$ we start outside the critical region and critical fluctuations can never develop. If we reduce the temperature this only drives system further away from the critical region (Figure 1). Thus renormalization group critical scaling effects are not seen for $\mathcal{R} \ll 1$. For this case the system will be in a Fermi-liquid-plus-disorder-corrections state.

We conclude that effects of criticality due to a possible quantum critical point will not be observed when $\mathcal{R} \ll 1$. In this region Fermi liquid theory with disorder corrections will determine the scattering amplitude γ_2. In the range of temperature currently accessible experimentally and for $\mathcal{R} \ll 1$ there is no indication of temperature dependent scaling of γ_2. Consequently, in the following we neglect any temperature dependence of the Fermi liquid parameters and take γ_2 to depend only on density.

If γ_2 at fixed density remains fixed at its starting value, the coefficient $\alpha(\gamma_2)$ does not evolve with T. It is then straightforward to integrate Eq. (2) from the starting energy of order $k_B T_\tau = \hbar/\tau$, where τ is the Drude elastic scattering time, down to the measurement temperature $k_B T$. The resulting equation is then simply,

$$\sigma(T) = \sigma_0 + \frac{e^2}{\pi h}\alpha(\gamma_2)\log\left(\frac{T}{T_\tau}\right), \tag{4}$$

where σ_0 is the Drude conductivity. Eq. (5) was first derived in the renormalization group approach, with $\alpha(\gamma_2)$ fixed, by Finkelstein [12, 14]. In this derivation the strong electron-electron interactions were treated non-perturbatively to all orders. We note that if logarithmic behaviour is observed at low temperature in the diffusive regime it is direct evidence of a lack of rescaling in that temperature range.

Interaction corrections at higher temperatures $k_B T > \hbar/\tau$, in the ballistic regime, to lowest order in $\mathcal{R}$ and to all orders in the electron-electron interaction have been studied using a Fermi-liquid approach [16]. However, for consideration of the approach to the ground state at low temperature, experimental data in the diffusive regime ($T/T_\tau \ll 1$) are required.

The simple expression Eq. (4) with constant $\alpha(\gamma_2)$ is applicable over a wide range of T, provided only that we are in the diffusive regime, that $\mathcal{R}$ is small, and that the correction $\delta\sigma$ is small relative to σ_0. For small T the correction to σ_0 in Eq. (4) is only the leading term of an infinite series. The growth of the logarithmic term at low temperature indicates that the higher order terms in the series cannot be neglected.

An interesting new phenomenon emerges from Eq. (4), a reentrant insulator. The reentrant insulator results from the change of sign of the fixed $\alpha(\gamma_2)$ as a function of density. At low densities where electron-electron interaction effects are strong the large γ_2 leads to a negative

$\alpha(\gamma_2)$ and $\mathcal{R}(T)$ will be "metallic-like". At high densities the electron-electron interaction effects are relatively weak, γ_2 becomes small so $\alpha(\gamma_2)$ will change sign and $\mathcal{R}(T)$ will be "insulator-like". This reentrant insulator phenomenon has been observed in both Si and GaAs [17, 18].

Figure 2 shows γ_2 for silicon as a function of density, determined from the spin susceptibility χ and the effective mass $m^\star$,

$$\chi/\chi_0 = (m^\star g^\star)/(m g_0)$$

$$\gamma_2 + 1 = g^\star/g_0 \qquad (5)$$

These have been measured for silicon by Pudalov *et al.* [19]. For $r_s < 5$ it is a good approximation to assume that the Dingle temperature T_D is constant since the change in the resistance over the studied temperature range is small.

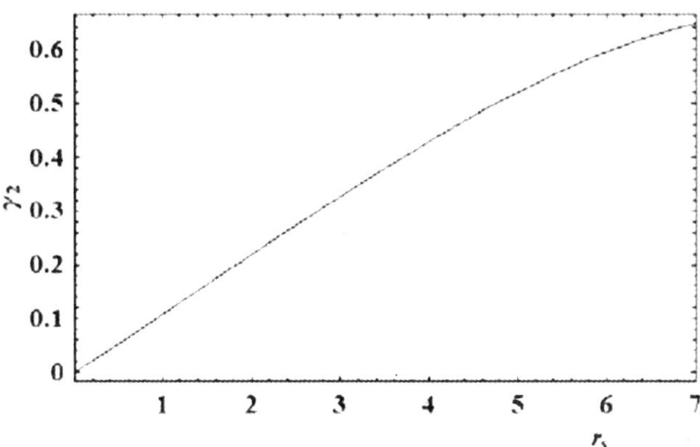

Figure 2. Calculated γ_2 for silicon.

The valley degeneracy for silicon is $n_v = 2$. The resulting $\alpha(\gamma_2)$ changes sign at $r_s = 3.2$ (Eq. (3)). Pudalov *et al.* [17] fitted the observed T dependence in $\delta\sigma$ in silicon at their lowest temperatures by the logarithmic form $\delta\sigma(T) = (e^2/h)C(n)\log(T)$. Their density dependent parameter $C(n)$ is simply related to the theoretical $\alpha(\gamma_2)$ by $C(n) = \alpha/\pi$. In Figure 3 we compare our calculated $\alpha(\gamma_2)$ with their measurements.

We see that theory and experiment are in good agreement except in the region $r_s \leq 3$. While it is not visible on the scale of this figure, Ref. Pudalov reports that the measured $C(n)$ does in fact change sign and become slightly positive. However our calculated $C(n)$ continues to grow with decreasing r_s, terminating in the high density limit at $C(r_s \rightarrow$

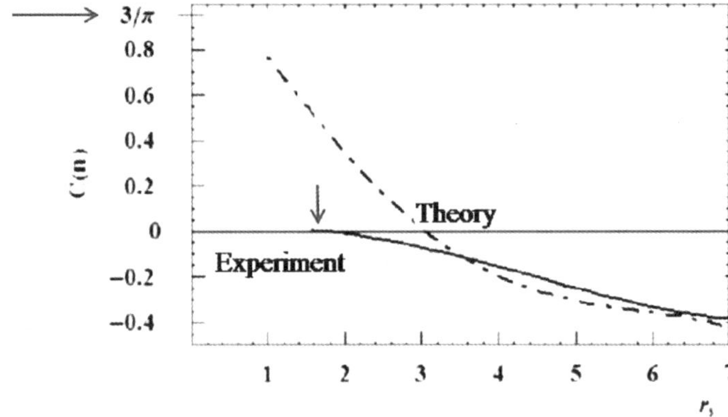

Figure 3. Dependence of prefactor $C(n)$ in silicon on density parameter r_s. Solid line: from measurements in Ref. Pudalov. Dashed line: calculated $C(n) \equiv \alpha(\gamma_2)/\pi$.

0) $= 3/\pi$ for $n_v = 2$. On the basis of current theory, one expects this high density limit to be precise. It is not clear why the experimental data at small r_s do not appear to extrapolate to this limit.

Thus for $\mathcal{R} \ll 1$ we are initially outside the quantum critical region and hence critical fluctuations cannot develop. If we reduce T this only drives system further away from the critical region. For this reason renormalization group critical scaling effects are not seen for $\mathcal{R} \ll 1$ and the system is in a Fermi-liquid-plus-disorder-corrections state. Experimental evidence of this can be found in the logarithmic behaviour observed at low temperature in the diffusive regime. In this region physical properties are controlled by the lowest order $\mathcal{R} \ll 1$ form of the renormalization group equation and not by the quantum critical point.

In a landmark event such as this, it is appropriate to add in a little speculation on other effects which have not yet been much considered, but which may become important in the future. As the density of electrons decreases, their average spacing can become comparable or larger than the distance separating them from the metal gate controlling their density. In this case coupling with the mobile charges in the metal gates becomes comparable in strength to the electron interactions within the layer. This has a significant effect on the effective screening between the electrons and may also lead to novel coupled states. At very low densities the spacing between the electrons can also be comparable with the width of the quantum well that confines the electrons to their layer. New dimerisation effects may then switch in, particularly for the Wigner crystal ground state.

ACKNOWLEDGMENTS. This research was supported by the Natural Sciences and Engineering Research Council of Canada and an Australian Research Council Grant.

References

[1] GORDON E. MOORE, Electronics **38** (1965), 8.

[2] S. V. KRAVCHENKO, W. E. MASON, G. E. BOWKER, J. E. FURNEAUX, V. M. PUDALOV and M. D'IORIO, Phys. Rev. B **51** (1995), 7038; S. V. KRAVCHENKO, D. SIMONIAN, M. P. SARACHIK, W. MASON and J. E. FURNEAUX, Phys. Rev. Lett. **77** (1996), 4938.

[3] D. SIMONIAN, S. V. KRAVCHENKO, M. P. SARACHIK and V. M. PUDALOV, Phys. Rev. Lett. **79** (1997), 2304.

[4] P. T. COLERIDGE, R. L. WILLIAMS, Y. FENG and P. ZAWADZKI, Phys. Rev. B **56** (1997), R12764.

[5] M. Y. SIMMONS, A. R. HAMILTON, M. PEPPER, E. H. LINFIELD, P. D. ROSE, D. A. RITCHIE, A. K. SAVCHENKO and T. G. GRIFFITHS, Phys. Rev. Lett. **80** (1998), 1292.

[6] N. F. MOTT and E. A. DAVIS, Electronic Properties of Non-Crystalline Materials, Clarendon, Oxford, 1971.

[7] See D. J. THOULESS and other reviewers in Ill-Condensed Matter, R. Balian, R. Maynard, and G. Toulouse (eds.), North-Holland, Amsterdam, 1979.

[8] E. ABRAHAMS, P. W. ANDERSON, D. C. LICCIARDELLO and T. V. RAMAKRISHNAN, Phys. Rev. Lett. **42** (1979), 673.

[9] B. L. ALTSHULER and D. L. MASLOV, Phys. Rev. Lett. **82** (1999), 145; B. L. ALTSHULER, G. W. MARTIN, D. L. MASLOV, V. M. PUDALOV, A. PRINZ, G. BRUNTHALER and G. BAUER, arXiv:cond-mat/0008005 v 1, unpublished.

[10] B. L. ALTSHULER, A. G. ARONOV and P. A. LEE, Phys. Rev. Lett. **44** (1980), 1288.

[11] H. FUKUYAMA, J. Phys. Soc. Jpn., **48** (1980), 2169.

[12] A. M. FINKELSTEIN, Zh. Eksp. Teor. Fiz. **84c** (1983), 168 [Sov. Phys. JETP **57** (1983), 97]; A. M. FINKELSTEIN, Z. Phys. B: Cond. Matt. **56** (1984), 189.

[13] C. CASTELLANI, C. DI CASTRO, P.A. LEE and M. MA, Phys. Rev. B **30** (1984), 527.

[14] A. M. FINKELSTEIN, Sov. Sci. Rev., Sect. A, **14** (1990), 3.

[15] B. L. ALTSHULER and A. G. ARONOV, Solid State Commun. **46** (1983), 429.

[16] G. ZALA, B. N. NAROZHNY and I. L. ALEINER, Phys. Rev. B **64** (2001), 214204.
[17] V. M. PUDALOV, G. BRUNTHALER, A. PRINZ and G. BAUER, JETP Lett. **68** (1998), 534.
[18] A. R. HAMILTON, M. Y. SIMMONS, M. PEPPER, E. H. LINFIELD, P. D. ROSE and D. A. RITCHIE, Phys. Rev. Lett. **82** (1999), 1542.
[19] V. M. PUDALOV, M. E. GERSHENSON, H. KOJIMA, N. BUTCH, E. M. DIZHUR, G. BRUNTHALER, A. PRINZ and G. BAUER, Phys. Rev. Lett. **88** (2002), 196404.

Physical nature of quantum entanglement

A. A. Klyachko and A. A. Shumovsky

In this paper, we discuss entanglement as a a manifestation of quantum fluctuations. We discuss a number of examples related to condensed matter physics, optics, and elementary particle physics.

1. Introduction

We dedicate this article to Professor Mario Tosi and to his unflagging interest in the enigmatic world of quantum physics, which is an inspirational example of everlasting scientific youth.

For decades, entanglement remained a mystery of quantum mechanics. The discussion was created by Einstein-Podolsky-Rosen (EPR) proposal of a thought experiment that appeared to demonstrate quantum mechanics to be an incomplete physical theory [1]. They claimed that predictions of quantum mechanics result from statistical distribution of classical hidden but real properties of the systems. In about thirty years, Bell found a way how to prove nonexistence of hidden variables in quantum mechanics [2]. Twenty years later the nonexistence of hidden variables was shown in experiment [3].

Straightway after EPR paper [1], Schrödinger [4] noticed information aspect of EPR paradox caused by nonlocality of multipartite EPR states and specific quantum correlations between spatially separated parts of the system. At the same time, Schrödinger found out the generality of the notion of entanglement, which he considered as a "characteristic trait of quantum mechanics" [4]. Fifty years later, protocols for the use of entanglement in quantum information processing [5–7] and quantum computing [8] were proposed. This discovery heralded the birth of quantum information science, which nowadays represents one of the hottest branches of physics, mathematics, and engineering. At present, quantum entanglement is considered to be the main physical resource of quantum information processing and quantum computing [9].

In spite of a great progress in theoretical and experimental investigation of entanglement and its practical implementations, some principle problems remain as dark as before. The very revelation of the physical nature of entanglement is among them.

The primordial definitions of quantum entanglement were mostly based on some secondary features of this physical phenomena such as nonseparability of quantum states of composite systems [10], violation of Bell's conditions [11], nonlocality [12] etc. All of them are valid for bipartite systems because of simple mathematical structure of states of those systems provided by the Schmidt decomposition [13], but fail for other systems. Generally speaking, they do not reflect the essence of entanglement as a physical phenomenon.

For example, the idea of nonlocality cannot be applied not only to a single-particle entanglement [14], but also to entanglement in quantum liquids and Bose-Einstein condensate. In the latter case, the requirement of nonlocality is completely meaningless because of the strong overlap of wave functions of individual particles [15]. Three-partite states of the so-called W-type [16] provide an example of nonseparable states that violate Bell's conditions [17] but do not manifest entanglement [18].

Probably, the most important fact which we are fully confident in is that entanglement is purely quantum phenomenon without any classical analogue. The poetic definition by Asher Peres states (for references, see [10]): *"Entanglement is a trick that quantum magicians use to produce phenomena that cannot be imitated by classical magicians"*.

Beginning with this fact, we should remember that the principle difference between the quantum and classical levels of description of nature consists in the definition of physical observables. In the former case, they are chosen to be Hermitian operators, acting in the Hilbert space of states of a system, while in the latter case observables are represented by c-numbers. As a result, a quantum observable $\mathcal{O}_i$ measured in a state $\psi \in \mathbb{H}$ manifests quantum fluctuations given by the variance

$$V_i(\psi) = \langle\psi|\mathcal{O}_i^2|\psi\rangle - \langle\psi|\mathcal{O}_i|\psi\rangle^2. \tag{1.1}$$

The quantum fluctuations are responsible for a number of physical phenomena such as Lamb shift, spontaneous emission, and quantum jumps (e.g., see [19]). They also define specific behavior of quantum liquids [20] and solids. According to Wigner [21], the variance (1.1) can be associated with so-called skew information, specifying a knowledge that can be extracted from macroscopic measurements of quantum observables about a state of quantum system.

The aim of this paper is to show that quantum entanglement represents a certain manifestation of quantum fluctuations. It builds upon our previous results [22–26]. We have arranged the paper as follows. In Sec. II we discuss the definition of a quantum mechanical system via states in the Hilbert space and their efficient operator representation. We define coherent and completely entangled states in terms of extremal properties of quantum fluctuations peculiar for the operator basis of the system. In Sec. III we illustrate the approach by examples of entangled states in quantum systems of different physical nature (condensed matter, atoms and photons, elementary particles). Finally, in Sec. IV we summarize the main conclusions of the work.

2. States and observables

Specifying a quantum mechanical system Q, we should determine a structure of the corresponding Hilbert space $\mathbb{H}_Q$ and an operator basis accessible for measurements.

Consider as an example a system that can be prepared in only three mutually orthogonal states $|\ell\rangle$ ($\ell = 0, 1, 2$), forming a basis of the three-dimensional Hilbert space. This system can be interpreted as a qutrit (an object considered within the quantum information in connection with ternary quantum logic [27]) if the efficient operator representation uses the Hermitian generators of the $su(3)$ algebra. This means that the basic set of observables that should be measured to specify a state of qutrit is provided by eight independent operators. If instead we choose and efficient operator representation defined by three spin-1 operators, the system, a "spin-qutrit", will have quite different physical properties.

From the physical point of view, the choice of basic observables is caused by the measurements we are able to perform over the system to characterize its state.

In general, the Lie algebra $\mathcal{L}$ of essential (basic) observables $\mathcal{O}_i$ is determined by the set of observables that we choose to perform measurement of states or, what is the same, by the Hamiltonians that are accessible for manipulations with the states. The choice of $\mathcal{L}$ determines the dynamic symmetry group $G = \exp(\mathcal{L})$ of the Hilbert space. For example, if we choose spin projection operators as the basis of $\mathcal{L} = su(2)$, the symmetry properties of the system are specified by the $SU(2)$ group. In the case of qutrit, the choice of $\mathcal{L} = su(3)$ causes the $SU(3)$ symmetry of the system, and so on.

To characterize a state $\psi \in \mathbb{H}_Q$, we should examine quantum fluctuations (1.1) of all observables, forming the basis of $\mathcal{L}$, in addition to mean values of those observables. The total amount of quantum fluctuations

$$V(\psi) = \sum_i V_i(\psi) \tag{2.2}$$

gives the total amount of knowledge about the state ψ that can be extracted from macroscopic measurement of all basic observables peculiar for the symmetry of the system Q. In a sense, (2.2) specifies the remoteness of the state ψ from "classical reality" provided by the measurement of classical observables (c-numbers), when $V_i = 0$ for all i and $\psi \in \mathbb{H}$.

For all quantum states, $V(\psi) > 0$. The minimal amount of total variance (2.2) corresponds to the so-called generalized coherent states [28]. It has been noticed in Ref. [22] that completely entangled states $\psi_{CE} \in \mathbb{H}_Q$ have the property

$$\forall i, \quad \langle \psi_{CE} | \mathcal{O}_i | \psi_{CE} \rangle = 0. \tag{2.3}$$

This property can be interpreted in the following way. Assume that the essential observables form a basis of compact Lie algebra $\mathcal{L}$. Then, there is the Casimir operator

$$\hat{C} \equiv \sum_i \mathcal{O}_i^2 = C \times \hat{1}, \tag{2.4}$$

It is easily seen now from Eqs. (1.1) and (2.2) that $V_{\max} = C$ provided by the condition (2.3). Thus,

$$V(\psi_{CE}) = \max_{\psi \in \mathbb{H}_Q} V(\psi). \tag{2.5}$$

This Eq. (2.5) can be used as a definition of completely entangled states in an arbitrary quantum system Q [25]. In fact, Eq. (2.5) represents a variational principle that clarifies the physical nature of complete entanglement as a *manifestation of quantum fluctuations of basic observables at their extreme*. In other words, the completely entangled states are maximally remote from the "classical reality" provided by a measurement of classical observables (c-numbers) free of quantum fluctuations.

Recall that entangled states of a system Q can be easily separated from the other states in $\mathbb{H}_Q$. In particular, there are certain local operations that can transform entangled states into completely entangled states and vice versa but they are not able either to destroy or to create entanglement. An example is provided by the so-called stochastic local operations assisted by classical communications (SLOCC) [16,29]. Mathematically SLOCC transformations amounts to action of complexified dynamic symmetry group $G^c = \exp(\mathcal{L} \otimes \mathbb{C})$. Within an approach based on the theory of geometric invariants it is possible to show that an arbitrary entangled state belongs to the class of semistable vectors (see Ref. [23]). The classification of states of three qubits (six-dimensional system with dynamic symmetry $SU(2) \times SU(2) \times SU(2)$ composed of three spin-$\frac{1}{2}$ particles) via geometrical invariants was done in Ref. [18].

It is wrong to assume that all entangled states have maximal amount of quantum fluctuations. For example, unentangled W-states of three qubits [16] manifest more quantum fluctuations than some entangled states [25] ($V(W) = 8 + 2/3$, while $V_{\max} = 9$). The point is that entangled states belong to a special class with respect to geometric invariants [18] and cannot be transformed into unentangled states by local operations [16].

Thus, definition of completely entangled states (2.5) can be used to determine all entangled states of a given system Q via SLOCC.

One of the benefits obtained from the variational principle (2.5) is a clue in the problem of stabilization of entanglement. The point is that practical realization of quantum information protocols requires a robust entanglement (complete entanglement with long enough life time). It follows from (2.5) that first we have to prepare a state with maximum amount of quantum fluctuations. Then, we should decrease the energy of the system up to a (local) minimum under the condition of conservation of total variance (2.2). Such a "mnimax" protocol leads to a stable entanglement. Possible realizations of this procedure in physical systems were discussed in [30].

3. Examples

The simplest example usually discussed in the context of entanglement is provided by a qubit system. A single qubit is defined to be a state from the two-dimensional Hilbert space $\mathbb{H}_2$ with the dynamic symmetry $SU(2)$ [31]. Irrespective of physical realization, this system is equivalent to spin $s = \frac{1}{2}$. The basis of essential observables is provided by the Pauli operators σ, forming an infinitesimal representation of the $s\ell(2, \mathbb{C})$ algebra, which is the complexification of the $su(2)$ algebra. The Casimir operator has the form $\hat{C} = 3 \times \hat{1}$. The SLOCC operations are given by the transformations from the $SL(2, \mathbb{C})$ group.

Denote the basis of a qubit by $|\ell\rangle$, $\ell = 0, 1$, where $\ell = 0$ ($\ell = 1$) corresponds to spin up (down). Following [32], we introduce coherent state of a qubit as follows

$$|\alpha\rangle = \exp(\alpha\sigma_+ - \alpha^*\sigma_-)|1\rangle, \quad \alpha \in \mathbb{C},$$

where $\sigma_+ = |0\rangle\langle 1|$ and $\sigma_- = |1\rangle\langle 1|$. It is easily seen that

$$|\alpha\rangle = e^{i \arg \alpha} \sin(|\alpha|)|0\rangle + \cos(|\alpha|)|1\rangle. \tag{3.6}$$

Since the action of Pauli operators on the base states $|\ell\rangle$ are defined as follows

$$\sigma_x \left\{ \begin{array}{l} |0\rangle \\ |1\rangle \end{array} = \right\{ \begin{array}{l} |1\rangle \\ |0\rangle \end{array}, \quad \sigma_y \left\{ \begin{array}{l} |0\rangle \\ |1\rangle \end{array} = \right\{ \begin{array}{c} i|1\rangle \\ -i|0\rangle \end{array}, \quad \sigma_z \left\{ \begin{array}{l} |0\rangle \\ |1\rangle \end{array} = \right\{ \begin{array}{c} |0\rangle \\ -|1\rangle \end{array},$$

the total variance (2.2) in coherent state (3.6) is $V_{coh} = 2$. This is the minimal value of the total variance of a single qubit. In the case of N qubits, the coherent state is defined as

$$|\alpha_1, \cdots, \alpha_N\rangle = \bigotimes_{j=1}^{N} |\alpha_j\rangle$$

with the total variance $V_{coh}^{(N)} = 2N$. In turn, a completely entangled state of N qubits is specified by the maximum value $V_{\max}^{(N)} = 3N$.

It follows from the condition (2.3) that completely entangled state of a single qubit is impossible. It is also seen from (2.3) that a general normalized state of two qubits

$$|\psi\rangle = \sum_{\ell,\ell'=0}^{1} \psi_{\ell\ell'} |\ell, \ell'\rangle, \quad |\ell, \ell'\rangle = |\ell\rangle \otimes |\ell'\rangle, \tag{3.7}$$

is completely entangled iff [24]

$$2|\det[\psi]| \equiv 2|\psi_{00}\psi_{11} - \psi_{01}\psi_{10}| = 1.$$

Here $[\psi]$ denotes the matrix of coefficients in (3.7). The so-called EPR and Bell states of two qubits obey this condition.

It was shown in [33] that any proper measure of entanglement should be an entangled monotone. The only entangled monotone for two qubits is given by the $\det[\psi]$. The conventional measure of two-qubit entanglement called concurrence is given by the expression [34]

$$\mathcal{C}(\psi) = 2|\det[\psi]|, \tag{3.8}$$

whose form coincides with the left-hand side in the above condition of complete two-qubit entanglement. The concurrence (3.8) can also be represented in terms of the total variance as follows

$$\mathcal{C}(\psi) = \sqrt{\frac{V(\psi) - V_{\min}}{V_{\max} - V_{\min}}}. \tag{3.9}$$

An unconditional advantage of measure (3.9) is that it deals with the physical quantities, which can be directly measured. In the case of mixed states of two qubits, the definition of concurrence is described in [34].

The multi-qubit completely entangled states can be determined via the variational principle (2.5) and equivalent condition (2.3) in similar fashion [24, 25].

Physical realizations of qubits include two-level atoms, photons with two polarizations, and spin-$\frac{1}{2}$ particles.

Much less obvious example is provided by a single spin-qutrit, i.e., by a spin-1 like system [26]. In this case, the three-dimensional Hilbert space $\mathbb{H}_3$ is spanned by vectors $|\ell\rangle$ ($\ell = 0, 1, 2$)

$$|0\rangle = \begin{pmatrix} 1 \\ 0 \\ 0 \end{pmatrix}, \quad |1\rangle = \begin{pmatrix} 0 \\ 1 \\ 0 \end{pmatrix}, \quad |2\rangle = \begin{pmatrix} 0 \\ 0 \\ 1 \end{pmatrix}, \tag{3.10}$$

and essential observables are given by the spin-1 operators

$$S_x = \begin{pmatrix} 0 & 1/\sqrt{2} & 0 \\ 1/\sqrt{2} & 0 & 1/\sqrt{2} \\ 0 & 1/\sqrt{2} & 0 \end{pmatrix}, \quad S_y = \begin{pmatrix} 0 & -i/\sqrt{2} & 0 \\ i/\sqrt{2} & 0 & -i/\sqrt{2} \\ 0 & i/\sqrt{2} & 0 \end{pmatrix},$$
$$S_z = \begin{pmatrix} 1 & 0 & 0 \\ 0 & 0 & 0 \\ 0 & 0 & -1 \end{pmatrix} \tag{3.11}$$

with the Casimir operator $S_x^2 + S_y^2 + S_z^2 = 2 \times \hat{1}$. A general pure state of such a system has the form

$$|\psi\rangle = \sum_{\ell=0}^{2} \psi_\ell |\ell\rangle, \quad \sum_{\ell=0}^{2} |\psi_\ell|^2 = 1. \tag{3.12}$$

Then, the use of condition (2.3) gives the following set of equations

$$\left.\begin{aligned} \mathrm{Re}(\psi_1\psi_2^*) + \mathrm{Re}(\psi_2\psi_3^*) &= 0 \\ \mathrm{Im}(\psi_1\psi_2^*) + \mathrm{Im}(\psi_2\psi_3^*) &= 0 \\ |\psi_1|^2 - |\psi_3|^2 &= 0 \end{aligned}\right\} \tag{3.13}$$

whose solutions determine the completely entangled states of a single spin-qutrit. In view of the normalization condition for the state (3.12), it follows from the last equation in (3.13) that

$$2|\psi_1|^2 + |\psi_2|^2 = 1, \quad |\psi_2| = |\psi_0|. \tag{3.14}$$

Neglecting the total phase of the state (3.12), we can put $\arg(\psi_1) = 0$. Then, for the first two equations in (3.13) we have the following solutions.

1. $\psi_1 = 1$, $\psi_0 = \psi_2 = 0$, which corresponds to the state

$$|\psi_{CE}^{(1)}\rangle = |1\rangle. \tag{3.15}$$

By construction, this is a completely entangled state. Using the language peculiar for the spin-1 systems, one can say that this is a state with zero projection of spin on the quantization axis.

2. $|\psi_1| = 0$, which, in view of (3.14), gives the state

$$|\psi_{CE}^{(\pm)}\rangle = \frac{1}{\sqrt{2}}(|0\rangle + |2\rangle). \tag{3.16}$$

It is seen that the state (3.15) and (3.16) are mutually orthogonal. Thus, they form a basis of completely entangled states in $\mathbb{H}_3$.

To clarify the physical meaning of the obtained results, let us consider now a physical system called "biphoton" (see Ref. [35] and references therein). A photon pair created by spontaneous resonance down-conversion and carrying correlations with respect to frequency, wave vector, moment of birth, and polarization is usually called *biphoton*. A pure state of biphoton with an arbitrary polarization can be represented in the space of states $\mathbb{H}_2 \otimes \mathbb{H}_2$ of two photons as follows

$$|\psi_{BP}\rangle = \psi_1|2,0\rangle + \psi_2|1,1\rangle + \psi_3|0,2\rangle, \tag{3.17}$$

where $|n_x, n_y\rangle$ denotes a state with n_x photons in the horizontal (x) polarization mode and n_y photons in the vertical (y) polarization mode, and

$$\sum |\psi_i|^2 = 1.$$

In the picture where both spatial modes of a biphoton are accessible, the state $|1,1\rangle$ in (1.1) is interpreted as a symmetric combination of states of two photons with orthogonal polarization (two qubits)

$$|1,1\rangle = \frac{1}{\sqrt{2}}(|\uparrow,\rightarrow\rangle + |\rightarrow,\uparrow\rangle). \tag{3.18}$$

Since the photons propagate together and can be separated neither in space nor in time, the biphoton can be considered as a single "particle", which can be observed in three "spin-1" states

$$|2,0\rangle = |0\rangle, \quad |1,1\rangle = |1\rangle, \quad |0,2\rangle = |2\rangle. \tag{3.19}$$

Thus, the state (3.17) is similar to the single spin-qutrit state (3.12).

In view of the analogy (3.19), the completely entangled state (3.15) of a single $SU(2)$-qutrit can now be interpreted as the completely entangled states of two qubits (3.18). In turn, the states in (3.16) can also be interpreted in terms of completely entangled two-qubit states.

We now note that Eq. (3.9) can also be used to measure the amount of entanglement carried by the single spin-qutrit state. This immediately follows from the coincidence between (3.9) and two-qubit concurrence (3.8) and interpretation of a single spin-qutrit in terms of two qubits.

Another example of a single spin-qutrit is provided by the orbital angular momentum of photons, which is equal to $L = 1$ in the case of Gauss-Laguerre beams [36].

One more example is given by the Cooper pairs in superfluid liquid ^{3}He. This Cooper pair has both spin an orbital angular momentum equal to one. In the so-called B-phase, both spin and orbital angular momentum of the Cooper pair are completely entangled. In other stable phases, they are separated and can be treated as independent "spin-1" objects.

An example of a real single-particle "spin-1" entanglement with respect to intrinsic degrees of freedom is provided by the π^0 meson, composed of "up" and "down" quarks in completely entangled state. Two other members of the π-meson triplet, namely π^+ and π^-, are prepared in unentangled states of quarks. Probably, strong quantum fluctuations in completely entangled state are responsible for quite short lifetime of π^0 meson.

4. Conclusions

We have discussed an approach to quantum entanglement based on consideration of quantum fluctuations of essential observables, specified by the symmetry properties of the system under consideration.

The generality of the approach makes it possible to describe entanglement of objects with very different physical nature. The examples provided by the orbital angular momentum of the Gauss-Laguerre photon, Cooper pairs in superfluid liquid ^{3}He, and π-mesons have been discussed in Sec. III do not exhaust all possibilities.

Besides that, we have shown that the amount of quantum fluctuations can be used as a measure of the amount of quantum entanglement in the form given by Eq. (3.9) that is valid for all pure states of single and bipartite systems. The problem of measure of entanglement in the case of mixed states deserves special consideration.

References

[1] A. EINSTEIN, B. PODOLSKY and N. ROSEN, Phys. Rev. **47** (1935), 777.

[2] J. S. BELL, Physics **1** (1964), 195; Rev. Mod. Phys. **38** (1966), 447.

[3] A. ASPECT, F. GRANGIER and G. ROGER, Phys. Rev. Lett. **49** (1982), 91.

[4] E. SCHRÖDINGER, Naturwisseschaften **23** (1935), 807; 823; 844; Proc. Cambridge Phil. Soc. **31** (1935), 555.

[5] C. H. BENNET and G. BRASSARD, In: IEEE Int. Conf. Computer, Systems, and Signal Processing, (1984), 175.

[6] A. EKERT, Phys. Rev. Lett. **67** (1991), 661.

[7] C. H. BENNETT, G. BRASSARD, G. CREPEAU, R. JOSA, A. PERES and W. K. WOOTERS, Phys. Rev. Lett. **70** (1993), 1895.

[8] P. W. SHOR, In: Proc. of the 35th Annual Symposium on the Foundations of Computer Science, S. Soldwasser (ed.), IEEE Computer Society Press, Los Alamoc, 1994.

[9] D. BOUWMEESTER, A. EKERT and A. ZEILINGER (eds.), "The Physics of Quantum Information", Springer, Berlin, 2000.

[10] D. BRUSS, J. Math. Phys. **43** (2002), 4237.

[11] A. V. BELINSKII and D. N. KLYSHKO, Phys. Usp. **36** (1993), 653; J. URFNIK, Phys. Rev. Lett. **88** (2002), 230406.

[12] S. POPESCU, Phys. Rev. Lett. **72** (1994), 797; R. CLIFTON and H. HALVORSON, Phys. Rev. A **61** (1999), 012108.

[13] A. EKERT and P. L. KNIGHT, Am. J. Phys. **63** (1995), 415; J. H. EBERLY, K. W. CHAN and C. K. LAW, In: Quantum Communication and Information Technologies, A. S. Shumovsky and V. I. Rupasov (eds.), Kluwer Academic, Dordrecht, 2003.

[14] S. M. TAN, D. F. WALLS and M. J. COLLETT, Phys. Rev. Lett. **66** (1991), 252; C. C. GERRY, Phys. Rev. A **53** (1996), 4583; A. BEIGE, B.-G. ENGLERT, C. KURTSIEFER and W. WEINFURTER, J. Phys. A **35** (2002), L407.

[15] A. J. LEGGETT, Rev. Mod. Phys. **73** (2001), 307.

[16] W. DÜR, G. VIDAL and J. I. CIRAC, Phys. Rev. A **62** (2000), 06231.

[17] A. CABELO, Phys. Rev. A **65** (2002), 032108.

[18] A. MIYAKE, Phys. Rev. A **67** (1003), 012108.

[19] M. O. SCULLY and M. S. ZUBAIRY, "Quantum Optics", Cambridge University Press, New York, 1997.

[20] N. H. MARCH and M. TOSI, "Introduction to Liquid State Physics", Oxford University Press, Oxford, 2002.

[21] E. P. WIGNER, Z. Physik **51** (1952), 262; E. P. WIGNER and M. M. YANASE, Proc. Nat. Acad. Sci. USA **19** (1963), 910.
[22] M. A. CAN, A. A. KLYACHKO and A. S. SHUMOVSKY, Phys. Rev. A **66** (2002), 02111.
[23] A. A. KLYACHKO, quant-ph/0206012.
[24] A. A. KLYACHKO and A. S. SHUMOVSKY, J. Opt. B: Quantum and Semiclassical Opt. **5** (2003), S322.
[25] A. A. KLYACHKO and A. S. SHUMOVSKY, J. Opt. B: Quantum and Semiclassical Opt. **6** (2004), S29.
[26] M. A. CAN, A. A. KLYACHKO and A. S. SHUMOVSKY, J. Opt. B: Quantum and Semiclassical Opt. **7** (2005), L1.
[27] H. BECHMAN-PASQUINUCCI and A. PERES, Phys. Rev. Lett. **85** (2000), 3313; D. BRUSS and C. MACCHIAVELLO, Phys. Rev. Lett. **88** (2002), 127901.
[28] A. PERELOMOV, "Generalized Coherent States and Their Applications", Springer, Berlin, 1986.
[29] C. H. BENNETT, S. POPESCU, D. ROHRLICH, J. A. SMOLIN and A. V. THAPALIYA, Phys. Rev. A **63** (2001), 012307; F. VERSTRAETE, J. DEHAENE, B. DE MOOR and H. VERSCHELDE, Phys. Rev. A **65** (2002), 052112.
[30] M. A. CAN, A. A. KLYACHKO and A. S. SHUMOVSKY, Appl. Phys. Lett. **81** (2002), 5072; Ö. CAKIR, M. A. CAN, A. A. KLYACHKO and A. S. SHUMOVSKY, Phys. Rev. A **68** (2003), 022305; Ö. CAKIR, M. A. CAN, A. A. KLYACHKO and A. S. SHUMOVSKY, J. Opt.B: Quantum and Semiclassical Opt. **6** (2004), S13.
[31] M. A. NIELSEN and L. L. CHUANG, "Quantum Computing and Quantum Information", Cambridge University Press, New York, 2000.
[32] J. M. RADECLIFFE, J. Phys.A, **4** (1971), 313.
[33] G. VIDAL, J. Mod. Opt. **47** (2000), 355; J. EISERT, K. AUDENART and M. B. PLENIO, J. Phys. A **36** (2003), 5605.
[34] S. HILL and W. K. WOOTERS, Phys. Rev. Lett. **78** (1997), 5022.
[35] M. V. CHECHOVA, L. A. KRIVITSKY, S. P. KULIK and G. A. MASLENNIKOV, Phys. Rev. A **70** (2004), 053801.
[36] A. VAZIRI, G. WEIHS and A. ZEILINGER, J. Opt. B **4** (2002), S47.

5

MANY BODY PHYSICS

Session 1
Chair
Bilal Tanatar

Session 2
Chair
Giuseppe Grosso

Collective excitations in double-layer electron systems

B. Tanatar

We present a short review of the many-body effects on the collective charge density excitations in double-layer electron systems. In particular, the long wavelength behavior of out-of-phase plasmon mode within theories including local-field factors is compared and contrasted with experiments. We point out the possible use of recently developed exact forms of exchange-correlation kernels to shed new light into this problem.

A double-layer electron system is a useful model to study interesting many-body effects arising from interlayer interactions [1]. Recent advances in the semiconductor growth technology make it possible to fabricate parallel layers of electron or hole gases in the form of double-quantum wells enabling the study of various correlation effects in such systems. The electron gas is characterized by a dimensionless coupling parameter $r_s = a/a_B$ where a and a_B are the average distance between the electrons and the Bohr radius, respectively. Within a layer, the correlation effects are important for $r_s > 1$ or when the layer density is decreased. When the layer separation distance d becomes comparable to a, the interlayer correlation effects also start to make their influence. Thus, compared to a single layer electron system we have much richer possibilities controlled by the parameters r_s and d/a. The experimental verification of the correlation effects come from transport measurements, photoluminescence and Raman spectroscopy type light scattering experiments. In particular, the collective charge and spin density excitations are accessible through inelastic light scattering experiments [2–5].

Collective mode dispersions for charge density excitations in a double-layer system have initially been theoretically studied within the random-phase approximation (RPA) [6]. It is established that the charges in the layers oscillate either in-phase or out-of-phase with respect to each other giving rise to two distinct collective modes. In these calculations it was assumed that the layer separation is large enough (or the barrier height is high enough) so that no charge transfer (i.e. tunneling) occurs between the layers. In this case, the long wavelength plasmon modes have the dispersion relations given by $\omega_+(q) \sim q^{1/2}$ for the in-phase mode and $\omega_-(q) \sim q$ for the out-of-phase mode. These collective excitations are also called "optical" and "acoustic" plasmons, respectively. The experimental observation of the plasmon modes in double-layer systems has generally confirmed the above picture. However, the experiments so far have not probed the large r_s region to clearly distinguish the correlation effects. There has been quite a body of theoretical work [7–11] that goes beyond the RPA on the collective excitations in double-layer systems with varying degrees of sophistication. Some of these approaches make completely differing predictions for the long-wavelength behavior of the out-of-phase plasmon mode as we shall discuss below.

With this background, in this paper we provide a short review of the collective charge excitations in a double-layer electron system. In particular, we concentrate on the exchange-correlation effects neglected by the random-phase approximation (RPA) treatment. Various theoretical approaches will be surveyed. We hope that our exposition will motivate the experimentalists to study the plasmon modes in double-layer systems at low density to settle some of the outstanding issues.

We consider an electron system confined in a double quantum-well structure consisting of symmetrical wells separated by a infinite barrier. In the limit of very thin quantum wells we obtain the idealized system of double-layer electron gas with no layer thickness. The intra and interlayer Coulomb interactions are given by $V_{11} = 2\pi e^2/\epsilon_s q$ and $V_{12}(q) = V_{11}(q)e^{-qd}$, respectively. The idealizations above may easily be relaxed in more realistic calculations, especially when detailed comparisons with experiments are warranted. The density of electrons is conveniently described by the dimensionless coupling parameter r_s defined as $r_s = 1/\sqrt{\pi n a_B^2}$, where $a_B = \hbar^2\epsilon_s/m^* e^2$ is the Bohr radius expressed in terms of the effective band mass and dielectric constant. The Fermi wave vector and r_s are related through $k_F a_B = \sqrt{2}/r_s$.

The collective charge density modes are obtained from the zeros of the determinant of the dielectric matrix constructed for a double layer system [4]. Within the RPA and in the case of equal layer densities the collective modes are obtained by the solution of the following equations.

$$1 - [V_{11}(q) + V_{12}(q)]\chi_0(q, \omega) = 0\,, \qquad (1)$$

and

$$1 - [V_{11}(q) - V_{12}(q)]\chi_0(q, \omega) = 0\,, \qquad (2)$$

where

$$\chi_0(q,\omega) = 2\sum_{\mathbf{k}} \frac{f_{\mathbf{k}+\mathbf{q}} - f_{\mathbf{k}}}{\omega - \epsilon_{\mathbf{k}+\mathbf{q}} + \epsilon_{\mathbf{k}} + i\eta} \tag{3}$$

is the noninteracting dynamic susceptibility of a single layer electron gas. $f_{\mathbf{k}}$ and $\epsilon_{\mathbf{k}} = \hbar^2 k^2/2m$ are the Fermi distribution function and single-particle energies, respectively. Making use of the explicit form of $\chi_0(q,\omega)$ in two-dimensions [12] the resulting plasmon modes are obtained to have the following dispersion relations

$$\omega_{\pm}(q) = q v_F \, \frac{B_{\pm}\sqrt{1+4/A_{\pm}}}{2} \tag{4}$$

where

$$B_{\pm} = C_{\pm}\frac{q^2}{q_{TF}k_F} + \frac{q}{k_F}\,, \tag{5}$$

$$A_{\pm} = 2C_{\pm}\frac{q^3}{q_{TF}k_F^2}\left(1 + \frac{qC_{\pm}}{2q_{TF}}\right), \tag{6}$$

and

$$C_{\pm} = \frac{1}{1 \pm e^{-qd}}\,. \tag{7}$$

In the above expressions $q_{TF} = 2/a_B$ denotes the Thomas-Fermi screening wave number. The full dispersion relations simplify in the long wavelength ($q \to 0$) limit to yield

$$\omega_+(q) \approx \sqrt{q q_{TF} v_F} \qquad \text{and} \qquad \omega_-(q) \approx q v_F \frac{1 + q_{TF}d}{\sqrt{1 + 2q_{TF}d}}\,, \tag{8}$$

for the in-phase and out-of phase modes, respectively. As noted previously, the out-of-phase plasmon mode within the RPA shows an acoustic behavior.

The experimental confirmation of these collective modes came recently from the inelastic Raman scattering measurements of Kainth *et al.* [4]. The main results of these investigations were that (i) RPA supplemented with Hubbard type exchange-correlation effects accounts for the observed dispersion relations, and (ii) the temperature dependence of the linewidth of the acoustic plasmon mode can be modeled within the same theoretical framework only partially (see below). It is to be noted that the layer densities in these experiments were such that $r_s \sim 1.3$ and therefore large deviations from the RPA are not expected.

The correlation effects in the double quantum-well system are most conveniently described by the local-field factors. The density-density response of a double-layer system has been calculated within the celebrated Singwi-Tosi-Sjölander-Land [13] (STLS) approach. STLS approximation improves upon the RPA by considering the local depletion of the charge density around any given electron. The short-range effects correlation effects neglected by the RPA are described by static (wave vector dependent) local-field factors $G_{ij}(q)$, $(i,\ j = 1,\ 2$ are layer indices) which follow from the assumption that the two-particle distribution function may be decoupled as a product of two one-particle distribution functions multiplied by the pair-correlation function. They enter the response functions through the replacement $V_{ij}(q) \to V_{ij}(q)[1 - G_{ij}(q)]$. The aforementioned decoupling scheme yields the following expressions for the local-field factors [8, 14]

$$G_{ij}(q) = -\frac{1}{n} \int \frac{d^2\mathbf{k}}{(2\pi)^2} \frac{V_{ij}(k)}{V_{ij}(q)} \frac{\mathbf{k}\cdot\mathbf{q}}{q^2} [S_{ij}(\mathbf{k}-\mathbf{q}) - \delta_{ij}] \tag{9}$$

where $S_{ij}(q)$ are the static structure factors, being on one hand the Fourier transform of pair-correlation functions and related on the other hand to the density-density response functions via the fluctuation-dissipation theorem. Collective modes calculated within the STLS scheme resemble qualitatively the RPA results except that they are lowered by the presence of the local-field factors. This follows from the fact that $G_{ij}(q) \sim q$ as $q \to 0$ and the out-of-phase plasmon mode retains its acoustic character. With regard to the experiments of Kainth *et al.* [4] which are at relatively high density STLS approximation gives plasmon dispersions close to the Hubbard approximation.

The damping of the plasmon modes is also of importance in determining the overall consistency of the various theories. Within the RPA and STLS approximation the plasmons are undamped except when they enter the particle-hole continuum (Landau damping). However, Kainth *et al.* [4] have measured temperature dependence of the damping of acoustic plasmons and found that they cannot be fully accounted by the static local-field approaches. This has led Tanatar and Davoudi [15] to develop the frequency dependent local-field factors for a double-layer system within the so-called quantum STLS (qSTLS) approach. Here, instead of the classical distribution functions the Wigner functions are used to derive explicit expressions for the frequency dependent $G_{ij}(q,\omega)$. The resulting damping rates calculated at $T = 0$ nevertheless agreed favorably with the observed values. The plasmon dispersions within the qSTLS were also similar to the static STLS results. Theoretical studies at lower electron densities indicate that plasmon energies of qSTLS approximation generally lie between the RPA and STLS results. To experimentally verify this, however, experiments at lower densities need to be performed. Recently, plasmon measurements in single layer electron systems at very low densities were accomplished [16]. Through an ingenious experimental setup plasmon dispersion at wave vectors $q \gtrsim k_F$ were recorded so that deviations from the RPA can be readily assessed. An interesting outcome is that the measured plasmon dispersion can be accounted for using a frequency dependent local-field factor, but not a purely static one [17]. It would then seem to be most interesting to explore a similar regime of q values in double-layer systems experimentally. Equally important would be to measure the plasmon dispersions at long wavelength ($q \to 0$) to discern the behavior of the out-of-phase plasmon mode.

In the quasi-localized charge (QLC) theory of Kalman and coworkers [10, 18, 19]. the ensuing out-of-phase plasmon dispersion has a completely different character. Rather than being acoustic in nature, at long-wavelength it exhibits a gap. The main difference between the STLS approach and that of QLC approximation is the way respective local-field factors are defined. In the case of QLC approximation the double-layer system approximately fulfills the third frequency moment sum rule [20]. The plasmon modes in the QLC approximation are obtained by solving [19]

$$1 - [V_{11}(q) \pm V_{12}(q) + D_{11}(q) \pm D_{12}(q)]\chi_0(q,\omega) = 0 \tag{10}$$

where

$$D_{11} = \frac{1}{n}\int \frac{d^2\mathbf{k}}{(2\pi)^2}\, V_{11}(k)\, \frac{(\mathbf{q}\cdot\mathbf{k})^2}{q^4}\, [S_{11}(\mathbf{k}-\mathbf{q}) - S_{11}(k)]$$
$$-\frac{1}{n}\int \frac{d^2\mathbf{k}}{(2\pi)^2}\, V_{12}(k)\, \frac{(\mathbf{q}\cdot\mathbf{k})^2}{q^4}\, S_{12}(q) \tag{11}$$

and

$$D_{12} = \frac{1}{n}\int \frac{d^2\mathbf{k}}{(2\pi)^2}\, V_{12}(k)\, \frac{(\mathbf{q}\cdot\mathbf{k})^2}{q^4}\, S_{12}(\mathbf{k}-\mathbf{q}) \tag{12}$$

related to the previously defined local-field factors by

$$G_{ij}(q) = -D_{ij}(q)/V_{11}(q).$$

The most interesting behavior of the local-field factors in the QLC approximation is that as $q \to 0$, $G_{ij} \sim 1/q$ which is in contrast to the linear in q behavior in the STLS approximation. The out-of-phase plasmon mode thus acquires a gap at long wavelengths where the gap energy depends on r_s and layer separation d.

Most recent work in this direction [19] uses the quantum Monte Carlo (QMC) data of static structure factors $S_{ij}(q)$ of a double-layer system to evaluate the local-field corrections. The gap in the out-of-phase plasmon persists down to small couplings r_s. Thus, the QLC theory does not go over to the RPA as $r_s \to 0$. Molecular dynamics (MD) simulations on these structures support the notion of a gapped excitation for the out-of-phase mode [21]. However, the MD method is classical by nature and may be missing some essential quantum mechanical ingredient in capturing the true behavior of collective excitations. QMC simulations, on the other hand, are geared more toward the static properties of quantum systems, thus give no direct information on the dynamical properties such as plasmon modes.

A critical comparison between the STLS and QLC theories with regard to the out-of-phase plasmon was elucidated by Ortner [22]. He has argued that the neglect of damping processes overestimates the role of correlations and the correct account of damping processes would lead to an absence of an energy gap in the acoustic plasmon dispersion. It is known that the damping due to disorder effects suppresses the plasmon dispersion so that it is conceivable that the RPA behavior is restored because of these competing effects. However, damping may also arise from multi-particle excitations and their effect on the collective modes in double-layer systems remains an open problem.

Lastly, we would like to point out to the recent developments in the study of the dynamical exchange-correlation kernels $f_{\rm xc}(q,\omega)$ of homogeneous electron systems which has also some bearing into the present problem. They are extensively used in the time dependent density functional theory and may be related to the familiar local-field factors through the relation

$$f_{{\rm xc},ij}(q,\omega) = -V(q)\, G_{ij}(q,\omega)\,. \tag{13}$$

Tosi and coworkers [23] and Qian and Vignale [24] in a series of papers have advanced our understanding of the frequency dependence of $f_{\rm xc}(q,\omega)$ by employing many-body methods and imposing known limiting forms for q and ω dependences. The exact asymptotic form (as $q \to 0$) of the exchange-correlation kernels in a double-layer system (in analogy to a spin resolved system considered by Qian and Vignale [24]) is given by

$$f_{{\rm xc},ij}(q,\omega) = \frac{A(\omega)}{q^2}\,\frac{(2\delta_{ij}-1)n^2}{4n_i n_j} + B_{ij}(\omega) + \mathcal{O}(q^2) \tag{14}$$

where $A(\omega)$ and $B_{ij}(\omega)$ are complex functions of frequency. They are determined from the known limiting behavior at high and low frequency. Comparison with the long wavelength forms of the local-field factors in the QLC approximation reveals the fact that in that approach $A(\omega)$ was treated as a constant (i.e. its value at $A(\infty)$ is used throughout). This may be the key to the prediction of QLC approximation that a gap exists for all densities and layer separations. Note that the frequency dependent local-field factors in the qSTLS approach do not reproduce the exact limiting form of $f_{xc,ij}(q, \omega)$. It would be most productive to explore these ideas to calculate the out-of-phase plasmon dispersion using the dynamic exchange-correlation kernels constructed for a double-layer system.

In summary, we have presented an overview of the present status of research on the many-body effects on collective excitation modes of a double-layer electron system. It appears that the nature of the out-of-phase plasmon mode in these structures is far from being settled. There are competing theoretical approaches predicting an acoustic or a gapped behavior at long wavelengths. We have briefly outlined how the respective dispersion relations are obtained in these theoretical schemes. We have also introduced the elements of recent developments in the exact exchange-correlation kernels which may be applied to double-layer systems. This may shed new light to the controversy over the out-of-phase plasmon in such structures. Needless to say there is great need in performing systematic experimental studies to measure the plasmon modes in double-layer systems with a view to probe the exchange-correlation effects.

ACKNOWLEDGMENTS. It is an immense pleasure to dedicate this paper to Professor Mario Tosi on the occasion of his 72nd birthday who has made many a contribution to the electron correlation effects. This work is supported by the Scientific and Technical Research Council of Turkey (TUBITAK) and the Turkish Academy of Sciences (TUBA). We thank INFM for its support and Professor F. Beltram for his hospitality in Pisa. We gratefully acknowledge fruitful discussions with Professor G. Vignale.

References

[1] C. B. HANNA, D. HAAS and J. C. DÍAZ-VÉLEZ, Phys. Rev. B **61**, 13 882 (2000). For an earlier review see T. ANDO, A. B. FOWLER and F. STERN, Rev. Mod. Phys. **54** (1981), 437.

[2] A. PINCZUK *et al.*, Phys. Rev. Lett. **56** (1986), 2092.

[3] G. FASOL, N. MESTERS, H. P. HUGHES, A. FISCHER and K. PLOOG, Phys. Rev. Lett. **56** (1986), 2517.

[4] D. S. Kainth *et al.*, Phys. Rev. B **57** (1998), 2065; J. Phys.: Condens. Matter **12** (2000), 439.
[5] A. S. Bhatti, D. Richards, H. P. Hughes and D. A. Ritchie, Phys. Rev. B **53** (1996), 11016.
[6] S. Das Sarma and A. Madhukar, Phys. Rev. B **23** (1981), 805; G. E. Santoro and G. F. Giuliani, *ibid.* **37** (1988), 937.
[7] C. Zhang and N. Tzoar, Phys. Rev. A **38** (1988), 5786.
[8] D. Neilson, L. Swierkowski, J. Szymanski and L. Liu, Phys. Rev. Lett. **71** (1993), 4035: L. Liu, L. Swierkowski, D. Neilson and J. Szymanski, Phys. Rev. B **53** (1996), 7923.
[9] A. Gold, Z. Phys. B **86** (1992), 193.
[10] B. Dong and X. L. Lei, J. Phys.: Condens. Matter **10** (1998), 7535.
[11] G. Kalman, V. Valtchinov and K. I. Golden, Phys. Rev. Lett. **82** (1999), 3124.
[12] F. Stern, Phys. Rev. Lett. **18** (1967), 546.
[13] K. S. Singwi, M. P. Tosi, R. H. Land and A. Sjölander, Phys. Rev. **176** (1968), 589.
[14] L. Zheng and A. H. MacDonald, Phys. Rev. B **49** (1994), 5522.
[15] B. Tanatar and B. Davoudi, Phys. Rev. B **63** (2001), 165328.
[16] C. F. Hirjibehedin, A. Pinczuk, B. S. Dennis, L. N. Pfeiffer and K. W. West, Phys. Rev. B **65** (2002), 161309.
[17] A. Yurtsever, V. Moldoveanu and B. Tanatar, Phys. Rev. B **67** (2003), 115308.
[18] K. I. Golden and G. J. Kalman, Phys. Plasmas **7** (2000), 14 and references therein.
[19] K. I. Golden, H. Mahassen, G. J. Kalman, G. Senatore and F. Rapisarda, preprint cond-mat/0410575 (unpublished).
[20] K. N. Pathak and P. Vashishta, Phys. Rev B **7** (1973), 3649.
[21] S. Ranganathan and R. E. Johnson, Phys. Rev. B **69** (2004), 085310.
[22] J. Ortner, Phys. Rev. B **59** (1999), 9870.
[23] R. Nifosi, S. Conti and M. P. Tosi, Phys. Rev. B **58** (1998), 12 758; S. Conti, R. Nifosi and M. P. Tosi, J. Phys.: Condens. Matter **9** (1997), L475.
[24] Z. Qian and G. Vignale, Phys. Rev. B **68** (2003), 195113; **65** (2003), 235121; Z. Qian, A. Constantinescu and G. Vignale, Phys. Rev. Lett. **90** (2003), 066402.

Theoretical approach for imaging currents in the integer quantum Hall effect

A. Cresti, G. Grosso, and G. Pastori Parravicini

The questions concerning where local currents flow in quantum Hall samples, which energy levels are involved, and why is the carrier transport so perfectly dissipationless, have received much attention and controversial interpretations. These problems are here addressed within the framework of the nonequilibrium Green's function formalism, which provides very accurate space and energy resolved current distributions in two-dimensional mesoscopic systems.

1. Introduction

On the eve of the world year of physics, it is an easy forecast that the Hall effects will receive a lot of attention from the scientific community. In 1879 Hall discovered the effect that bears his name [1] by measuring the electrical potential acting perpendicularly to a current flow in a magnetic field. The far reaching relevance of the discovery can be better appreciated considering that in Hall's time the electron was not yet identified. About one century later, when the classical Hall effect had reached a mature age and routine applications, unheralded arrived the discovery of the integer quantum Hall effect [2]; this was soon followed by the discovery of the fractional Hall effect [3], bringing the evidence of carriers with fractional electron charge [4].

The integer quantum Hall effect occurs in two-dimensional electron gases in strong magnetic fields and very low temperatures. The most spectacular feature is the occurrence of plateaus in the Hall resistance at the exact values $R_H = h/(e^2 i)$, with i integer number, and concomitant vanishing of the longitudinal resistance. The extraordinary experimental accuracy of the von Klitzing constant is about 10^{-10}, and the value $h/e^2 = 25812.807\Omega$ is the presently adopted standard of resistance [5].

The quantization aspects of the integer Hall effect were explained on the basis of the gauge arguments of Laughlin [6] and Halperin [7] for closed cylindrical or annular topologies, respectively. However the gauge arguments, because of their generality and topological assumptions, remain elusive for what concerns the spatial and energy distribution of currents in the actual experiments, usually performed on open planar bars (or also Corbino disks). Elusive remain also standard transport measurements, since the observed resistances have universal values, independent from the microscopic properties of the system. Thus it is not surprising that, twenty five years after the experimental discovery, there is still in the literature considerable debate and disagreement on the problem of current distribution in Hall devices, with a number of authors believing that current flows prevalently along edge-states channels, and others believing that current flows throughout the whole device at or near bulk Landau levels. Besides the issue of current profiles, at stake is the much more ambitious aim of the microscopic interpretation of the quantum Hall effect.

On the last years ingenious imaging techniques [8–10] and theoretical investigations [11–18] have been pursued for understanding how the microscopic currents are partitioned between edge and bulk channels in quantum Hall regime. The purpose of this paper is to illustrate a theoretical scheme for imaging of currents based on the tight-binding Keldysh Green's function formalism [17,18], which is a most valuable tool for accurate description of energy and space resolved persistent and transport currents in quantum magneto-transport.

2. General expression of current profiles

The tight-binding approach is one of the most useful tools for investigating equilibrium and nonequilibrium properties of solids. The implementation of the basic concepts of the nonequilibrium Keldysh formalism [19] in tight-binding systems has been pioneered by Caroli *et al.* [20] and is now well established in the literature [21,22]. A distinctive feature of the tight-binding framework is its flexibility to model electronic states in different classes of materials, at the desired degree of accuracy and sophistication. Different topologies of mesoscopic systems, can be studied from planar two-dimensional systems, to cylindrical nanotubes, Bethe lattices (for dendrimers), Aharonov-Bohm loop geometries. However the most remarkable advantage of the use of this method is the treatment of nonequilibrium magneto-transport. In the tight-binding representation, magnetic fields are accurately taken into account via phase factors in the off-diagonal matrix elements; moreover, the crossover from the orthogonal class symmetry to the unitary class symmetry or to the extreme chiral regime can be followed with accuracy and simplicity [17,18].

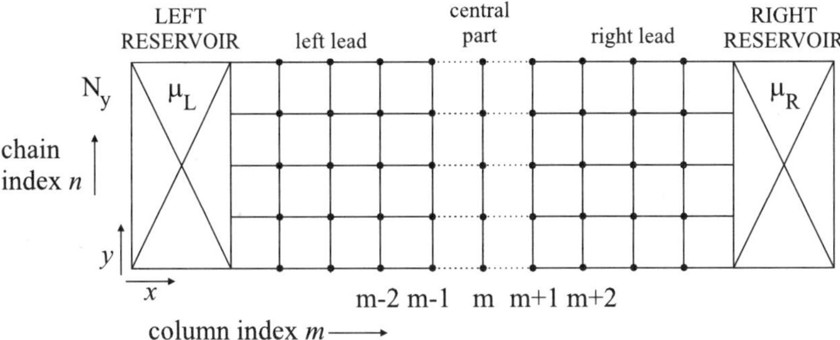

Figure 1. Schematic representation of a two-dimensional device for the calculation of current distributions. The central device is restricted to the single column m. The left part and the right part of the structure are in contact with the particle reservoirs kept at the chemical potentials μ_L and μ_R respectively.

The mesoscopic device we consider is schematically represented by a square lattice of parameter a, with a single orbital per site, diagonal energies E_{mn} and nearest neighbor interactions ($t < 0$), as shown in Figure 1. The open electronic system with rectangular shape is arbitrarily extended in the x-direction, width $W = (N_y - 1)a$ in the y-direction, threaded by a uniform perpendicular magnetic field along the z-direction.

The electron Hamiltonian in the site representation is

$$H = \sum_{mn} E_{mn} c^{\dagger}_{mn} c_{mn} + t \sum_{mn} [c^{\dagger}_{m,n+1} c_{mn} + c^{\dagger}_{mn} c_{m,n+1}]$$

$$+ t \sum_{mn} [e^{-i2\pi\beta n} c^{\dagger}_{mn} c_{m+1,n} + e^{i2\pi\beta n} c^{\dagger}_{m+1,n} c_{mn}] \qquad (2.1)$$

where $c^{\dagger}_{mn}$, c_{mn} denote creation and annihilation operators corresponding to the orbital ϕ_{mn} centered at the site (ma, na); the Peierls phase is given by $\beta = \Phi_p(B)/\Phi_0$ where $\Phi_p(B) = Ba^2$ is the magnetic flux through the elementary plaquette and $\Phi_0 = hc/e$ is the flux quantum. For an ideal defect-free sample, the diagonal matrix elements E_{mn} are taken equal. Different types of impurities and disorder can be represented by appropriate choice of site independent terms; in the Anderson model of disorder, the diagonal elements E_{mn} are randomly chosen in the energy range $[-w, w]$.

Utilizing the Keldysh formalism [19], we have elaborated formal expressions of the spatial distribution of microscopic currents flowing through each bond of the lattice. With the assumption $\mu_L \leq \mu_R$, carriers are injected from the right reservoir to the left reservoir (while current flows in the opposite direction). The microscopic current between the $(m-1)$-th column and m-th column takes the expression [17]

$$I_{m-1\,n,mn} = I^{(eq)}_{m-1\,n,mn} + I^{(neq)}_{m-1\,n,mn} , \tag{2.2}$$

where the *equilibrium* (or *persistent*) component is

$$I^{(eq)}_{m-1\,n,mn} = \frac{2(-e)}{h} \int dE\,(-f_L) \tag{2.3}$$

$$2\Re <\phi_{mn}|G^R \Sigma^{R(left)} - \Sigma^{R(left)} G^R|\phi_{mn}> ,$$

and the *nonequilibrium* (or *transport* or *applied*) component is

$$I^{(neq)}_{m-1\,n,mn} = \frac{2(-e)}{h} \int dE\,(f_L - f_R) \tag{2.4}$$

$$2\Im <\phi_{mn}|G^R \Gamma^{(right)} G^A \Sigma^{A(left)}|\phi_{mn}> .$$

In the above equations G, Σ, Γ denote Green's functions, self-energies and linewidth matrices of rank N_y; f_L and f_R are the Fermi-Dirac occupation functions for the left and the right lead respectively. The states with energy below μ_L are fully occupied, and contribute to persistent currents (although the net flow between electrodes vanishes); the states of energy between the two chemical potentials contribute to the net transport current. With an eye at the integrand in Eq. (2.3) and Eq. (2.4), it is convenient to consider the equilibrium and the transport current per unit energy given by

$$i^{(eq)}_{m-1\,n,mn} = i_0 2\Re <\phi_{mn}|G^R \Sigma^{R(left)} - \Sigma^{R(left)} G^R|\phi_{mn}> \tag{2.5}$$

and

$$i^{(neq)}_{m-1\,n,mn} = i_0 2\Im <\phi_{mn}|G^R \Gamma^{(right)} G^A \Sigma^{A(left)}|\phi_{mn}> . \tag{2.6}$$

The quantity

$$i_0 = \frac{2e}{h} = \frac{2e^2}{eh} = 2 \times \frac{1}{25812.807} \frac{1}{\mathrm{e(V/A)}} = 7.748092 \times 10^{-5}\mathrm{A/eV}$$

represents the natural unit of energy resolved current. The actual transport current through any given bond is obtained by integrating the energy resolved current between the two chemical potentials μ_L and μ_R; for the actual persistent current the integration extends to all energies below μ_L.

3. Chiral regime, conductance quantization and dissipationless edge channels

In the absence of magnetic fields the Hamiltonian (2.1) is invariant under time reversal symmetry and the persistent currents (2.3) rigorously vanish. In the presence of magnetic fields, the time reversal symmetry is broken and persistent currents flow in the device even if kept at thermal equilibrium with unbiased reservoirs at the same chemical potential $\mu_L = \mu_R$. Thus chiral regime, i.e. complete spatial separation of left and right moving carriers, becomes possible in sufficiently strong magnetic fields. In fact, the quantum Hamiltonian (2.1) automatically takes into account the driving Lorentz force toward the lower (upper) edge for left (right) moving carriers. Within the tight-binding Keldysh formalism, it is straightforward to examine the occurrence of the chiral regime in the ideal or disordered sample described by the model Hamiltonian (2.1). At the (arbitrarily) chosen column m one considers the matrix of rank N_y given by

$$M(E) = G^R(E)\Gamma^{(right)}(E)G^A(E)\Gamma^{(left)}(E).$$

It is possible to demonstrate that the eigenvalues m_i of the matrix M are confined in the interval [0,1]; when these eigenvalues are either zero or one the transport current is noiseless and the chiral regime is at work. Since the trace of M gives the total currents per unit energy in units of i_0 (or equivalently the total differential conductance in units of $\sigma_0 = ei_0$), it follows that conductance is perfectly quantized in the chiral regime.

The chiral regime also entails the semilocal character of the retarded (or advanced) Green's functions, because of the gapped and gapless density-of-states in the interior and at the periphery of the sample, respectively. The clockwise semilocal character of the retarded Green's function for magnetic field in the positive z-direction, is related to the Lorentz force included in the electron Hamiltonian. The behavior is evident from the numerical results reported in Figure 2 for the case of a perfect quantum wire, constituted by 101 chains, threaded by a uniform magnetic field. This property is robust against disorder, and the retarded Green's function remains semilocal in the backward direction in the lower edge and in the forward direction in the upper edge also in the presence of a moderate disorder.

We can now show that the semilocal property of the Green's function entails perfect equilibration of edge potentials and hence dissipationless carrier transport. Consider in fact an arbitrary part of the two-dimensional device from column 1 to column M. The lesser Green's function within

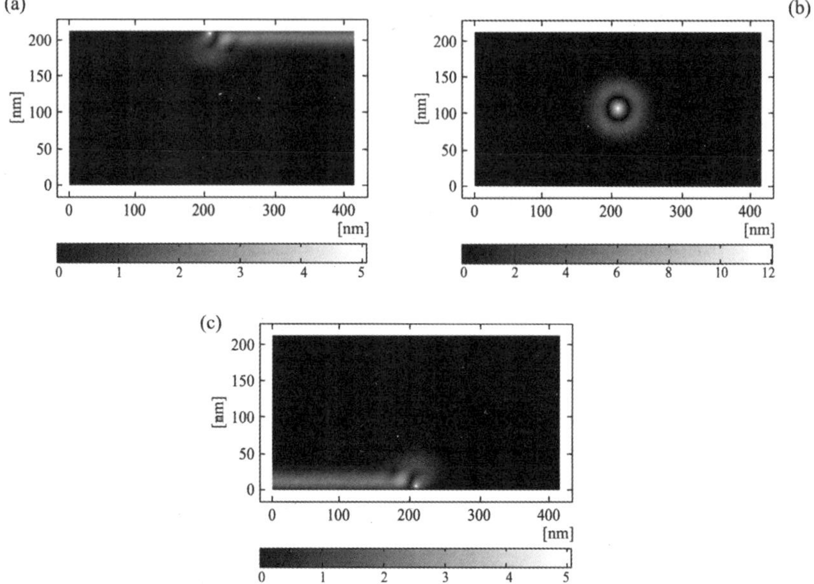

Figure 2. Modulus of the retarded Green's function, $|G^R(E)|$, for the propagation from a chosen initial site to other sites, at the energy $E = 12$ meV. The defect-free sample, with magnetic energy $\hbar\omega_c^* = 8.2$ meV is described in the text. (a) The starting site is in the upper edge and electron propagation can occur only from the left to the right. (b) The starting site is in the bulk region, the Green's function rapidly decreases to zero and no propagation is possible. (c) The starting site is in the lower edge and electron propagation can occur only from the right to the left. The unit on the grey scale is eV^{-1}.

this region is determined by the kinetic equation

$$G^< = G^R \Sigma^{<(leads)} G^A = G^R \Sigma^{<(left)} G^A + G^R \Sigma^{<(right)} G^A.$$

The clockwise and counter-clockwise semilocal character of the retarded and advanced Green's functions, entails the equality

$$G^< = \begin{cases} G^R \Sigma^{<(left)} G^A & \text{upper region} \\ G^R \Sigma^{<(right)} G^A & \text{lower region.} \end{cases}$$

This demonstrates the perfect thermodynamic equilibration of the upper and lower conductive lanes at the chemical potentials μ_L and μ_R of the left reservoir and right reservoir respectively.

The numerical calculations given below show that the currents in Hall devices flow in the bulk or edge regions of the sample depending on specific situations, such as carrier density, geometry, Hall potential, impurities etc.

4. Edge and bulk currents in model Hall devices

The sample used to verify numerically the above concepts is a quantum wire, arbitrarily extended in the x-direction, width equal to hundred unit cells, threaded by a perpendicular field $B = 5$ Tesla. By choosing $t = -0.125$ eV in the model Hamiltonian (2.1) and effective mass of the two-dimensional electron gas $m^* = 0.067\, m_e$, typical of GaAs-AlGaAs heterosctructure, we have for the lattice parameter $a = 2.11$ nm. The magnetic energy is $\hbar\omega^* = 8.2$ meV, the magnetic length is $\ell_0 = (\hbar c/eB)^{(1/2)} \approx 10$ nm, and the Peierls phase is $\beta(B) = \Phi(B)/\Phi_0 \approx 1/100$.

This mesoscopic system is studied in the ideal defect-free situation and in the presence of Anderson type disorder, in the presence or absence of built in Hall fields. The energy band structure shows the presence of flat Landau levels, localized in the bulk region, and edge levels, with opposite group velocity at the boundaries of the sample.

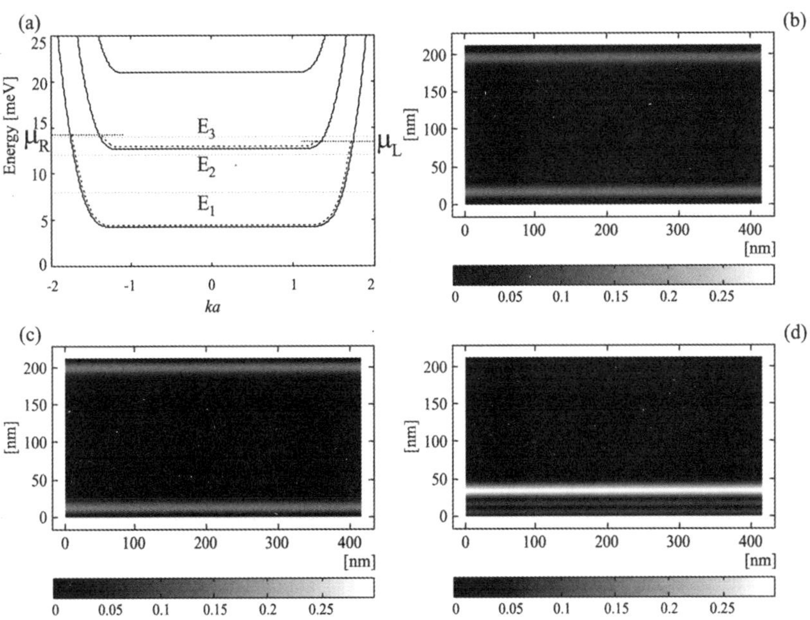

Figure 3. (a) Band structure of the perfect system described in the text, threaded by a magnetic field of 5 T. The electron occupation, schematically indicated by dots, refers to the case $\mu_R \approx \mu_L \approx 14$ meV; (b) persistent current per unit energy at the energy $E_1 = 8$ meV; (c) persistent current per unit energy at the energy $E_2 = 12$ meV; (d) transport current per unit energy at the energy $E_3 = 14$ meV. The unit on the grey scale is i_0.

In Figure 3 we show the current per unit energy at several chosen energies in the defect-free sample described in the text. The chosen value of

μ_L=14 meV is between the second and the third Landau level, and μ_R is chosen above it by an infinitesimal quantity. In Figures 3(b) and 3(c) it is shown the spatial distribution of energy-resolved persistent currents at the $E_1 = 8$ meV and $E_2 = 12$ meV. In Figures 3(b) and 3(c) persistent currents flow on the periphery of the sample in opposite directions (in the figure only the absolute values of the bond currents are reported). At $E_3 = \mu_R$, the transport current flows only on the lower edge. Persistent currents per unit energy in each of the two edges, as well as the transport currents per unit energy in Figure 3(d) are perfectly quantized because of chirality.

In Figure 4 we show the current per unit energy in the defect free sample, assuming the values $\mu_L = 14$ meV and $\mu_R = 18$ meV, and including the built in Hall potential of 4 mV. In Figures 4(b), 4(c) and 4(d) are reported the spatial distributions of currents per unit energy at the same energies considered in the previous Figure 3. It is seen that the persistent currents in Figure 4(c) change significantly both in intensity and space distribution; the transport currents remain near the lower edge of the sample and are perfectly quantized to 2 i_0.

The Figure 5(a) should be directly compared with Figure 4(c), the only difference being the introduction of an Anderson type disorder of 20 meV strength in the region extending from 100 to 300 nm. Figures 5(b), 5(c), 5(d) report sections of Figure 5(a) at $x = 100, 150, 200$ nm respectively. The presence of disorder influences the spatial distribution of currents. The influence also occurs for the transport current, which however remains perfectly quantized.

5. Conclusions

In this paper we have considered a theoretical approach for imaging transport currents and persistent currents in two-dimensional Hall devices, based on the Keldysh nonequilibrium Green's function formalism and the tight-binding representation of the electronic states. From the formal structure of the basic equations (2.3) and (2.4) of Section II, the following general features can be drawn: (i) both edge states and bulk states contribute to the local currents in Hall devices, although only the states in the energy range between the two chemical potentials contribute to the observed lead-to-lead transport current, (ii) in the chiral regime the currents per unit energy are quantized, (iii) the equilibration of conductance channels is related to the clockwise or counter-clockwise semilocal character of the retarded and advanced Greens' functions in magnetic fields, (iv) the stabilization mechanism of Hall plateaus is due to the peculiar behavior of scatterers that, in the chiral regime, do not change the total

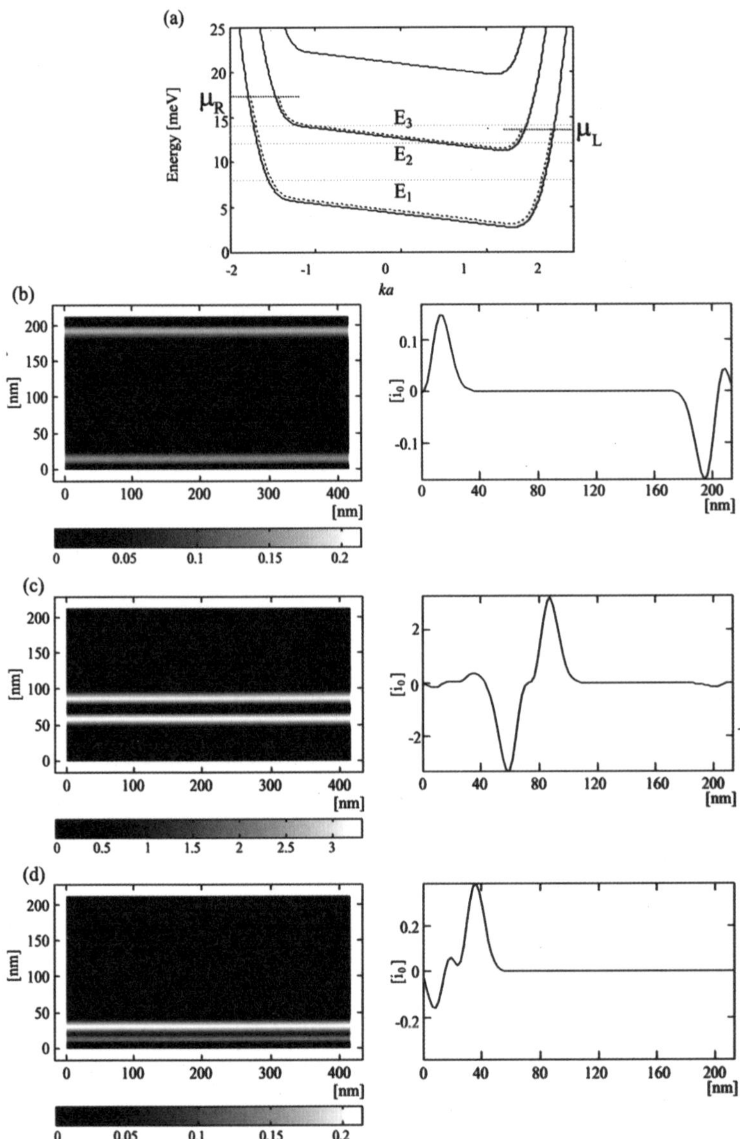

Figure 4. (a) Band structure of the perfect wire described in the text; the electron occupation, schematically indicated by dots, refers to the case $\mu_L = 14$ meV and $\mu_R = 18$ meV; (b) persistent current per unit energy at the energy $E_1 = 8$ meV; (c) persistent current per unit energy at the energy $E_2 = 12$ meV; (d) transport current per unit energy at the energy $E_3 = 14$ meV. The unit on the grey scale is i_0 (and only absolute values of currents are reported). In order to clarify the sign of the local current direction, a plot of the current through a generic section is reported near each map.

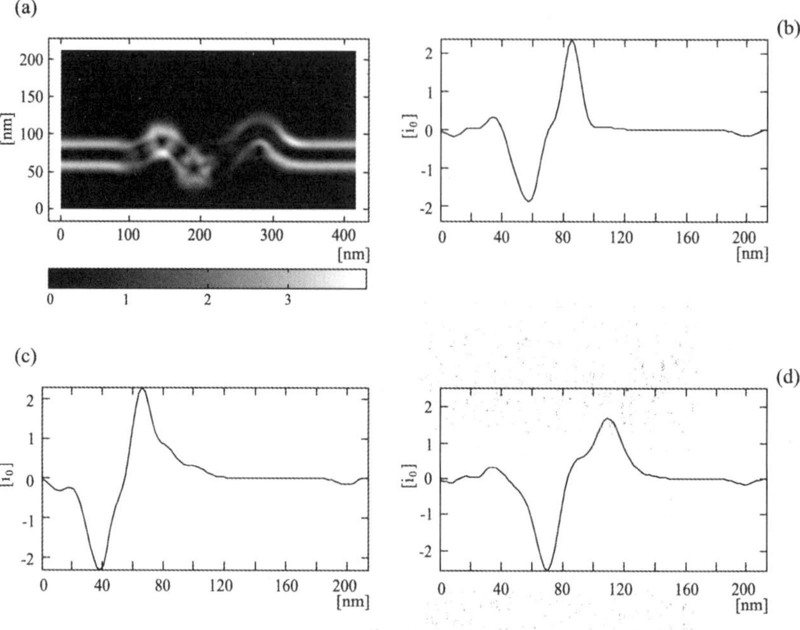

Figure 5. (a) Persistent current per unit energy at the energy $E_1 = 12$ meV with $\mu_L = 14$ meV and $\mu_R = 18$ meV when an Anderson type disorder of 20 meV is added in the central zone; the unit on the grey scale is $i_0 \times 10^{-3}$; (b) section at $x = 100$ nm; (c) section at $x = 150$ nm; (d) section at $x = 200$ nm.

net current flowing through the device, no matter how strongly they modify the local currents.

Many other appealing aspects of the quantum Hall effect have been investigated in the literature, from an experimental and theoretical point of view. The nonequilibrium Keldysh tight-binding formalism for the calculation of space and energy resolved current profiles, emerges as precious tool in this restless area of research.

ACKNOWLEDGMENTS. This paper has been written in the occasion of the meeting held at Scuola Normale Superiore to honor Prof. Mario Tosi. It is a pleasure to express our deep appreciation for his outstanding professional activity, that has benefited not only direct coworkers but also so many friends and colleagues.

References

[1] E. H. HALL, Am. J. of Math. **2** (1879), 287.
[2] K. VON KLITZING, G. DORDA and M. PEPPER, Phys. Rev. Lett. **45** (1980), 494.
[3] D. C. TSUI, H. L. STORMER and A. C. GOSSARD, Phys. Rev. Lett. **48** (1982), 1559.
[4] R. B. LAUGHLIN, Phys. Rev. Lett. **18** (1983), 1395.
[5] B. JECKELMANN and B. JEANNERET, Rep. Prog. Phys. **64** (2001), 1603.
[6] R. B. LAUGHLIN, Phys. Rev. B **23** (1981), 5632.
[7] B. I. HALPERIN, Phys. Rev. B **25** (1982), 2185.
[8] B. J. LEROY, J. Phys.: Condens. Matter **15** (2003), R1835.
[9] S. CHAKRABORTY, I. J. MAASILTA, S. H. TESSMER and M. R. MELLOCH, Phys. Rev. B **69** (2004), 073308.
[10] J. HUELS, J. WEIS, J. SMET, K. VON KLITZING and Z. R. WASILEWSKY, Phys. Rev. B **69** (2004), 085319.
[11] P. STREDA, J. KUCERA and A. H. MACDONALD, Phys. Rev. Lett. **59** (1987), 1973.
[12] D. B. CHKLOVSKII, B.I. SHKLOVSKII and L. I. GLAZMAN, Phys. Rev. B **46** (1992), 4026.
[13] M. BÜTTIKER, Phys. Rev. B **38** (1988), 9375.
[14] T. CHRISTEN and M. BÜTTIKER, Phys. Rev. B **53** (1996), 2046.
[15] S. KOMIYAMA and H. HIRAY, Phys. Rev. B **54** (1996), 2067.
[16] T. TANIGUCHI, Physics Letters A **279** (2001), 81.
[17] A. CRESTI, R. FARCHIONI, G. GROSSO and G. PASTORI PARRAVICINI, Phys. Rev. B **68** (2003), 075306; J. Phys.: Condens. Matter, **15** (2003), L377.
[18] A. CRESTI, G. GROSSO and G. PASTORI PARRAVICINI, Phys. Rev. B **69** (2004), 233313.
[19] L. V. KELDYSH, Soviet Physics JEPT **20** (1965), 1018.
[20] C. CAROLI, R. COMBESCOT, P. NOZIERES and D. SAINT-JAMES, J. Phys. C.: Solid State Phys. **4** (1971), 916; J. Phys. C.: Solid State Phys. **5** (1972), 21.
[21] D. K. FERRY and S. M. GOODNICK, "Transport in Nanostructures", Cambridge, Cambridge University Press, 1997.
[22] R. LAKE, G. KLIMECK, R. C. BOWEN and D. JOVANOVIC, J. Appl. Phys. **81** (1997), 7845.

The Bose-Hubbard model
- from Josephson Arrays to optical lattices -

R. Fazio

1. Introduction

The physics of strongly interacting physical systems has been under scrutiny of the physics community since many decades. Despite all the experimental and theoretical efforts made so far, there are still several important questions that are lacking of a complete understanding. The need to explore the physics of these systems has stimulated the realization of artificial structures that could reproduce accurately the properties of strongly interacting quantum systems and that could allow to access the various regimes by a tuning of some externally controlled parameters.

In this note I would like to briefly review the very succesful story of "home-made" strongly interacting bosons. The first experiments were realized with Josephson Junction Arrays (JJAs) in the submicron regime at the University of Delft in the middle of the '80s. Since then quantum-JJAs have been extensively studied both experimentally and theoretically. Many intereasting properties of the phase diagram and of the quantum dynamics of the topological excitations has been uncovered so far. Probably the most notable experimental finding is the discovery of a zero temperature superconductor-insulator phase transition. The vast body of knowledge accumulated with the studies of JJAs has found, in the middle of '90s, a natural development in the area of ultracold quantum gases [1]. It was first proposed theoretically and the experimentally verified that bosons loaded in an optical lattice could also undergo a superfluid-insulator phase transition at zero temperature. As I will review briefly below, the superconductor-insulator observed in JJAs and the superfluid-insulator in optical lattices transition have the same origin. They stem from the interplay between the tendency to global coherence, due to the hopping of bosonic particles, with the localization induced by the strong local repulsion. However, despite the strong similarities, these two model systems allow for the investigation of different observables. Josephson arrays are studied by means of transport measurements while a free expansion once the trap is realesed or scattering of light are the most common tools used in optical lattices.

This paper is organized as follows, I will first introduced the Bose-Hubbard model, a paradigm model for studying strongly interacting bosons and discuss its phase diagram. I will then give a very brief introduction to Josephson arrays and optical lattices and show how to describe those systems by means of the Bose-Hubbard model.

It is a great pleasure to write a contribution to celebrate Mario Tosi's birthday. He made many important contributions in several different areas of condensed matter physics. His constant interest and attention to the study of new physical problems is of great example for all of us.

2. The Bose-Hubbard model

The Bose-Hubbard model proposed for the first time by Fisher *et al.* [2] is defined as

$$H = \frac{1}{2}\sum_{ij} n_i U_{ij} n_j - \mu \sum_i n_i - \frac{t}{2}\sum_{\langle ij\rangle} b_i^\dagger b_j + \text{h.c.} \qquad (2.1)$$

Here, $b_i^\dagger$, b_i are the creation and annihilation operators for bosons in the i-th site and $n_i = b_i^\dagger b_i$ is the number operator. The coupling U_{ij} describes the interaction between bosons, μ is the chemical potential, and t the hopping matrix element. The range and the nature of the boson interaction depends on the system considered. It is on-site in the case of optical lattices while it may be long range in the case of Josephson junction arrays. The $< ... >$ brackets mean that the sum is restricted to nearest neighbors.

A qualitative understanding of the zero temperature phase diagram can be obtained by considering the two limiting cases in which one of the two coupling energies (U or t) is largest. If the hopping term is dominant, the lowest energy state ia a condensate of bosons delocalized through the whole lattice. If, on the contrary, the interaction energy is dominant, each site has a well defined number of bosons in the ground state and, in order to put an extra boson on a given site, one has to overcome an energy gap of the order of U, the system is therefore a Mott insulator. There is a critical value of the ratio t/U below which, even at zero temperature, the system becomes insulating. The transition to the insulating state is due to the strong local repulsion which prevent the establishing of global coherence.

For $t = 0$ the Mott gap E_G (I consider for simplicity the case of on-site repulsion $U_{ij} = U\delta_{ij}$) is given by

$$E_G = U(2n+1) - \mu \quad .$$

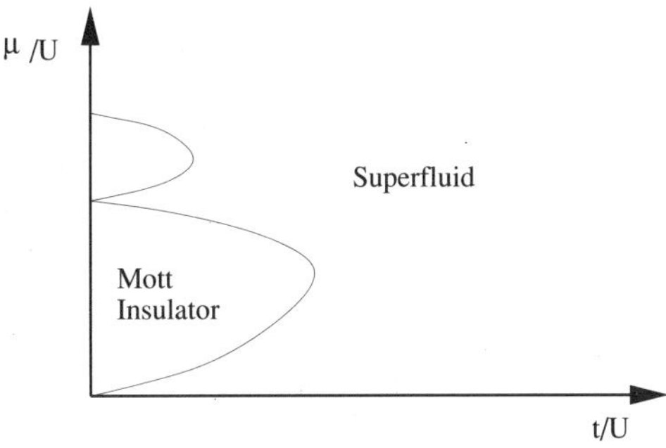

Figure 1. Zero temperature phase diagram of the Bose-Hubbard model. Inside the lobes the system is in a Mott insulator state. The different lobes (only the first two are drawn in the figure) have different fillings.

It is evident that at certain values of the chemical potential, the gap shrinks to zero. For these values the energy to add (or remove) a particle in a given site is zero. This implies that local repulsion is unable to suppress the condensation and in the phase diagram the superfluid phase extends to arbitrary small values of the hopping coupling.

A sketch of the phase diagram is depicted in Figure 1. It consists in a series of Mott-insulating lobes. The nature of the superfluid-insulator transition depends on the the position on the lobe at which the boundary is crossed. At the tip of the lobe the universiality class is that of a $d + 1$ classical XY model. The dynamical critical exponent $z = 1$. Away from the tip, where the particle-hole symmetry is broken, the dynamical critical exponent $z = 2$.

The Bose-Hubbard model was succesfully applied to Josephson arrays, granular superconductors, ultra-thin amorphus films and more recently to optical lattices. I will discuss only the case of arrays and optical lattices where disorder is virtually absent or it can be introduced in a controlled way.

3. Josephson Junction Arrays

A Josephson array consists of a regular network of superconducting islands weakly coupled by tunnel junctions. The first artificially fabricated JJAs were realized twenty years ago at IBM [3] as part of their effort to develop an electronics based on superconducting devices. Soon after their realization Josephson arrays were intensively investigated to explore a wealth of classical phenomena [4–6]. Josephson arrays proved to be an ideal model system in which classical phase transitions, frustration effects, classical vortex dynamics, non-linear dynamics and chaos could be studied in a controlled way.

Josephson arrays in the quantum regime were first realized in Delft in the group of J.E. Mooij [8]. The possibility to enter this regime was due to the impressive progresses in lythography that allowed to realize submicron junctions with a high (of the order of few kΩ) tunneling resistance. Array's parameters (associated to the shape of the islands, the thickness of the oxide barrier,...) can be made uniform across the whole array. The largest samples realized consist of about 10^4 junctions.

What is the appropriate model to describe a JJA? The coupling strength between adjacent islands is determined by the Josephson energy E_J. Quantum effects in Josephson arrays come into play whenever the charging energy (associated with non-neutral charge configurations of the islands) is comparable with the Josephson coupling (the physics associated with charging effects in single normal and superconducting junctions has been reviewed in Refs. [9]). In addition, as mentioned before, the junction resistance should be of the order of (or larger than) the quantum of resistance $R_Q = h/4e^2$ The electrostatic energy can be determined once the capacitance matrix C_{ij} and the gate voltages (if present) are known. Generally one only considers the junction capacitance C and the capacitance to the ground C_0. In the case of square lattices the capacitance matrix has the form $C_{ii} = C_0 + 4C$, $C_{ij} = -C$ (if i, j nearest neighbors) and zero in all other cases. Consequently the charging energy (for two charges placed in islands i and j of coordinates $\mathbf{r}_i$ and $\mathbf{r}_j$ respectively) is given by

$$U_{ij} = \frac{e^2}{2} \int \frac{d\mathbf{k}}{4\pi^2} \frac{e^{i\mathbf{k}\cdot(\mathbf{r}_i - \mathbf{r}_j)}}{C_0 + 2C(1 - \cos k_x) + 2C(1 - \cos k_y)} \quad , \tag{3.2}$$

The charging interaction increases logarithmically up to distances of the order of the screening length $\lambda \sim \sqrt{C/C_0}$ and then dies out exponentially. A characteristic energy scale tha characterizes the strength of electrostatic interaction is given by $E_C = e^2/2C$.

The previous considerations leads to the following Hamiltonian descibing Cooper pair tunneling in superconducting quantum networks. This model is frequently called the Quantum Phase Model and is its most general form it is given by:

$$H = \frac{1}{2} \sum_{i,j} (q_i - q_\text{x})\, U_{ij}\, (q_j - q_\text{x}) - E_J \sum_{<i,j>} \cos\left(\phi_i - \phi_j\right) \ . \tag{3.3}$$

The first term in the Hamiltonian is the charging energy; the second is due to the Josephson tunneling, $Q_i = 2eq_i$ is the net charge on the i-

th island. Quantum mechanics enters through the commutator between charge and phase operators

$$[q_i, e^{i\phi_j}] = \delta_{ij} e^{i\phi_j}$$

An external gate voltage V_x gives the contribution to the energy via the induced charge $q_x = \sum_j C_{ij} V_x / 2e$.

The two contributions in the Hamiltonian of Eq. (3.3) favor different types of ground states. The Josephson energy tends to establish phase coherence which can be achieved if supercurrents flow through the array. On the other hand the charging energy favors charge localization on each island and therefore tends to suppress superconducting coherence. This interplay becomes evident if one recalls the Josephson relation (which here can be obtained at the operator level by calculating the Heisenberg equation of motion for the phase)

$$\frac{d\phi_i}{dt} = \frac{2e}{\hbar} V_i = \frac{2e}{\hbar} C_{ij}^{-1} Q_j \tag{3.4}$$

A constant (in time) charge on the islands implies strong fluctuations in the phases. On the other hand phase coherence leads to strong fluctuations in the charge.

The properties of Josephson arrays in the quantum regime are reviewed in [10]. Here I simply recall some characteristics of the superconductor-insulator transition. The transition is measured by looking at the behaviour of the resistance as a function of temperature for different values of the ratio E_J/E_C. A schematic plot is given in Figure 2. Those sample for which the Josephson coupling is dominant undergo a transition at a finite temperature to a superconducting state. Below this critical temperature the array is globally coherent. However on lowering the ratio E_J/E_C for some critical value the resistance increases on lowering the temperature. This upturn of the resistance signals the existence of an insulator state at zero temperature. The striking fact is that the whole array is insulating despite the fact that each island is still superconducting!

The connection between the Bose-Hubbard model and the Quantum Phase Model is easily seen by writing the field b_i in terms of its amplitude and phase and by subsequently approximating the amplitude by its average (the mapping becomes more accurate as the average number of bosons per sites increases). This procedure leads to the identification

4. Optical Lattices

Following the work of Jaksch *et al.* [11], optical lattices have been suggested as concrete realization of the Bose-Hubbard model. The experimental test of the Mott insulator-superfluid transition by Greiner *et al.*

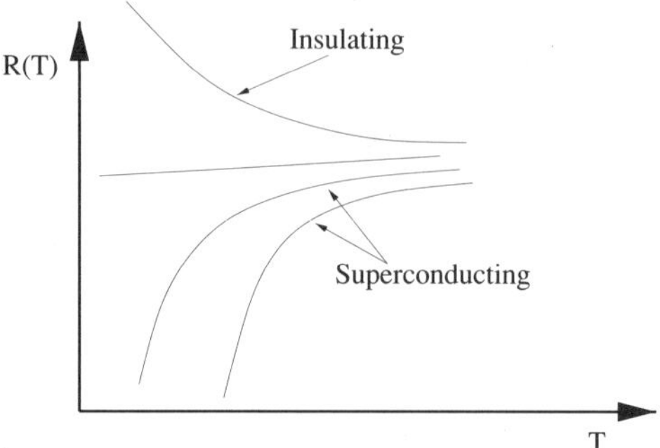

Figure 2. A schematic plot of the resistance of a Josephson array as a function of temperature for different samples. At the critical value of E_J/E_C the system becomes insulating

Bose-Hubbard		Josephson array
b_i	$\longrightarrow$	$e^{-i\phi_i}$
$< n > t$	$\longrightarrow$	E_J
n_i	$\longrightarrow$	q_i
μ	$\longrightarrow$	$\sum_j U_{ij} q_{x,j}$

[12] has paved the way to study strongly correlated phenomena in trapped cold atomic gases. An optical lattice can be realized by using counter-propagating laser fields. At large detuning, they provide a conservative potential of the form (I consider here the two-dimensional case)

$$V_p(\mathbf{r}) = V_0(\sin^2 ax + \sin^2 ay)$$

where V_0, the strength of periodic potential is controlled by the intensity of the lasers whereas the lattice constant a by means of their wavelenght.

In the following I briefly recall the leading steps, following the original derivation in the paper by Jacksh *et al.*, that lead to the mapping onto the Bose-Hubbard model. A system of interacting bosons which is subject to a periodic potential $V_p(\mathbf{r})$ and to a trapping potential $V_{ext}(\mathbf{r})$ and interacting via a potential $V_{int}(|\mathbf{r}-\mathbf{r}'|)$ is described by the Hamiltonian

$$\mathcal{H}=\int d^2\mathbf{r}\, \hat{\Psi}^\dagger(\mathbf{r}) \left[\frac{\mathbf{p}^2}{2M} + V_p(\mathbf{r}) + V_{ext}(\mathbf{r})\right] \hat{\Psi}(\mathbf{r}) \\ +\frac{1}{2}\int d^2\mathbf{r} \int d^2\mathbf{r}'\, \hat{\Psi}^\dagger(\mathbf{r})\hat{\Psi}^\dagger(\mathbf{r}')V_{int}(|\mathbf{r}-\mathbf{r}'|)\hat{\Psi}(\mathbf{r}')\hat{\Psi}(\mathbf{r})\,. \quad (4.5)$$

Here $\hat{\Psi}(\mathbf{r})$ and $\hat{\Psi}^\dagger(\mathbf{r})$ are field operators satisfying the commutation relations $[\hat{\Psi}(\mathbf{r}), \hat{\Psi}^\dagger(\mathbf{r}')] = \delta^{(3)}(\mathbf{r}-\mathbf{r}')$.

It is convenient to decompose the field operator $\hat{\Psi}(\mathbf{r})$ in the following Wannier basis,

$$\hat{\Psi}(\mathbf{r}) = \sum_{n,\mathbf{r}_i} w_n(\mathbf{r})\hat{b}_n(\mathbf{r}_i)\,, \tag{4.6}$$

where $\hat{b}_n(\mathbf{r}_i)$ ($\hat{b}_n^\dagger(\mathbf{r}_i)$) destroys (creates) a boson around the lattice site $\mathbf{r}_i$ in the n-th band described by the Wannier function $w_n(\mathbf{r})$. If the external fields are sufficiently weak and in the absence of band crossings, the band index n can be taken to be a constant of the motion. Under these assumptions Eq. (4.6) can be restricted to the lowest band maps the continuous model (4.5) onto the Bose-Hubbard Hamiltonian, with the coupling parameters given by

$$t = -\int d^2\mathbf{r}\; w^*(\mathbf{r}-\mathbf{r}_i)\left[-\hbar^2\nabla_{\mathbf{r}}^2/2M + V_{\text{lat}}(\mathbf{r})\right] w(\mathbf{r}-\mathbf{r}_j)\,, \tag{4.7}$$

$$U_{ij} = \int d^2\mathbf{r} \int d^2\mathbf{r}'\; |w(\mathbf{r}-\mathbf{r}_i)|^2\, V_{\text{int}}(|\mathbf{r}-\mathbf{r}'|)\, |w(\mathbf{r}'-\mathbf{r}_j)|^2 \tag{4.8}$$

and

$$\mu = \int d^2\mathbf{r}\; V_{\text{ext}}(\mathbf{r})|w(\mathbf{r}-\mathbf{r}_i)|^2. \tag{4.9}$$

The long-range phase coherence of cold bosons in an optical lattice can be directly tested by observing a multiple matter-wave interference pattern after ballistic expansion when all external trapping potentials are switched off in a time-of-flight measurement. Phase-coherent matter waves originating from different lattice site overlap and interfere with each other. Narrow peaks in the momentum distribution, that are due to the periodicity of the lattice, and a constant macroscopic phase difference across the lattice sites become visible [13–15].

The momentum distribution function is given by

$$n(\mathbf{k}) = \langle \hat{\Psi}^\dagger(\mathbf{k})\hat{\Psi}(\mathbf{k})\rangle \tag{4.10}$$

where $\hat{\Psi}(\mathbf{k})$ is the Fourier transform of the field operator. Using the previous expression for the field operator restricted to the lowest band) we thus find

$$n(\mathbf{k}) = \sum_{\mathbf{r}_i,\mathbf{r}_j} \langle \hat{b}^\dagger(\mathbf{r}_i)\hat{b}(\mathbf{r}_j)\rangle\; \exp\left[i\mathbf{k}\cdot(\mathbf{r}_i-\mathbf{r}_j)\right] w^*_{\mathbf{r}_i}(\mathbf{k})w_{\mathbf{r}_j}(\mathbf{k}) \tag{4.11}$$

An example of the result of a fre exapnsion of a condensate in an optical lattice is given in Figure 3. The interference pattern, appearing because of the global coherence of the lattice, is described by the fringe structure.

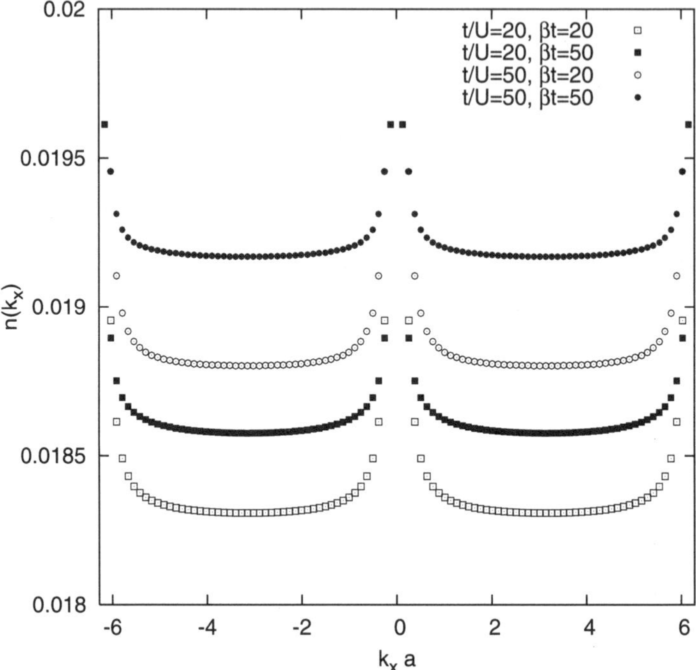

Figure 3. Momentum distribution along the x direction for a Bose condensate loaded into an optical lattice in the superfluid phase. The values of the ratio t/U and βt are listed in the figure. The various curves are shifted for clarity.

5. Conclusions

This brief note was intended as a reminder of the connection between two apparently very different systems, Josephson arrays and bosons in optical lattices. Both systems are described by the Bose-Hubbard model. Presently, the research in optical lattices is experiencing a very lively athmosphere. In this context I would like to mention only two among the many possible directions of study. Optical lattices with spinor condensates [16] present a very rich phase diagram that is presently intesively investigated. Another interesting possibility which is emerging recently is the study of frustration in optical lattices [17].

ACKNOWLEDGMENTS. I am enjoying a very fruitful collaboration on the physics of optical lattices with the group of Mario Tosi. I take this opportunity to expecially thank Marco Polini, Mario Tosi and Patrizia Vignolo for very fruitful discussions. I also would like to thank Marco Polini for the help in preparing the figures.

References

[1] A. MINGUZZI, S. SUCCI, F. TOSCHI, M. P. TOSI and P. VIGNOLO, Phys. Rep. **395** (2004), 223.

[2] M. P. A. FISHER, B. P. WEICHMAN, G. GRINSTEIN and D. S. FISHER, Phys. Rev. B **40** (1989), 546.

[3] R. F. VOSS and R. A. WEBB, Phys. Rev. B **25** (1982), 3446; R. A. WEBB, R. F. VOSS, G. GRINSTEIN and P. M. HORN, Phys. Rev. Lett. **51** (1983), 690.

[4] M. GOLDMAN and S. A. WOLF (eds.)., "Percolation, Localization, and Superconductivity", NATO ASI **108**, 1986.

[5] J. E MOOIJ and G. SCHÖN (eds.), "Coherence in Superconducting Networks", Physica **B 152**, 1988, pp. 1-302.

[6] H. A. CERDEIRA and S. R. SHENOY (eds.), "Josephson Junction Arrays", Physica **B 222**, 1996, pp. 253-406.

[7] C. GIOVANELLA and M. TINKHAM (eds.), "Macroscopic Quantum Phenomena and Coherence in Superconducting Networks", World Scientific, Singapore, 1995.

[8] L. J. GEERLIGS, M. PETERS, L. E. M. DE GROOT, A. VERBRUGGEN and J. E. MOOIJ, Phys. Rev. Lett. **63** (1989), 326.

[9] G. SCHÖN and A. D. ZAIKIN, Phys. Rep. **198** (1990), 237.

[10] R. FAZIO and H. VAN DER ZANT, Phys. Rep. **355** (2001), 235.

[11] D. JAKSCH, C. BRUDER, J. I. CIRAC, C. W. GARDINER and P. ZOLLER, Phys. Rev. Lett. **81** (1998), 3108.

[12] M. GREINER *et al.*, Nature **415** (2002), 39.

[13] L. PITAEVSKII and S. STRINGARI, Phys. Rev. Lett. **87** (2001), 180402.

[14] A. CUCCOLI, *et al.*, Phys. Rev. A **64** (2001), 061601.

[15] R. ROTH and K. BURNETT, Phys. Rev. A **67** (2003), 031602(R).

[16] E. DEMLER and F. ZHOU, Phys. Rev. Lett. **88** (2002), 163001.

[17] M. POLINI, R. FAZIO, A. H. MACDONALD and M. P. TOSI, cond-mat/0501387

Many-body effective mass enhancement in a two-dimensional electron liquid

M. Polini[1]

Motivated by a large number of recent magnetotransport studies we have revisited the problem of the microscopic calculation of the many-body effective mass in an unpolarized two-dimensional ($2D$) electron liquid (EL). We report extensive calculations over a broad range of electron densities. In this respect we critically examine the relative merits of the on-shell approximation, commonly used in weak-coupling situations, *versus* the actual self-consistent solution of the Dyson equation. We show that already for $r_s \simeq 3$ and higher, a solution of the Dyson equation proves necessary in order to obtain a well behaved effective mass. Finally, we also show that our theoretical results for a quasi-$2D$ EL, free of any adjustable fitting parameters, are in good qualitative agreement with recent measurements in GaAs/AlGaAs heterostructures.

1. Introduction

An EL on a uniform neutralizing background is used as the reference system in most realistic calculations of electronic structure in condensed-matter physics [1]. At zero temperature there is only one relevant parameter for an unpolarized, disorder-free EL in the absence of quantizing magnetic fields and spin-orbital coupling: the usual Wigner-Seitz density parameter $r_s = (\pi n_{2D} a_B^2)^{-1/2}$, n_{2D} being the average density and a_B the Bohr radius in vacuum. In a medium $a_B \to a_B^\star = \hbar^2 \bar{\kappa}/(m_b e^2)$ with $\bar{\kappa}$ and m_b appropriate dielectric constant and bare band mass respectively. Understanding the many-body aspects of this model has attracted continued interest for many decades [2–4]. The EL, unlike systems of classical particles, behaves like an ideal paramagnetic gas at high density ($r_s \ll 1$)

[1] Based on a work done together with Reza Asgari, Bahman Davoudi, Gabriele F. Giuliani, Mario P. Tosi, and Giovanni Vignale.

and like a Wigner solid at low density [5] ($r_s \gg 1$). In the intermediate density regime, which is relevant in $3D$ to conduction electrons in simple metals and in $2D$ to electrons in an inversion layer of a Si metal-oxide-semiconductor field-effect transistor (MOSFET) or in an AlGaAs/GaAs quantum well, perturbative techniques are clearly not effective owing to the lack of a small expansion parameter. One has to take recourse to approximate semi-analytical methods, a number of which have been reviewed in Refs. [3] and [4], or to Quantum Monte Carlo (QMC) simulation methods [6, 7].

Among the methods designed to deal with the intermediate density regime, of particular interest for its physical appeal and elegance is Landau's phenomenological theory [8]. The basic idea of Landau's theory is that the low-lying excitations of a system of interacting Fermions with repulsive interactions can be constructed starting from the low-lying states of a noninteractig Fermi gas by adiabatically switching-on the interaction between particles. This procedure allows to establish a one-to-one correspondence between the eigenstates of the ideal system and the approximate eigenstates of the interacting one. Landau called such single-particle excitations of an interacting Fermi-liquid "quasiparticles" (QP's). He wrote the excitation energy of the Fermi-liquid $E[n_{\mathbf{p}}]$ as a functional of the QP distribution function $n_{\mathbf{p}}$ in terms of the isolated quasiparticle energy $\mathcal{E}_{\mathbf{p}}$ and of the QP-QP interaction function $f_{\mathbf{p},\mathbf{p}'}$. The latter can in turn be used to obtain various physical properties of the system, such as the compressibility and the spin-susceptibility.

One of the implications of Landau's theory is the fact that the QP mass $m^\star$ is renormalized by electron-electron interactions [2]: $m^\star \neq m$, m being the bare electron mass. In a translationally-invariant system the current $\mathbf{j}_{\mathbf{p}}$ carried by a single excited QP of momentum $\mathbf{p}$ is controlled only by the bare mass, $\mathbf{j}_{\mathbf{p}} = \mathbf{p}/m$. On the other hand, the QP group velocity $\mathbf{v}_{\mathbf{p}}$ is instead defined by $\mathbf{v}_{\mathbf{p}} = \nabla_{\mathbf{p}}\mathcal{E}_{\mathbf{p}}$. In an isotropic system $\mathbf{v}_{\mathbf{p}}$ is parallel to $\mathbf{p}$ and the relation $\mathbf{v}_{\mathbf{p}} = \mathbf{p}/m^\star$ defines the QP effective mass. Thus $\mathbf{j}_{\mathbf{p}} \neq \mathbf{v}_{\mathbf{p}}$, the reason being that due to interactions the moving QP tends to drag part of the electronic medium along with it producing an extra current [2, 3].

The QP effective mass is a measurable quantity. The most direct way to determine $m^\star$ would be a measurement of the low-temperature heat capacity $C_V(T)$. It is in fact remarkable that electron-electron interaction effects enter $C_V(T)$ only through $m^\star$: the Landau interaction functions $f_{\mathbf{p},\mathbf{p}'}$ are not explicitly invoked [2, 3]. These type of experiments are exceedingly challenging and have not yet been realized with high precision (see Ref. [26] below and references therein). A powerful alternative tool to access experimentally the QP effective mass (and other Fermi-liquid parameters) is to analyze quantum Shubnikov-de Haas (SdH) oscillations of the magnetoresistance [9–14].

The contents of the paper are described briefly as follows. In Sect. 2 we present the theoretical background [15] referring the reader to the original works for a detailed derivation. We proceed in Sect. 3 to present some illustrative numerical results for $m^\star$ in a strictly-$2D$ EL and a comparison between our theory and some recent experimental results in GaAs/AlGaAs heterostructures. Finally, Sect. 4 concludes the paper.

2. Many-body effective mass enhancement

The aim of this section is to present a brief summary of the theory we have used for the retarded QP self-energy $\Sigma_{\rm ret}(\mathbf{k},\omega)$ of a paramagnetic $2D$ EL (that we have summarized in Eqs. (2.1) and (2.3) below) from which we have calculated $m^\star$. The formal justification of Eqs. (2.1) and (2.3), which essentially rests on both diagrammatic perturbation theory and on the so-called renormalized Hamiltonian approach [16], can be found in the original works [17,18], in Ref. [3], and in Ref. [15]. For later purposes we introduce the Fermi wave number $k_F = (2\pi n_{\rm 2D})^{1/2}$, the Fermi energy $\varepsilon_F = \hbar^2 k_F^2/(2m)$ and the quantity $\xi_{\mathbf{k}} = \hbar^2\mathbf{k}^2/(2m) - \varepsilon_F$ which is the single-particle energy $\varepsilon_{\mathbf{k}} = \hbar^2\mathbf{k}^2/(2m)$ measured from ε_F.

The retarded QP self-energy $\Sigma_{\rm ret}(\mathbf{k},\omega)$ is written as the sum of two terms, $\Sigma_{\rm ret}(\mathbf{k},\omega) = \Sigma_{\rm SX}(\mathbf{k},\omega) + \Sigma_{\rm CH}(\mathbf{k},\omega)$, where the first term is called "screened-exchange" (SX) and the second term is called "Coulomb-hole" (CH). The frequency ω is measured from $\varepsilon_F/\hbar$.

The SX contribution is given by

$$\Sigma_{\rm SX}(\mathbf{k},\omega) = -\int \frac{d^2\mathbf{q}}{(2\pi)^2}\,\frac{v_{\mathbf{q}}}{\varepsilon(\mathbf{q},\omega-\xi_{\mathbf{k}+\mathbf{q}}/\hbar)}\,\Theta(-\xi_{\mathbf{k}+\mathbf{q}}/\hbar)\,. \tag{2.1}$$

Here $v_{\mathbf{q}} = 2\pi e^2/q$ is the $2D$ Fourier transform of the bare Coulomb interaction e^2/r, $\Theta(x)$ is the step function and $\varepsilon(\mathbf{q},\omega)$ is a screening function originating from effective Kukkonen-Overhauser (KO) interactions [19], $\varepsilon^{-1}(\mathbf{q},\omega) = 1 + v_{\mathbf{q}}[1-G_+(\mathbf{q})]^2\chi_{\rm C}(\mathbf{q},\omega) + 3v_{\mathbf{q}}G_-^2(\mathbf{q})\chi_{\rm S}(\mathbf{q},\omega)$. Here the charge and spin response functions $\chi_{\rm C,S}(\mathbf{q},\omega)$ are determined by the spin-symmetric and spin-antisymmetric *static* [20] local-field factors $G_+(\mathbf{q})$ and $G_-(\mathbf{q})$ [22],

$$\chi_{\rm C,S}(\mathbf{q},\omega) = \frac{\chi_0(\mathbf{q},\omega)}{1 - v_{\mathbf{q}}[\mathcal{A}_{\rm C,S} - G_\pm(\mathbf{q})]\chi_0(\mathbf{q},\omega)}\,. \tag{2.2}$$

Here $\mathcal{A}_{\rm C} = 1$, $\mathcal{A}_{\rm S} = 0$, and $\chi_0(\mathbf{q},\omega)$ is the Lindhard response function of a noninteracting $2D$ EL [23].

The CH contribution to the retarded self-energy is given by

$$\Sigma_{\rm CH}(\mathbf{k},\omega) = -\int \frac{d^2\mathbf{q}}{(2\pi)^2}\, v_{\mathbf{q}} \int_0^{+\infty} \frac{d\Omega}{\pi}\, \frac{\Im m[\varepsilon^{-1}(\mathbf{q},\Omega)]}{\omega - \xi_{\mathbf{k}+\mathbf{q}}/\hbar - \Omega + i0^+}\,. \tag{2.3}$$

The real part of the retarded self-energy $\Re e\,\Sigma_{\rm ret}(\mathbf{k},\omega)$ can be readily obtained from Eqs. (2.1) and (2.3).

Once $\Re e\,\Sigma_{\rm ret}(\mathbf{k},\omega)$ is known, the QP excitation energy $\delta\mathcal{E}_{\rm QP}(\mathbf{k})$, which is the QP energy measured from the chemical potential μ of the interacting EL, can be calculated by solving self-consistently the Dyson equation

$$\delta\mathcal{E}_{\rm QP}(\mathbf{k}) = \xi_{\mathbf{k}} + \Re e\,\Sigma^{\rm R}_{\rm ret}(\mathbf{k},\omega)\big|_{\omega=\delta\mathcal{E}_{\rm QP}(\mathbf{k})/\hbar}\,, \tag{2.4}$$

where $\Re e\,\Sigma^{\rm R}_{\rm ret}(\mathbf{k},\omega) = \Re e\,\Sigma_{\rm ret}(\mathbf{k},\omega) - \Sigma_{\rm ret}(k_F,0)$. For later purposes we introduce at this point the so-called on-shell approximation (OSA). This amounts to approximating the QP excitation energy by calculating $\Re e\,\Sigma^{\rm R}_{\rm ret}(\mathbf{k},\omega)$ in Eq. (2.4) at the frequency $\omega = \xi_{\mathbf{k}}/\hbar$ corresponding to the single-particle energy, that is

$$\delta\mathcal{E}_{\rm QP}(\mathbf{k}) \simeq \xi_{\mathbf{k}} + \Re e\,\Sigma^{\rm R}_{\rm ret}(\mathbf{k},\omega)\big|_{\omega=\xi_{\mathbf{k}}/\hbar}\,. \tag{2.5}$$

The effective mass $m^\star$ can be calculated from the QP excitation energy by means of the relationship

$$\frac{1}{m^\star} = \frac{1}{\hbar^2 k_F}\, \frac{d\delta\mathcal{E}_{\rm QP}(k)}{dk}\bigg|_{k=k_F}\,. \tag{2.6}$$

As remarked above, $\delta\mathcal{E}_{\rm QP}(k)$ may be calculated either by solving self-consistently the Dyson equation (2.4) or by using the OSA in Eq. (2.5). In what follows the identity

$$\begin{aligned}\frac{d\Re e\,\Sigma^{\rm R}_{\rm ret}(k,\omega(k))}{dk} &= \partial_k \Re e\,\Sigma^{\rm R}_{\rm ret}(k,\omega)\big|_{\omega=\omega(k)} \\ &\quad + \partial_\omega \Re e\,\Sigma^{\rm R}_{\rm ret}(k,\omega)\big|_{\omega=\omega(k)}\, \frac{d\omega(k)}{dk}\end{aligned} \tag{2.7}$$

will be used, $\omega(k)$ being an arbitrary function of k.

Using Eqs. (2.6) and (2.7) with $\omega(k) = \delta\mathcal{E}_{\rm QP}(k)/\hbar$ we find that the effective mass $m^\star_{\rm D}$ calculated within the Dyson scheme is given by

$$\frac{m^*_{\rm D}}{m} = \frac{Z^{-1}}{1 + (m/\hbar^2 k_F)\, \partial_k \Re e\,\Sigma^{\rm R}_{\rm ret}(k,\omega)\big|_{k=k_F,\omega=0}}\,. \tag{2.8}$$

The renormalization constant Z, that measures the discontinuity of the momentum distribution function at the Fermi surface, is given by $Z^{-1} = 1 - \hbar^{-1}\partial_\omega \Re e \Sigma^{\rm R}_{\rm ret}(k,\omega)|_{k=k_F,\omega=0}$. The normal Fermi-liquid assumption translates mathematically into the inequality $0 < Z \leq 1$. This implies $\partial_\omega \Re e \Sigma^{\rm R}_{\rm ret}(k,\omega)|_{k=k_F,\omega=0} \leq 0$. Thus we see that the effective mass $m^\star_{\rm D}$ can diverge at a *finite* value of r_s by one of two mechanisms: (i) $\partial_\omega \Re e \Sigma^{\rm R}_{\rm ret}(k,\omega)|_{k=k_F,\omega=0}$ going to minus infinity at some finite value of r_s [24]; (ii) $\partial_k \Re e \Sigma^{\rm R}_{\rm ret}(k,\omega)|_{k=k_F,\omega=0}$ going to $-\hbar^2 k_F/m$ at some finite value of r_s.

Neither possibility is realized in our calculation: the first is barred *a priori* by the fact that the analytic expression for the frequency derivative of $\Sigma^{\rm R}_{\rm ret}$ is always finite at finite r_s; the second is found *a posteriori* not to occur since the momentum derivative of $\Sigma^{\rm R}_{\rm ret}$ is positive up to the largest r_s considered (see below).

On the other hand, using Eqs. (2.6) and (2.7) with $\omega(k) = \xi_{\bf k}/\hbar$ we find that the effective mass $m^\star_{\rm OSA}$ within the OSA is given by

$$\frac{m^\star_{\rm OSA}}{m} = \frac{1}{1 + (m/\hbar^2 k_F)\,\partial_k \Re e \Sigma^{\rm R}_{\rm ret}(k,\omega)\big|_{k=k_F,\omega=0} + (1 - Z^{-1})}\,. \tag{2.9}$$

Of course, Eq. (2.9) is a valid approximation to the effective mass in the weak coupling limit, as can be seen by expanding Eq. (2.8) for small values of $\Sigma^{\rm R}_{\rm ret}$: however its application becomes problematic at large values of r_s. In particular, we see that because Z decreases monotonically with increasing r_s, there must necessarily be a critical value of r_s for which the denominator of Eq. (2.9) vanishes and $m^\star_{\rm OSA}$ diverges. A recent paper by Zhang and Das Sarma [25] infers from this fact a true divergence of the effective mass within the RPA. In our view, however, this must be considered an artifact of Eq. (2.9). The unphysical character of the behavior $m^\star_{\rm OSA} \to \infty$ is revealed by the fact that the divergence is driven by a negative but finite value of $\partial_\omega \Re e \Sigma^{\rm R}_{\rm ret}(k_F, 0)$, whereas we know, from the general analysis given above, that a genuine divergence would have to be driven either by an infinite $\partial_\omega \Re e \Sigma^{\rm R}_{\rm ret}(k_F, 0)$ or by a negative $\partial_k \Re e \Sigma^{\rm R}_{\rm ret}(k_F, 0)$ becoming equal to $-\hbar^2 k_F/m$. We conclude that there is no evidence, within the present theory, for a physically relevant divergence of the effective mass.

3. Numerical results

We turn now to a presentation of our main numerical results for $m^\star/m$. In all figures the labels "RPA", "G_+" and "$G_+ \& G_-$" refer to three possible choices for the local-field factors: "RPA" refers to the case in which local-field factors are not included, "G_+" to the case in which $G_-({\bf q})$

is set to zero (*i.e.* spin-density fluctuations are not allowed), and finally "G_+&G_-" refers to the full theory including both charge- and spin-density fluctuations.

In Figure 1 we show our numerical results for $m^\star_{\rm D}$ and $m^\star_{\rm OSA}$. The effective mass enhancement is substantially smaller in the Dyson-equation calculation than in the OSA, the reason being that a large cancellation occurs between numerator and denominator in Eq. (2.8). In both calculations the combined effect of charge and spin fluctuations is to enhance the effective mass over the RPA result, whereas the opposite effect is found if only charge fluctuations are included – a manifestly incorrect result that neglects the spinorial nature of the electron. For completeness we have also included in Figure 1 the variational QMC results of Kwon *et al.* [7]. The reader should bear in mind that the effective mass is not a ground-state property and thus its evaluation by the QMC technique is quite delicate, as it involves the construction of excited states. There clearly is quantitative disagreement between our "best" theoretical results (the "G_+&G_-/D" predictions) and the QMC data.

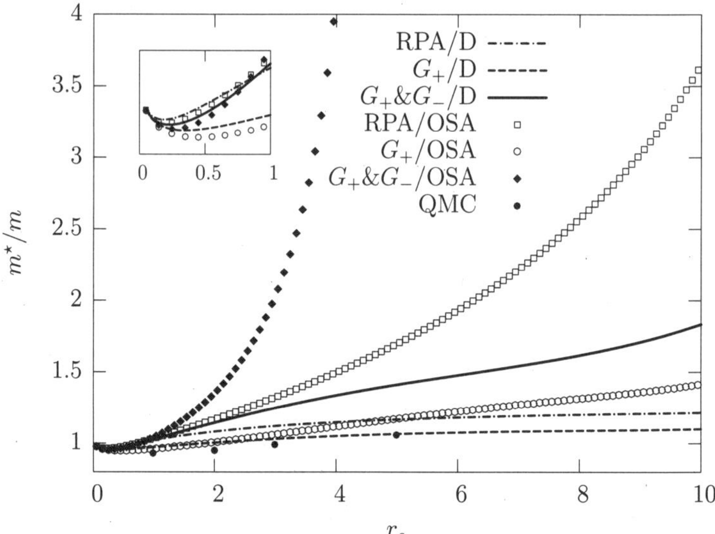

Figure 1. Effective mass enhancement as a function of r_s for $0 \leq r_s \leq 10$. The inset shows an enlargement of the results for $r_s \leq 1$. The lines show the results from Eq. (2.8), while the symbols (except for the dots) are from Eq. (2.9). The QMC data (dots) are from Ref. [7].

In Figure 2 we show the behavior of the two terms in the denominator of Eq. (2.9) as functions of r_s. This figure clearly shows how a spurious divergence can arise in $m^\star_{\rm OSA}$: for instance, within the RPA the denominator in Eq. (2.9) has a zero at $r_s \simeq 15.5$ (see the inset in Figure 2). Our numerical evidence, within the three theories we have studied, is that indeed (i) $\partial_\omega \Re e \Sigma^{\rm R}_{\rm ret}(k_F, 0)$ is negative as it should for a normal Fermi-liquid, and monotonically increasing in absolute value as a function of r_s; and (ii) $\partial_k \Re e \Sigma^{\rm R}_{\rm ret}(k_F, 0)$ is positive and monotonically increasing too. Within the theory outlined in Ref. [15], which uses as a key ingredient the KO effective screening function $\varepsilon(\mathbf{q}, \omega)$, the effect of a charge-only local field is to shift this divergence to higher values of r_s, while the opposite occurs upon including both charge and spin fluctuations. For instance, within the "$G_+\&G_-$/OSA" theory the divergence occurs near $r_s = 5$.

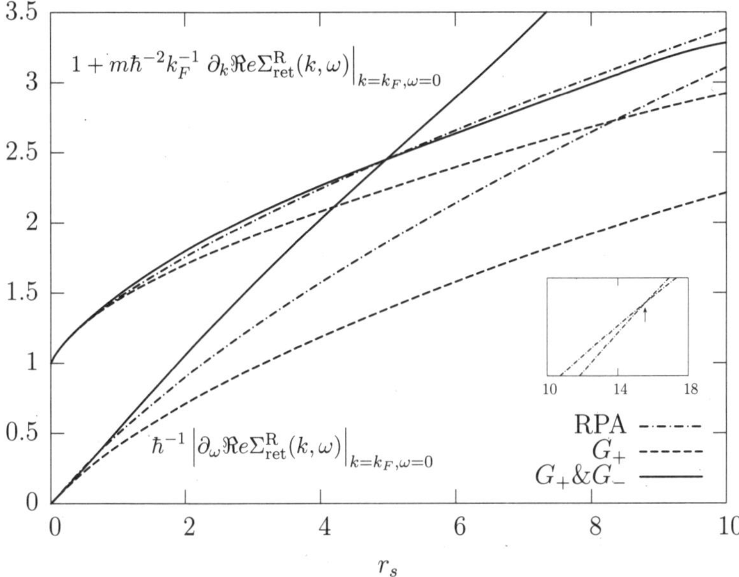

Figure 2. Illustrating the divergence of the effective mass within the OSA. The three curves starting from unity at $r_s = 0$ refer to the quantity $1 + m\hbar^{-2}k_F^{-1}\partial_k \Re e \Sigma^{\rm R}_{\rm ret}(k_F, 0)$, and the other three curves to $\hbar^{-1}|\partial_\omega \Re e \Sigma^{\rm R}_{\rm ret}(k_F, 0)| = Z^{-1} - 1$. The intersection of two lines with the same line-style in the two sets of curves corresponds to a zero in the denominator of Eq. (2.9) and thus to a divergence in $m^*_{\rm OSA}$. The inset shows this divergence occurring within the RPA at $r_s \simeq 15.5$.

We now pass to illustrate how our theory compares with experiments. A full analysis of the published data for the effective mass of carriers in Si-MOSFET's [9, 10] would require a more complete theoretical study, mainly to account for the two-valley nature of the material. We will focus here instead on the experimental results of Tan *et al.* [26] in a GaAs/AlGaAs heterojunction-insulated gate field-effect transistor (HIGFET) of exceedingly high quality. A quantitative comparison between theory and experiment would also require a refined treatment of a series of effects such as those due to (i) the detailed band-structure of the host semiconductor, (ii) disorder, and (iii) finite temperature [27]. Note also that the SdH measurements of Tan *et al.* [26] are performed in a small but obviously *finite* magnetic field B and, in general, the ground-state of an EL in a small finite B field can be profoundly different from that at $B = 0$ (see *e.g.* Ref. [28]). In our zero-field calculations we have: (i) included band-structure effects only through the GaAs band mass $m_{\rm b} = 0.067\, m$; (ii) neglected possible effects due to disorder because the concentration of background impurities n_i in the HIGFET used in Ref. [26] has been estimated [29] to be $n_i \approx 5 \times 10^{12}\,{\rm cm}^{-3}$ which is a very low number; and (iii) neglected thermal effects even though in Ref. [26] the temperature of the dilution refrigerator was kept relatively high ($100 \lesssim T \lesssim 400\,{\rm mK}$) to avoid the quantum Hall regime in which the SdH oscillations become non-sinusoidal. A simple inspection shows however that, assuming that for the highest temperature $T = 400\,{\rm mK}$ the EL is in thermal equilibrium with the refrigerator, the ratio of thermal to Fermi energy is quite small even at the lowest densities, *e.g.* $k_B T/\varepsilon_F \approx 0.03$ for $r_s = 6$.

We thus restrict our analysis solely to the effect of finite sample thickness, by discussing how a softened Coulomb potential modifies $m^\star$ against the strictly-$2D$ results shown in Figure 1. The expectation is that the QP effective mass will be noticeably smaller when a softened Coulomb interaction is at work.

We have thus recalculated $m^\star$ after renormalizing the bare Coulomb potential by means of a form factor to take into account the finite width of the EL in the HIGFET used in Ref. [26]. The appropriate renormalized potential is given by $V_{\bf q} = v_{\bf q} \mathcal{F}(qd)/\bar{\kappa}$, where

$$\mathcal{F}(x) = \left(1 + \frac{\kappa_{\rm ins}}{\kappa_{\rm sc}}\right) \frac{8 + 9x + 3x^2}{16(1+x)^3} + \left(1 - \frac{\kappa_{\rm ins}}{\kappa_{\rm sc}}\right) \frac{1}{2(1+x)^6}\,, \qquad (3.10)$$

with $d = [\hbar^2 \kappa_{\rm sc}/(48\pi m_{\rm b} e^2 n^\star)]^{1/3}$ representing an effective width of the quasi-$2D$ EL [30]. Here $\kappa_{\rm ins} = 10.9$ and $\kappa_{\rm sc} = 12.9$ are the dielectric constants of the insulator and of the space charge layer, $\bar{\kappa}$ is their average and $n^\star = n_{\rm depl} + 11 n_{\rm 2D}/32$, the depletion layer charge density $n_{\rm depl}$ being essentially zero (see Ref. 18 of Ref. [29]) in the experiments of Ref. [26]. Note that the renormalized potential does not contain any adjustable fitting parameter. The results that we obtain with the softened potential $V_{\bf q}$ are shown in Figure 3 against the experimental results of Tan *et al.* [26]. A *caveat* to keep in mind is that we have used the same local-field factors as for a strictly-$2D$ EL [22] in the lack of a better choice. Thus the results labeled by "G_+" and "$G_+\&G_-$" in Figure 3 contain the effect of finite thickness only through the renormalization of the Coulomb potential. We believe that the explicit dependence of $G_\pm(\mathbf{q})$ on the finite width of the $2D$ EL should not change the results of Figure 3 in a substantial manner.

As it appears from Figure 3, a quite satisfactory qualitative agreement exists between theory and experiment, considering the assumptions we have made to study the quasi-$2D$ EL in the HIGFET of Ref. [26]. Nevertheless, in detail the discrepancies are quite substantial. In particular both the RPA and "$G_+\&G_-$" theories, which give rather similar results, underestimate the considerable effective mass enhancement at strong-coupling, *i.e.* for $r_s \gtrsim 5$. A comparison in the weak-coupling regime $r_s \lesssim 1$ would be very helpful but unfortunately is not possible due to the lackness of experimental data.

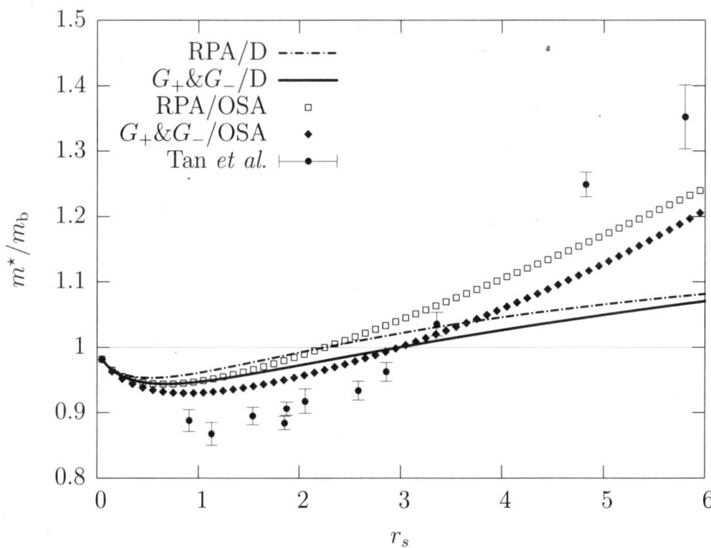

Figure 3. Effective mass enhancement for a $2D$ EL confined in a GaAs/AlGaAs triangular quantum well of the type used in Refs. [13,26]. The theoretical results are shown in the same notation of Figure 1. The experimental results (Tan *et al.*) are from Ref. [26].

4. Conclusions

In summary, we have revisited the problem of the microscopic calculation of the many-body effective mass enhancement in a $2D$ EL. We have performed a systematic study based on the many-body local-fields theory, taking advantage of the results of the most recent QMC calculations of the static charge and spin response of the EL expressed through static local-field factors. We have presented results for the effective mass enhancement over a wide range of electron densities. In this respect we have critically examined the merits of the OSA *versus* the Dyson-equation calculation. Depending on the local-field factors, the OSA predicts a spurious divergence of the effective mass at strong coupling and a solution of the Dyson equation is therefore necessary in order to obtain the correct value of the effective mass within Fermi-liquid theory. The comparison with the experimental data of Ref. [26] shows good qualitative agreement but substantial discrepancies especially at strong-coupling calling for further theoretical work and computer simulations.

ACKNOWLEDGMENTS. We would like to gratefully acknowledge the early contributions to this project by E. Strepparola. We are indebted to G. Simion for pointing out to us the existence of an ultraviolet divergence (see Ref. [15] for more details) in the calculation of the self-energy within the static many-body local fields theory. We would also like to thank J. Zhu, M. Shayegan, Y.-W Tan, and K. Vakili for sharing with us their considerable physical insight and their experimental results. S. Moroni provided useful clarifications concerning the QMC results. This work was partially supported by MIUR through the PRIN2001 and PRIN2003 programs.

References

[1] D. M. CEPERLEY, Nature **397** (1999), 386.
[2] D. PINES and P. NOZIÉRES, “The Theory of Quantum Liquids”, W. A. Benjamin, Inc., New York, 1966.
[3] G. F. GIULIANI and G. VIGNALE, “Quantum Theory of the Electron Liquid”, Cambridge University Press, Cambridge, 2005.
[4] K. S. SINGWI and M. P. TOSI, In: Solid State Physics, **36**, H. Ehrenreich, F. Seitz and D. Turnbull (eds.), Academic, New York, 1981, p. 177.
[5] E. P. WIGNER, Phys. Rev. **46** (1934), 1002.
[6] B. TANATAR and D. M. CEPERLEY, Phys. Rev. B **39** (1989), 5005; S. MORONI, D. M. CEPERLEY and G. SENATORE, Phys. Rev. Lett. **69** (1992), 1837; F. RAPISARDA and G. SENATORE, Austr. J. Phys. **49** (1996), 161; D. VARSANO, S. MORONI and G. SENATORE, Europhys. Lett. **53** (2001), 348; C. ATTACCALITE, S. MORONI, P. GORI-GIORGI and G. BACHELET, Phys. Rev. Lett. **88** (2002), 256601.
[7] Y. KWON, D. M. CEPERLEY and R. M. MARTIN, Phys. Rev. B **50** (1994), 1684.

[8] L. D. LANDAU, Sov. Phys. JEPT **3** (1957), 920.
[9] A. A. SHASHKIN, S. V. KRAVCHENKO, V. T. DOLGOPOLOV and T. M. KLAPWIJK, Phys. Rev. Lett. **87**, 086801 (2001); Phys. Rev. B **66** (2002), 073303.
[10] V. M. PUDALOV, M. E. GERSHENSON, H. KOJIMA, N. BUTCH, E. M. DIZHUR, G. BRUNTHALER, A. PRINZ and G. BAUER, Phys. Rev. Lett. **88** (2002), 196404.
[11] E. TUTUC, S. MELINTE and M. SHAYEGAN, Phys. Rev. Lett. **88** (2002), 036805.
[12] H. NOH, M. P. LILLY, D. C. TSUI, J. A. SIMMONS, E. H. HWANG, S. DAS SARMA, L. N. PFEIFFER and K. W. WEST, Phys. Rev. B **68** (2003), 165308.
[13] J. ZHU, H. L. STORMER, L. N. PFEIFFER, K. W. BALDWIN and K. W. WEST, Phys. Rev. Lett. **90** (2003), 056805.
[14] K. VAKILI, Y. P. SHKOLNIKOV, E. TUTUC, E. P. DE POORTERE and M. SHAYEGAN, Phys. Rev. Lett. **92** (2004), 226401.
[15] R. ASGARI, B. DAVOUDI, M. POLINI, G. F. GIULIANI, M. P. TOSI and G. VIGNALE, Phys. Rev. B **71** (2005), 045323.
[16] D. R. HAMANN and A. W. OVERHAUSER, Phys. Rev. **143** (1966), 183.
[17] T. K. NG and K. S. SINGWI, Phys. Rev. B **34** (1986), 7738 and 7743.
[18] S. YARLAGADDA and G. F. GIULIANI, Solid State Commun. **69** (1989), 677; S. YARLAGADDA and G. F. GIULIANI, Phys. Rev. B **40** (1989), 5432; S. YARLAGADDA and G. F. GIULIANI, *ibid.* **49** (1994), 7887 and 14188.
[19] C. A. KUKKONEN and A. W. OVERHAUSER, Phys. Rev. B **20** (1979), 550.
[20] Although the local-field factors are frequency-dependent quantities, we have made the common, and to a certain extent uncontrolled, approximation of neglecting their frequency dependence. Recent studies [21] have explored such a dependence in the long-wavelength limit $\mathbf{q} \to 0$, but clearly the knowledge of the full dependence on wave number is necessary for correctly carrying out the type of calculations that we are interested in this work. A possible role of dynamical exchange-correlation effects is currently under study.
[21] R. NIFOSÌ, S. CONTI and M. P. TOSI, Phys. Rev. B **58** (1998), 12758; Z. QIAN and G. VIGNALE, *ibid.* **65**, 235121 (2002) and **68** (2003), 195113.
[22] B. DAVOUDI, M. POLINI, G. F. GIULIANI and M. P. TOSI, Phys. Rev. B **64** (2001), 153101 and 233110.
[23] F. STERN, Phys. Rev. Lett. **18** (1967), 546.

[24] In this case the normal Fermi-liquid assumption breaks down and a singular behavior of $m^{\star}_{\rm D}$ could be interpreted as a quantum phase transition of the $2D$ EL to a non-Fermi-liquid state.
[25] Y. ZHANG and S. DAS SARMA, Phys. Rev. B **71**, 045322 (2005)
[26] Y. -W. TAN, J. ZHU, H. L. STORMER, L. N. PFEIFFER, K. W. BALDWIN and K. W. WEST, Phys. Rev. Lett. **94**, 016405 (2005).
[27] S. DAS SARMA, V. M. GALITSKI and Y. ZHANG, Phys. Rev. B **69** (2004), 125334.
[28] A. A. KOULAKOV, M. M. FOGLER and B. I. SHKLOVSKII, Phys. Rev. Lett. **76** (1996), 499; R. MOESSNER and J. T. CHALKER, Phys. Rev. B **54** (1996), 5006; M. P. LILLY, K. B. COOPER, J. P. EISENSTEIN, L. N. PFEIFFER and K. W. WEST, Phys. Rev. Lett. **82** (1999), 394; R. R. DU, D. C. TSUI, H. L. STORMER, L. N. PFEIFFER, K. W. BALDWIN and K. W. WEST, Solid State Commun. **109** (1999), 389; K. B. COOPER, M. P. LILLY, J. P. EISENSTEIN, L. N. PFEIFFER and K. W. WEST, Phys. Rev. B **60** (1999), R11285; F. D. M. HALDANE, E. H. REZAYI and K. YANG, Phys. Rev. Lett. **85** (2000), 5396.
[29] S. DE PALO, M. BOTTI, S. MORONI and G. SENATORE, Phys. Rev. Lett. **94**, 226405 (2005).
[30] T. ANDO, A. B. FOWLER and F. STERN, Rev. Mod. Phys. **54** (1982), 437.

On the two dimensional electron liquid in the presence of spin-orbit coupling

G. F. Giuliani and S. Chesi

A spin-orbit coupling of the Rashba type has profound effects on the spin degrees of freedom of a two dimensional electron liquid. In this system a prominent role is played by the local electron spin quantization axis in momentum space. We find that a useful device for the classification of the corresponding many-electron determinantal states is provided by the concept of generalized chirality. This allows us to systematically study the Hartree-Fock scenario for homogeneous, isotropic states. The ensuing relative phase diagram contains both chiral unpolarized and chiral spin polarized states which are characterized by peculiar distributions of the local electron spin quantization axis in momentum space.

1. Introduction

Great advances in electronic spin control have lately been achieved by taking advantage of the spin-orbit interaction of the Rashba [1, 2] or Dresselhaus [3] type experienced by the carriers in suitably prepared low dimensional electronic systems. Recent examples include spin current injection by magnetic focusing [4] and elastic reflection from a sharp boundary [5]. It is our aim to present here a number of results of a study of the effects of the interplay between the electron-electron interaction and the Rashba spin-orbit in a two dimensional electron liquid [6], a system believed to provide an accurate model for the description of the physics at hand. This problem is still far from having been properly explored and understood. A more detailed analysis can be found in reference [7].

2. The generalized chirality

We begin by considering the non interacting hamiltonian [1,2]:

$$\hat{H}_0 = \hat{K}_0 + \hat{R}_0 = \frac{\hat{\mathbf{p}}^2}{2m} + \alpha\,(\hat{\sigma}_x \hat{p}_y - \hat{\sigma}_y \hat{p}_x)\,, \qquad (2.1)$$

where the electronic motion is limited to the (x,y) plane and, by definition, the spin-orbit coupling constant α is assumed positive. This model hamiltonian is designed to describe the spin-orbit interaction induced under appropriate conditions by the confining potential.

It is readily found that the eigenstates of $\hat{H}_0$ are plane waves with spin oriented in the plane of motion in a direction perpendicular to the wave vector $\mathbf{k}$, i.e.

$$\varphi_{\mathbf{k},\pm}(\mathbf{r}) = \frac{e^{i\mathbf{k}\cdot\mathbf{r}}}{\sqrt{2L^2}} \begin{pmatrix} \pm 1 \\ i e^{i\phi_{\mathbf{k}}} \end{pmatrix} , \tag{2.2}$$

where the $\pm$ sub index on the left hand side labels positive or negative *chiral states*. The corresponding eigenvalues are

$$\epsilon_{\mathbf{k}\pm} = \frac{\hbar^2 \mathbf{k}^2}{2m} \mp \hbar\alpha\, k\,. \tag{2.3}$$

Non interacting many electron states of the system consist of single Slater determinants made out of the states of Eq. (2.2). Two topologically different yet compact possible ways of occupying such states are shown in Figure 1. The top panel represents a low density situation in which only one energy branch is compactly occupied, while the bottom panel depicts a high density regime in which both energy branches are compactly occupied. These states are completely determined by the geometrical parameters k_{in}, k_{out}, k_- and k_+ which are defined in Figure 1. We will refer to these two types of occupations as the basic compact occupation schemes. While it is clear that giving the electron density n (as for instance expressed in terms of the parameter $r_s = 1/\sqrt{\pi a_B^2 n})$ completely determines the state of lowest energy, the excited states of compact occupation necessitate in general two independent parameters for their classification. The physical significance of the second parameter is determined next. We notice first that all the states are characterized by a chiral polarization (or chirality) defined as

$$\chi_0 = \frac{n_+ - n_-}{n_+ + n_-} , \tag{2.4}$$

where $n_\pm$ is the number density of right handed and left handed electrons with $n = n_+ + n_-$. It is then clear that while all the high density states of this type can be uniquely classified by means of the value of their chirality $\chi_0 = \frac{k_+^2 - k_-^2}{k_+^2 + k_-^2} \leq 1$, the corresponding low density states are not amenable to such a classification since for all of them $\chi_0 = 1$. We find that, for a given density, an elegant way to classify all the different basic compact occupations is provided by the generalized chirality χ defined as:

$$\chi = \begin{cases} \chi_0 & \text{for } 0 < \chi_0 < 1\,, \\ \dfrac{k_{out}^2 + k_{in}^2}{k_{out}^2 - k_{in}^2} & \text{for } \chi_0 = 1\,. \end{cases} \tag{2.5}$$

Once the density and the generalized chirality are given the geometrical parameters characterizing the basic compact occupation schemes are immediately determined by the relations

$$k_{\pm} = \sqrt{2\pi n(1 \pm \chi)} \qquad \text{for } 0 < \chi < 1 \tag{2.6}$$

$$k_{out(in)} = \sqrt{2\pi n(\chi \pm 1)} \qquad \text{for } \chi \geq 1\,. \tag{2.7}$$

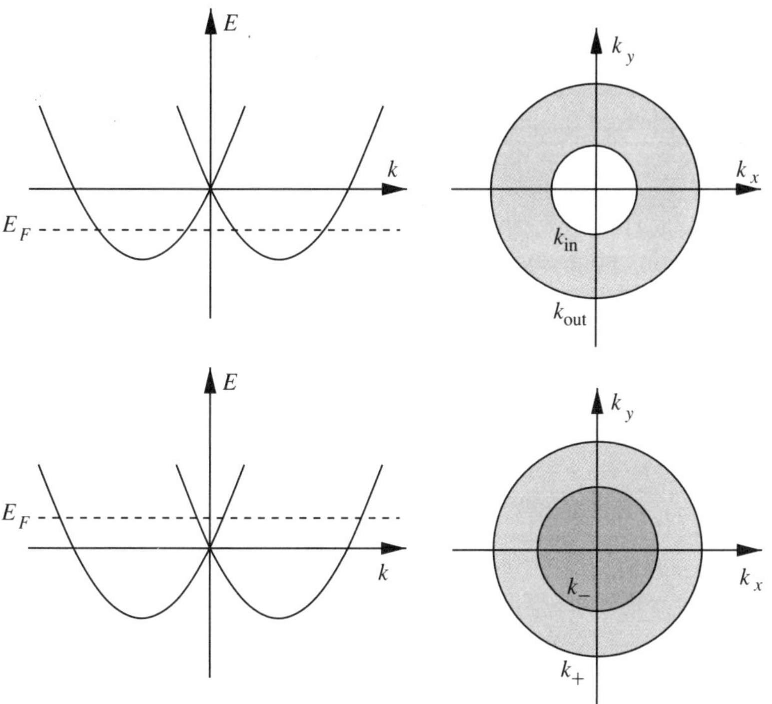

Figure 1. Two different ways of occupying non interacting chiral states in **k** space.

3. Interacting system: Mean field theory

The fully interacting problem is described by the many-body hamiltonian

$$\hat{H} = \sum_i \hat{H}_{0i} + \frac{1}{2}\sum_{i\neq j} \frac{e^2}{|\hat{\mathbf{r}}_i - \hat{\mathbf{r}}_j|} + \hat{H}_B \,, \tag{3.8}$$

where the last term represents the contributions stemming from the existence of a homogeneous neutralizing background. A number of properties of this system have been studied in reference [8] within a limited approach based on a first order perturbative expansion in terms of an *ad hoc* static screened interaction.

Limiting our study to the case of homogeneous, isotropic solutions, an interesting class of mean field states is represented by the Slater determinants formed occupying the single particle states obtained as the result of the following Bogolubov transformation:

$$\begin{pmatrix} \hat{b}^\dagger_{\mathbf{k}+} \\ \hat{b}^\dagger_{\mathbf{k}-} \end{pmatrix} = \begin{pmatrix} \cos\frac{\beta_\mathbf{k}}{2} & e^{i\gamma_\mathbf{k}}\sin\frac{\beta_\mathbf{k}}{2} \\ e^{-i\gamma_\mathbf{k}}\sin\frac{\beta_\mathbf{k}}{2} & -\cos\frac{\beta_\mathbf{k}}{2} \end{pmatrix} \begin{pmatrix} \hat{a}^\dagger_{\mathbf{k}\uparrow} \\ \hat{a}^\dagger_{\mathbf{k}\downarrow} \end{pmatrix} , \tag{3.9}$$

where the initial operators $\hat{a}_{\mathbf{k}\sigma}$ correspond to the standard plane wave spinors with spin quantization axis taken along the direction $\hat{z}$ perpendicular to the plane of motion. For a given $\mathbf{k}$, this transformation simply affects a rotation of the spin quantization axis from $\hat{z}$ to an arbitrary orientation $\hat{s}_\mathbf{k}$. The geometry of this rotation as well as the definition of the corresponding polar and azimuthal angles $\gamma_\mathbf{k}$ and $\beta_\mathbf{k}$ are provided in Figure 2. Notice that the choice $\gamma_\mathbf{k} = \phi_\mathbf{k} + \frac{\pi}{2}$ ($\phi_\mathbf{k}$ being the angle between $\mathbf{k}$ and the x-axis) and $\beta_\mathbf{k} = \frac{\pi}{2}$ leads to the eigenstates of $\hat{H}_0$ of Eq. (2.2).

At this point we evaluate the expectation value of the exact hamiltonian $\hat{H}$ over any of these determinantal states obtaining the following expression:

$$E_{HF}[n_{\mathbf{k}\pm}, \hat{s}_\mathbf{k}] = \sum_{\mathbf{k},\mu} \frac{\hbar^2\mathbf{k}^2}{2m} n_{\mathbf{k}\mu} - \hbar\alpha \sum_{\mathbf{k},\mu} \mu\, k\, \hat{\phi}_\mathbf{k}\cdot\hat{s}_\mathbf{k}\, n_{\mathbf{k}\mu} + \tag{3.10}$$

$$-\frac{1}{2L^2}\sum_{\mathbf{k},\mathbf{k}',\mu,\mu'} v_{\mathbf{k}-\mathbf{k}'}\,\frac{1+\mu\mu'\,\hat{s}_\mathbf{k}\cdot\hat{s}_{\mathbf{k}'}}{2}\, n_{\mathbf{k}\mu}n_{\mathbf{k}'\mu'} \,,$$

where the indices μ and μ' are summed over the values $\pm$. Clearly E_{HF} represents the (total) Hartree-Fock (HF) energy of this particular class of states and is a functional of the occupation numbers $n_{\mathbf{k}\pm}$ and of the orientation of the (wave vector space) local spin quantization axis unit

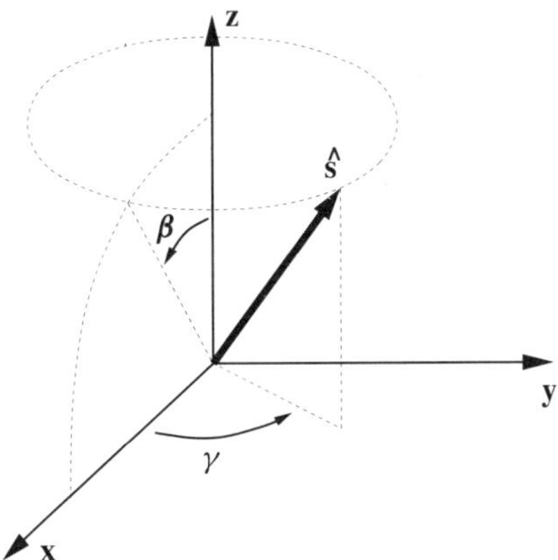

Figure 2. Geometry of the Bogolubov transformation.

vector $\hat{s}_{\mathbf{k}}$. It is still to be assessed which of these states correspond to actual solutions of the HF problem [9]. This we do by means of a standard Wick theorem based decoupling and diagonalization of the hamiltonian $\hat{H}$ expressed in terms of the operators $\hat{b}_{\mathbf{k}\pm}$.

As it turns out, the analysis is greatly simplified by setting $\gamma_{\mathbf{k}} = \phi_{\mathbf{k}} + \frac{\pi}{2}$ [10]. The resulting self-consistent equation involving the remaining functions $\beta_{\mathbf{k}}$ and $n_{\mathbf{k}\pm}$ reads:

$$\tan\beta_{\mathbf{k}} = \frac{\frac{1}{2L^2}\sum_{\mathbf{k}',\mu'} n_{\mathbf{k}'\mu'} v_{\mathbf{k}-\mathbf{k}'}\ \mu' \sin\beta_{\mathbf{k}'} \cos(\phi_{\mathbf{k}} - \phi_{\mathbf{k}'}) + \hbar\alpha k}{\frac{1}{2L^2}\sum_{\mathbf{k}',\mu'} n_{\mathbf{k}'\mu'} v_{\mathbf{k}-\mathbf{k}'}\ \mu' \cos\beta_{\mathbf{k}'}} \tag{3.11}$$

Although many possible solutions for the functions $n_{\mathbf{k}\pm}$ and $\beta_{\mathbf{k}}$ can in principle be found, it must be kept in mind that not all of them correspond to minima of the HF energy. Even restricting our analysis to wave vector space occupations of the basic compact types (and therefore to spatially isotropic states), solving the HF problem is not trivial. In this situation a simplifying feature is represented by the fact that, when the wave vector dependence of the function β is properly rescaled, the solutions of Eq. (3.11) have the form $\beta_{\mathbf{k}} = \bar{\beta}(\frac{|\mathbf{k}|}{\sqrt{2\pi n}})$, with $\bar{\beta}$ independent of the electron density, being completely determined by the generalized chirality χ. We find that given the coupling constant α, the problem always admits at least the solution $\beta_{\mathbf{k}} = \frac{\pi}{2}$ corresponding to a paramagnetic state constructed out of the eigenfunctions of the non interacting problem posed by $\hat{H}_0$. Moreover, for certain values of χ, a second (polarized) solution also exists for which $\beta_{\mathbf{k}} \neq \frac{\pi}{2}$. This will be discussed below.

Once a set of HF solutions has been established, the putative ground state can be found by minimization of the energy E_{HF} over the set.

Within our approach, the lowest energy state must be accordingly found optimizing the value of χ, i.e.

$$\mathcal{E}_{HF}(\bar{\alpha}, r_s) = \min_\chi[\mathcal{E}(\bar{\alpha}, r_s, \chi)] , \tag{3.12}$$

where we have introduced the dimensionless parameter $\bar{\alpha} = \frac{\hbar\alpha}{e^2}$ and have used the notation $\mathcal{E}_{HF}$ to represent an HF energy per particle in Ry units.

Finally, for each HF state, the corresponding single particle energies can be calculated from the expression

$$\begin{aligned}\epsilon_\mu(\mathbf{k}) = & \frac{\hbar\mathbf{k}^2}{2m} - \frac{1}{2L^2}\sum_{\mathbf{k}',\mu'} v_{\mathbf{k}-\mathbf{k}'} n_{\mathbf{k}'\mu'} \\ & -\mu\hbar\alpha k\,\hat{\phi}_{\mathbf{k}}\cdot\hat{s}_{\mathbf{k}} - \frac{\mu}{2L^2}\sum_{\mathbf{k}',\mu'}\mu' n_{\mathbf{k}'\mu'} v_{\mathbf{k}-\mathbf{k}'}\hat{s}_{\mathbf{k}'}\cdot\hat{s}_{\mathbf{k}} ,\end{aligned} \tag{3.13}$$

which is easily derived from Eq. (3.10).

As already noted, a possible set of unpolarized HF solutions have $\beta_{\mathbf{k}} = \frac{\pi}{2}$ and are simply obtained by any circularly symmetric occupation of the states (2.2) of the non interacting hamiltonian. Following the general procedure outlined above, the total energy is readily calculated and can be expressed in the interesting following form:

$$\begin{aligned}\mathcal{E}^{(\text{unpol})}(\bar{\alpha}, r_s, \chi) = & \mathcal{K}(r_s, \chi) + \mathcal{R}^{(\text{unpol})}(\bar{\alpha}, r_s, \chi) + \mathcal{E}_x^{(\text{unpol})}(r_s, \chi) \\ = & \frac{\tilde{\mathcal{K}}(\chi)}{r_s^2} + \frac{\bar{\alpha}\,\tilde{\mathcal{R}}^{(\text{unpol})}(\chi)}{r_s} + \frac{\tilde{\mathcal{E}}_x^{(\text{unpol})}(\chi)}{r_s} ,\end{aligned} \tag{3.14}$$

where $\tilde{\mathcal{K}}^{(\text{unpol})}(\chi)$ and $\tilde{\mathcal{R}}^{(\text{unpol})}(\chi)$ have simple closed form expressions [7]. The dependence of the exchange energy product $\tilde{\mathcal{E}}_x^{(\text{unpol})}(\chi) = r_s\mathcal{E}_x^{(\text{unpol})}(r_s, \chi)$, on the other hand, can be obtained from a simple quadrature. A plot of this function is provided in Figure 3. It displays an interesting minimum for $\chi \simeq 0.9147$ and behaves like $\tilde{\mathcal{E}}_x(\chi) \simeq -\frac{\sqrt{2}\log\chi}{\pi\sqrt{\chi}}$ for large χ.

Since the spin axis orientation $\hat{s}_{\mathbf{k}}$ of the non interacting state and that of the HF unpolarized states of this type do coincide, the only difference is due to momentum space repopulation. This is exemplified by the fact that, for a given density, the value of $\chi_{\min}$ corresponding to the lowest energy differs in the two cases. This situation is displayed in Figure 6. As one would expect, the deviation is larger at lower density.

Polarized states are obtained as nontrivial solutions of Eq. (3.11). Typical results for the function $\bar{\beta}$ are plotted versus $y = \frac{|\mathbf{k}|}{\sqrt{2\pi n}}$ in Figure 4.

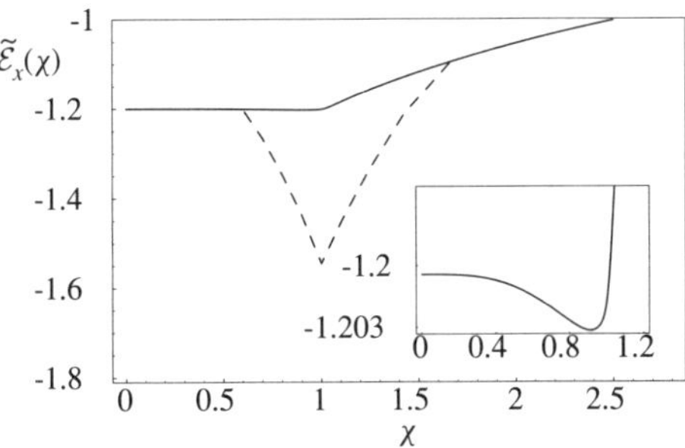

Figure 3. Dependence of $\tilde{\mathcal{E}}_x^{(\mathrm{unpol})}(\chi)$ (see Eq. (3.14) in the text) and of $\tilde{\mathcal{E}}_x^{(\mathrm{pol})}(\bar{\alpha}, \chi)$ (see Eq. (3.16)) on the generalized chirality χ. For the polarized case $\bar{\alpha} = 0.2$. In the inset, a particular of $\tilde{\mathcal{E}}_x^{(\mathrm{unpol})}(\chi)$ is shown.

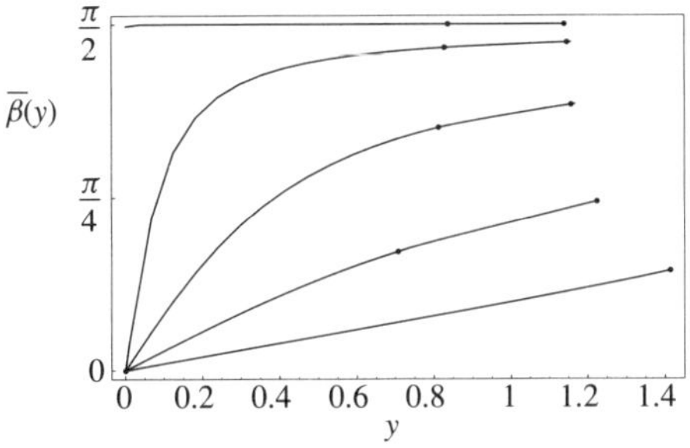

Figure 4. Plot of the complete $\bar{\beta}(y)$ (where $y = \frac{|\mathbf{k}|}{\sqrt{2\pi n}}$) for $\bar{\alpha} = 0.1$. The values of the generalized chirality are (from the top) $\chi = 0.3,\ 0.315,\ 0.34,\ 0.5,\ 1$. The dots mark the $[\sqrt{1-\chi}, \sqrt{1+\chi}]$ intervals.

Notice that as χ decreases towards an $\bar{\alpha}$ dependent critical value, the curves converge to the unpolarized solution. This corresponds to the fact that polarized solutions of Eq. (3.11) only exist for χ in a finite range of values.

Since $\beta_{\mathbf{k}}$ differs from $\frac{\pi}{2}$, these states have spin quantization axes $\hat{s}_{\mathbf{k}}$ lying outside of the plane of motion and therefore not only display an intriguing momentum space spin texture but also have a net fractional spin polarization that can be calculated from the relation:

$$p\,(\bar{\alpha}, \chi) = \frac{2\,\langle S_z \rangle}{\hbar\, N} = \int_{\sqrt{|1-\chi|}}^{\sqrt{|1+\chi|}} y \cos\bar{\beta}(y)\,\mathrm{d}y\,. \tag{3.15}$$

Note that in this case the fractional spin polarization does not coincide with the fraction of electrons with unpaired spin, as it is the case when spin-orbit is absent.

The energy per particle in Ry units can be calculated also in this case and can be cast in the following form:

$$\mathcal{E}^{(\mathrm{pol})}(\bar{\alpha}, r_s, \chi) = \mathcal{K}(r_s, \chi) + \mathcal{R}^{(\mathrm{pol})}(\bar{\alpha}, r_s, \chi) + \mathcal{E}_x^{(\mathrm{pol})}(\bar{\alpha}, r_s, \chi)$$
$$= \frac{\tilde{\mathcal{K}}(\chi)}{r_s^2} + \frac{\tilde{\mathcal{R}}^{(\mathrm{pol})}(\bar{\alpha}, \chi)}{r_s} + \frac{\tilde{\mathcal{E}}_x^{(\mathrm{pol})}(\bar{\alpha}, \chi)}{r_s}, \tag{3.16}$$

where $\tilde{\mathcal{R}}^{(\mathrm{pol})}(\bar{\alpha}, \chi)$ and $\tilde{\mathcal{E}}_x^{(\mathrm{pol})}(\bar{\alpha}, \chi)$ are expressed in terms of integrals involving $\bar{\beta}(y)$ [7]. A plot of $\tilde{\mathcal{E}}_x^{(\mathrm{pol})}(\bar{\alpha}, \chi)$ is provided in Figure 3 where it can be compared to the corresponding unpolarized result of Eq. (3.14). Again, for each density and spin-orbit strength, the ground state energy $\mathcal{E}_{HF}^{(\mathrm{pol})}(\bar{\alpha}, r_s)$ is determined by minimization with respect to χ. From numerical evaluations we find that, similarly to the $\alpha = 0$ case, the polarized ground state is obtained when $\chi_{\min} = 1$.

4. Relative phase diagram

A relative phase diagram within the space of the homogeneous, isotropic compactly occupied states described in the previous sections can be obtained by comparing the lowest energy in the various phases. The dependence of $\mathcal{E}_{HF}(\bar{\alpha}, r_s)$ and $\chi_{\min}$ on the density at particular values of $\bar{\alpha}$ is plotted in Figures 5 and 6. The ensuing scenario is as follows: the gas is unpolarized both at reasonably low and at high values of the density while a spin polarized state exists in a finite range of r_s. While the high density transition is the analog of the Bloch transition that occurs in absence of

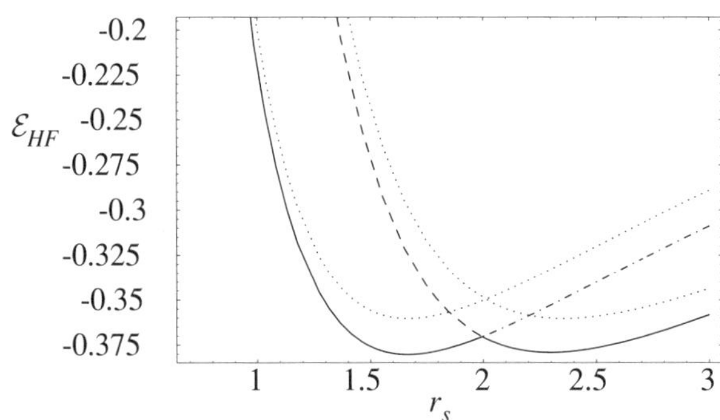

Figure 5. Plot of the total energy per particle in Ry units as function of the density parameter r_s ($\bar{\alpha} = 0.1$). Dot-dashed curve: unpolarized ground state; dashed curve: maximally polarized ($\chi = 1$) ground state; dotted curves: corresponding results for the familiar $\bar{\alpha} = 0$ case.

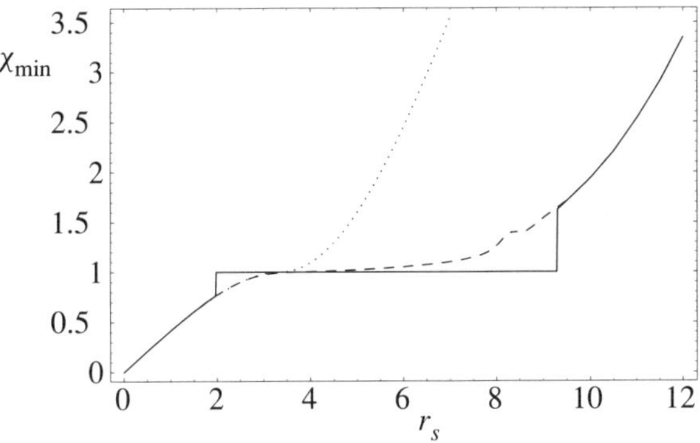

Figure 6. Value of the chirality $\chi_{\min}$ of the homogeneous, isotropic HF state of lowest energy as function of r_s (solid line). The flat region with $\chi = 1$ corresponds to the polarized state while outside of this density range the system is unpolarized. The dashed and dotted lines respectively represent the values appropriate to the unpolarized state and the non interacting one in the region where the polarized state lies lowest. For illustration purposes we have here chosen $\bar{\alpha} = 0.3$.

spin-orbit [9], the low density transition has no analog in absence of spin-orbit interaction. This simple phase diagram is also shown in Figure 7 in the $(r_s, \bar{\alpha})$ plane. One expects that as the density is further lowered, the system will make a transition into a Wigner crystal like state. It must be however noticed that the actual sequence of the transitions depends on the value of $\bar{\alpha}$. It is in fact possible, when $\bar{\alpha}$ is very small, for Wigner crystallization to preempt the reentrant unpolarized state transition [11].

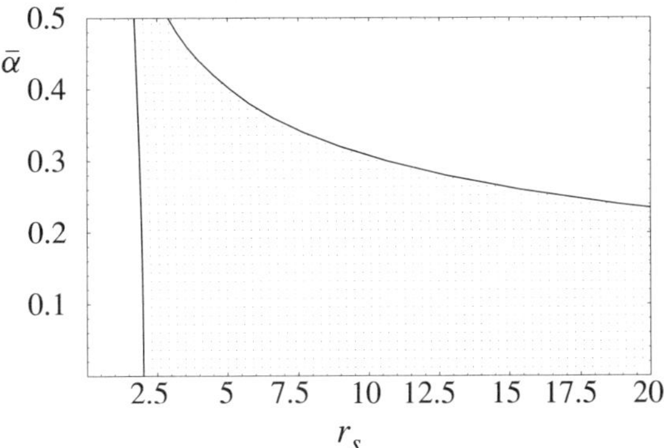

Figure 7. Phase diagram in the $(r_s, \bar{\alpha})$ plane. In the shaded region the gas is polarized.

Finally, Figure 8 shows the fractional polarization $p(\bar{\alpha}, 1)$, calculated from Eq. (3.15). This quantity is independent of r_s. Notice that, even if the state is maximally polarized (having $\chi = 1$), the fractional spin polarization drops to zero for large values of the spin-orbit coupling $\bar{\alpha}$.

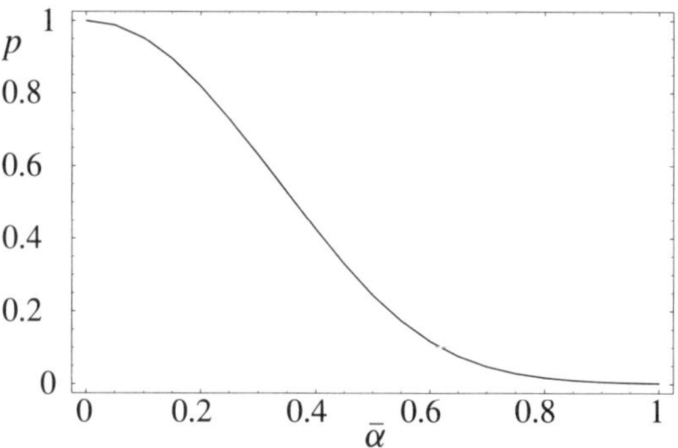

Figure 8. Plot of the polarization $p\,(\bar{\alpha}, 1)$ as function of $\bar{\alpha}$.

5. Conclusions

The concept of generalized chirality allows one to classify a vast class of compactly occupied determinantal states of a two dimensional electron liquid in the presence of Rashba spin-orbit. This enormously facilitates the study of the homogeneous, isotropic HF solutions for the problem. Our results show that within the HF theory the system admits both unpolarized and polarized states. The latter present a non trivial spin texture in momentum space. While the high density relative phase diagrams for these states closely patterns that, well known, of the same system in the absence of spin-orbit coupling, for suitably large values of the Rashba coupling constant $\bar{\alpha}$ a reentrant transition to the unpolarized state appears. The interplay of this transition and the phenomenon of Wigner crystallization still presents an untackled challenge.

While understanding the physical aspects of these spatially homogeneous states can be expected to prove reasonably useful, similarly to the standard electron liquid case, one can show that a class of inhomogeneous HF solutions of lower energy of the charge- and spin-density-wave type exists [12]. Ultimately the whole physical picture must be analyzed allowing for correlation effects beyond HF. While such a program is rather formidable we would like to conclude our contribution by stating that a possible avenue for the inclusion of correlations is suggested by a classical analogue of the problem that can be constructed starting from a direct inspection of the fundamental Eq. (3.10). This developments must await future consideration.

ACKNOWLEDGEMENTS. The authors would like to thank George Simion, Leonid Rokhinson and Giovanni Vignale for useful discussions.

References

[1] E. I. RASHBA, Sov. Phys.-Solid State **2** (1960), 1109.
[2] Y. A. BYCHKOV and E. I. RASHBA, JETP Lett. **39** (1984), 78; J. Phys. C **17** (1984), 6039.
[3] G. DRESSELHAUS, Phys. Rev. **100** (1955), 580.
[4] L. P. ROKHINSON, V. LARKINA, Y. B. LYANDA-GELLER, L. N. PFEIFFER and K. W. WEST, Phys. Rev. Lett. **93** (2004), 146601.
[5] H. CHEN, J. J. HEREMANS, J. A. PETERS, A. O. GOVOROV, N. GOEL, S. J. CHUNG and M. B. SANTOS, Appl. Phys. Lett. **86** (2005), 032113.
[6] By a trivial spin rotation, the present analysis can be applied also to the case in which the spin-orbit interaction is of the Dresselhaus type.
[7] S. CHESI and G. F. GIULIANI, to be published.
[8] G.-H. CHEN and M. E. RAIKH, Phys. Rev. B **59** (1999), 5090; Phys. Rev. B **60** (1999), 4826.
[9] G. F. GIULIANI and G. VIGNALE, "Quantum Theory of the Electron Liquid", Cambridge University Press, Cambridge, 2005.
[10] It is possible to prove that this choice of the polar angle $\gamma_{\mathbf{k}}$ minimizes the total energy.
[11] The effects of the spin-orbit interaction on the Wigner crystallization (which in absence of spin-orbit coupling occurs for $r_s \geq 34$) are to date unexplored. The fact that the spin-orbit energy scales with r_s in the same way as the Coulomb interaction does present an intriguing problem.
[12] G. SIMION and G. F. GIULIANI, to be published.

Nella stessa collana

E. DE GIORGI, F. COLOMBINI, L. C. PICCININI, *Frontiere orientate di misura minima e questioni collegate,* 1972.

C. MIRANDA, *Su alcuni problemi di geometria differenziale in grande per gli ovaloidi,* 1973.

G. PRODI, A. AMBROSETTI, *Analisi non lineare,* 1973.

C. MIRANDA, *Problemi di esistenza in analisi funzionale,* 1975 (esaurito/out of print).

I. T. TODOROV, M. MINTCHEV, V. B. PETKOVA, *Conformal Invariance in Quantum Field Theory,* 1978.

A. ANDREOTTI, M. NACINOVICH, *Analytic Convexity and the Principle of Phragmén-Lindelöf,* 1980.

S. CAMPANATO, *Sistemi ellittici in forma divergenza. Regolarità all'interno,* 1980.

Topics in Functional analysis 1980-81, Contributors: F. STROCCHI, E. ZARANTONELLO, E. DE GIORGI, G. DAL MASO, L. MODICA, 1981.

G. LETTA, *Martingales et intégration stochastique,* 1984.

OLD AND NEW PROBLEMS IN FUNDAMENTAL PHYSICS, Meeting in honour of GIAN CARLO WICK, 1986.

Interaction of Radiation with Matter, A Volume in honour of ADRIANO GOZZINI, 1987.

M. MÉTIVIER, *Stochastic Partial Differential Equations in Infinite Dimensional Spaces,* 1988.

Symmetry in Nature, A Volume in honour of LUIGI A. RADICATI DI BROZOLO, 2 voll., 1989 (ristampa 2005).

Nonlinear Analysis, A Volume in honour of GIOVANNI PRODI, 1991 (esaurito/out of print).

C. LAURENT-THIÉBAUT, J. LEITERER, *Andreotti-Grauert Theory on Real Hypersurfaces,* 1995.

J. ZABCZYK, *Chance and Decision. Stochastic Control in Discrete Time,* 1996 (ristampa 2000).

I. EKELAND, *Exterior Differential Calculus and Applications to Economic Theory,* 1998 (2000).

Electrons and Photons in Solids, A Volume in honour of FRANCO BASSANI, 2001.

J. ZABCZYK, *Topics in Stochastic Processes,* 2004.

TOUZI N., *Stochastic Control Problems, Viscosity Solutions and Application to Finance,* 2004.

Nuova serie

1. HIGHLIGHTS IN THE QUANTUM THEORY OF CONDENSED MATTER
   A symposium to honour Mario Tosi on his 72nd birthday, 2005.

Fotocomposizione "CompoMat" Loc. Braccone, 02040 Configni (RI) Italy
Finito di stampare nel mese di novembre 2005
dalla Nuova Grafica 86, Via Montieri 1/D, Roma